W0079476

MOLECULAR PHENOMENA IN BIOLOGICAL SCIENCES

TOPICS IN
MOLECULAR ORGANIZATION AND ENGINEERING

Honorary Chief Editor:

W. N. LIPSCOMB *(Harvard, U.S.A.)*

Executive Editor:

Jean MARUANI *(Paris, France)*

Editorial Board:

Henri ATLAN *(Jerusalem, Israel)*
Sir Derek BARTON *(Texas, U.S.A.)*
Christiane BONNELLE *(Paris, France)*
Paul CARO *(Meudon, France)*
Stefan CHRISTOV *(Sofia, Bulgaria)*
I. G. CSIZMADIA *(Toronto, Canada)*
P-G. DE GENNES *(Paris, France)*
J-E. DUBOIS *(Paris, France)*
Manfred EIGEN *(Göttingen, Germany)*
Kenishi FUKUI *(Kyoto, Japan)*
Gerhard HERZBERG *(Ottawa, Canada)*

Alexandre LAFORGUE *(Reims, France)*
J-M. LEHN *(Strasbourg, France)*
P-O. LÖDWIN *(Uppsala, Sweden)*
Patrick MacLEOD *(Massy, France)*
H. M. McCONNELL *(Stanford, U.S.A.)*
C. A. McDOWELL *(Vancouver, Canada)*
Roy McWEENY *(Pisa, Italy)*
Ilya PRIGOGINE *(Brussels, Belgium)*
Paul RIGNY *(Saclay, France)*
Ernest SCHOFFENIELS *(Liège, Belgium)*
R. G. WOOLLEY *(Nottingham, U.K.)*

Molecules in Physics, Chemistry, and Biology

Volume 4

Molecular Phenomena in Biological Sciences

Edited by

JEAN MARUANI
Centre de Mécanique Ondulatoire Appliquée,
Laboratoire de Chimie Physique,
CNRS and University of Paris, France.

Springer-Science+Business Media, B.V.

Library of Congress Cataloging in Publication Data

CIP

Molecules in physics, chemistry, and biology.

(Topics in molecular organization and engineering)
Includes bibliographies and indexes.
Contents: v. 1. General introduction to molecular
sciences --   -- v. 4. Molecular phenomena in
biological sciences.
1. Molecules.   2. Molecular biology.   I. Maruani,
Jean, 1937-   . II. Series.
QC173.M645   1988   539'.6   88-6811

ISBN 978-94-010-7022-5   ISBN 978-94-009-1173-4 (eBook)
DOI 10.1007/978-94-009-1173-4

All Rights Reserved

© 1989 Springer Science+Business Media Dordrecht
Originally published by Kluwer Academic Publishers, Dordrecht, The Netherlands in 1989
Softcover reprint of the hardcover 1st edition 1989

No part of the material protected by this copyright notice
may be reproduced or utilized in any form or by any means,
electronic or mechanical including photocopying, recording or by
any information storage and retrieval system, without written
permission from the copyright owner

Table of Contents

MOLECULAR PATHOLOGY

TOPICS IN BIOMOLECULAR PHYSICS

MOLECULAR NEUROBIOLOGY AND SOCIOBIOLOGY

Introduction to the Series

The Series 'Topics in Molecular Organization and Engineering' was initiated by the Symposium 'Molecules in Physics, Chemistry, and Biology', which was held in Paris in 1986. Appropriately dedicated to Professor Raymond Daudel, the symposium was both broad in its scope and penetrating in its detail. The sections of the symposium were: 1. The Concept of a Molecule; 2. Statics and Dynamics of Isolated Molecules; 3. Molecular Interactions, Aggregates and Materials; 4. Molecules in the Biological Sciences, and 5. Molecules in Neurobiology and Sociobiology. There were invited lectures, poster sessions and, at the end, a wide-ranging general discussion, appropriate to Professor Daudel's long and distinguished career in science and his interests in philosophy and the arts.

These proceedings have been arranged into eighteen chapters which make up the first four volumes of this series: Volume I, 'General Introduction to Molecular Sciences'; Volume II, 'Physical Aspects of Molecular Systems'; Volume III, 'Electronic Structure and Chemical Reactivity'; and Volume IV, 'Molecular Phenomena in Biological Sciences'. The molecular concept includes the logical basis for geometrical and electronic structures, thermodynamic and kinetic properties, states of aggregation, physical and chemical transformations, specificity of biologically important interactions, and experimental and theoretical methods for studies of these properties. The scientific subjects range therefore through the fundamentals of physics, solid-state properties, all branches of chemistry, biochemistry, and molecular biology. In some of the essays, the authors consider relationships to more philosophic or artistic matters.

In Science, every concept, question, conclusion, experimental result, method, theory or relationship is always open to reexamination. Molecules do exist! Nevertheless, there are serious questions about precise definition. Some of these questions lie at the foundations of modern physics, and some involve states of aggregation or extreme conditions such as intense radiation fields or the region of the continuum. There are some molecular properties that are definable only within limits, for example, the geometrical structure of non-rigid molecules, properties consistent with the uncertainty principle, or those limited by the neglect of quantum-field, relativistic or other effects. And there are properties which depend specifically on a state of aggregation, such as superconductivity, ferroelectric (and anti), ferromagnetic (and anti), superfluidity, excitons, polarons, etc. Thus, any molecular definition may need to be extended in a more complex situation.

Chemistry, more than any other science, creates most of its new materials. At least so far, synthesis of new molecules is not represented in this series, although the principles of chemical reactivity and the statistical mechanical aspects are

Jean Maruani (ed.), Molecules in Physics, Chemistry, and Biology, Vol. IV, xi–xii.
© 1989 *by Kluwer Academic Publishers.*

included. Similarly, it is the more physico-chemical aspects of biochemistry, molecular biology and biology itself that are addressed by the examination of questions related to molecular recognition, immunological specificity, molecular pathology, photochemical effects, and molecular communication within the living organism.

Many of these questions, and others, are to be considered in the Series 'Topics in Molecular Organization and Engineering'. In the first four volumes a central core is presented, partly with some emphasis on Theoretical and Physical Chemistry. In later volumes, sets of related papers as well as single monographs are to be expected; these may arise from proceedings of symposia, invitations for papers on specific topics, initiatives from authors, or translations. Given the very rapid development of the scope of molecular sciences, both within disciplines and across disciplinary lines, it will be interesting to see how the topics of later volumes of this series expand our knowledge and ideas.

WILLIAM N. LIPSCOMB

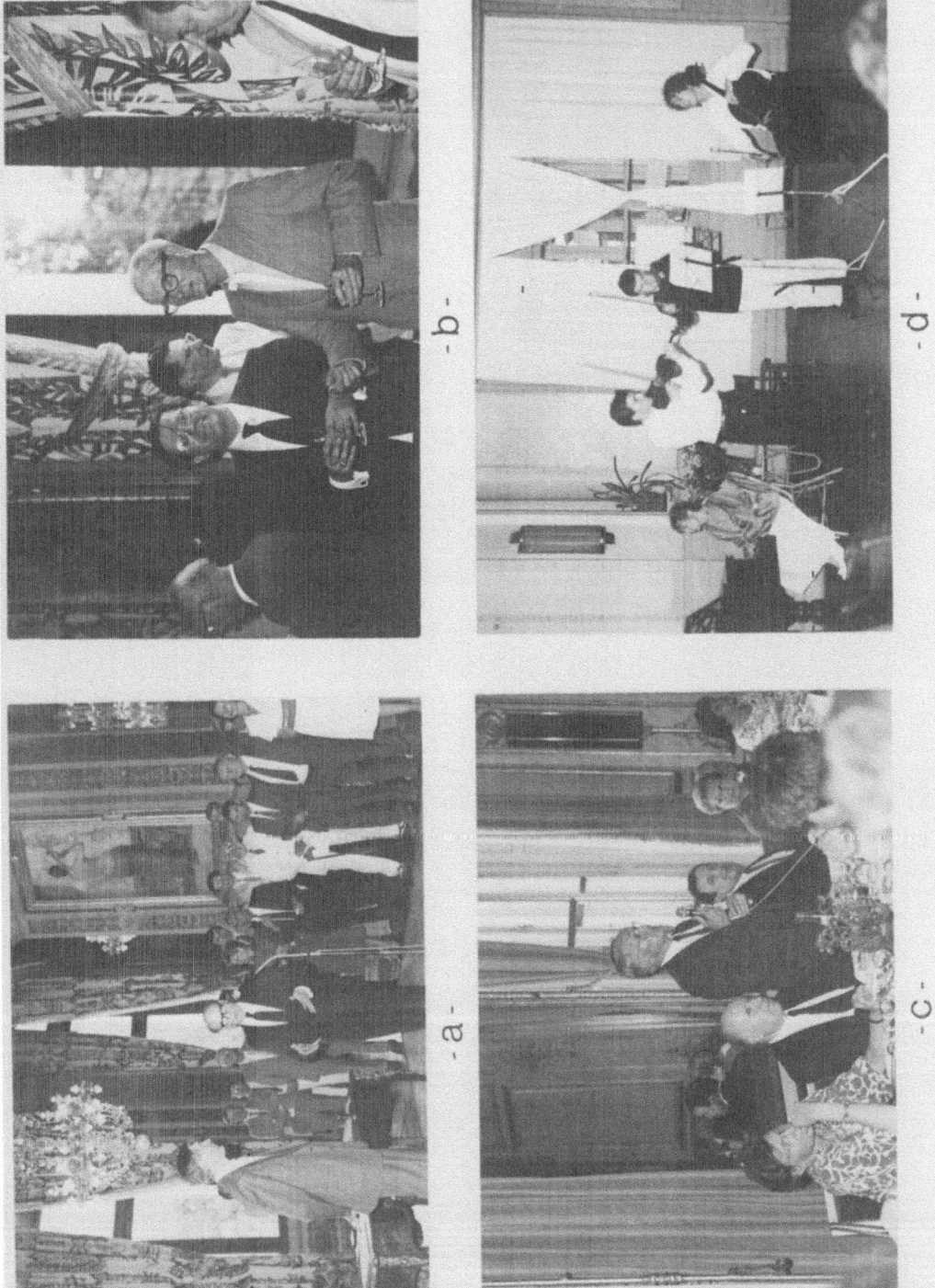

- a -

- b -

- c -

- d -

Preface to *Molecules in Physics, Chemistry, and Biology*

When we decided to organize an International Symposium dedicated to Professor Daudel, a question arose: on which themes should such a Symposium bear? After having reviewed all the themes on which Professor Daudel has worked during his long career, Imre Csizmadia and myself were somewhat at a loss; these themes ranged from Atomic Physics to Molecular Biology, with a stress on Theoretical Chemistry. Then I recalled a conversation I had in 1968, when I was in Vancouver, with Harden McConnell, on leave from Stanford. I asked him why he had switched to Biology; he answered: "I'm often asked this question. But I don't feel I've ever switched to Biology. I have always been interested in molecules, just molecules: in Physics, Chemistry, and Biology". I felt this flash of wit would make a perfect title for a Symposium dedicated to Professor Daudel, who has also been interested in molecules in Physics, Chemistry, and Biology, but from a theoretical viewpoint.

However, when it came to preparing a content appropriate to this title, we ended up with a several-page program, which defined what could have been some kind of an advanced-study institute, involving most of Physical Chemistry and parts of Molecular Biology. We announced the Symposium on that pluridisciplinary basis and then started receiving answers from invited speakers and proposals for communications. While classifying the letters, it appeared to us that a few key themes had emerged, which seemed likely to constitute 'hot topics' of the Molecular Sciences in the late 1980's and early 1990's. Indeed there are fashions in Science too, whether these are induced by the natural development of the sciences or by economic or cultural constraints. Afterwards we did our best to fill

---

LEGENDS TO THE PHOTOGRAPHS OF PLATE A
(Photographs by Miss Cristina Rusu)

— a — Minister of Research Alain Devaquet (on the left) awarding the Golden Medal of the City of Paris to Professor Raymond Daudel (on the right) in Paris City Hall. In the background, from left to right: Jean-Marie Lehn, William Lipscomb (between Devaquet and Daudel), Bernard Pullman, Jacques-Emile Dubois, Georges Lochak (all three wearing spectacles), Ernest Schoffeniels.

— b — William Lipscomb and Jean Maruani chatting after the ceremony. Also on the picture: Bernard Pullman (left), Jacques-Emile Dubois (center), Paul Caro (right).

— c — Senator Louis Perrein opening the closing banquet in the Senate House. From left to right: Alberte Pullman, Raymond Daudel, Jean-Pierre Changeux, Nicole D'Aggagio, Stefan Christov, Christiane Bonnelle.

— d — Composer and pianist Marja Maruani-Rantanen and Jean-Yves Metayer's string trio *I Solisti Europa* performing for participants in the Concordia Hotel.

Jean Maruani (ed.), Molecules in Physics, Chemistry, and Biology, Vol. IV, xv–xix.
© 1989 *by Kluwer Academic Publishers.*

what seemed to be gaps in the consistency of the emerging program. The main lines of the resulting program are recalled by Professor Lipscomb in his Introduction to the Series.

The Symposium gathered about 200 people, with interests ranging from the History and Philosophy of the Molecular Concept to Molecular Neurobiology and Sociobiology. A few social events were arranged, in order to help bring together participants with different interests, who otherwise would have tended to miss sessions not belonging to their own specialty. Miss Cristina Rusu recorded these oecumenical moments in photographs, a few of which are shown in Plate A.

During the nine months following the Symposium, I managed to gather together about 70% of the invited papers and 30% selected posters, as well as a few contributions not presented during the Symposium but expected to complete the Proceedings. The authors were requested to submit 'advanced-review' papers, including original material, and most of the manuscripts were refereed. The resulting arrangement of the topics is outlined in Table 1. In spite of the variety of the topics, there is a definite unity in the arrangement. This results from the specificity of the Molecular Sciences, which arises from the particular role played by the molecular concept in Science. In the hierarchy of structures displayed by Nature, molecules, supermolecules and macromolecules are situated just between atoms (which define the chemical elements) and proteins (which define biological

TABLE 1

Vol. I — General Introduction to Molecular Sciences
Part 1 — papers 01—03: History and Philosophy of the Molecular Concept
Part 2 — papers 04—06: Evolution and Organization of Molecular Systems
Part 3 — papers 07—11: Modelling and Esthetics of Molecular Structures

Vol. II — Physical Aspects of Molecular Systems
Part 1 — papers 12—13: Mathematical Molecular Physics
Part 2 — papers 14—15: Relativistic Molecular Physics
Part 3 — papers 16—17: Molecules in Space
Part 4 — papers 18—21: Small Molecular Structures
Part 5 — papers 22—25: Nonrigid and Large Systems
Part 6 — papers 26—28: Molecular Interactions
Part 7 — papers 29—33: Theoretical Approaches to Crystals and Materials

Vol. III — Electronic Structure and Chemical Reactivity
Part 1 — papers 34—40: Density Functions and Electronic Structure
Part 2 — papers 41—45: Structure and Reactivity of Organic Compounds
Part 3 — papers 46—49: Theoretical Approaches to Chemical Reactions

Vol. IV — Molecular Phenomena in Biological Sciences
Part 1 — papers 50—51: Biomolecular Evolution
Part 2 — papers 52—53: Biomolecular Chirality
Part 3 — papers 54—55: Topics in Molecular Pathology
Part 4 — papers 56—58: Topics in Biomolecular Physics
Part 5 — papers 59—63: Molecular Neurobiology and Sociobiology

specificity). In Physical Chemistry, indeed, there are thermodynamic, spectroscopic and diffraction data specifically related to molecular structure and dynamics.

Among the questions which arise in the Molecular Sciences, one may stress the following.

— How can a molecule be strictly defined with respect to the constitutive atoms, on the one hand, and the molecular gas, liquid, or solid, on the other? — Use of Topology and Fuzzy-Set Theory, Quantum and Statistical Mechanics, Effective Hamiltonian Operators and Reduced Density Matrices, X-ray and Neutron Diffraction, UV and IR Spectroscopy, etc. ('Molecular Phenomenology and Ontology').

— While hydrogen and helium constitute together 99% of the total mass of the natural elements (with, thank God! traces of heavier elements, including carbon), is molecular complexity a unique feature of the Earth or is it deeply related to the very structure of our Universe? Were Life and Man built into Nature or are they merely accidents? ('Molecular Cosmology and Evolution').

— What are the origin, nature and transfer of the information content packed in a molecular system? How can molecular information be extracted by the modelling of molecular structures? How can levels of information ordering be defined and what are the relations between the information on simple substructures and that on complex superstructures? Can the higher levels of organization and functioning be understood in purely physicochemical terms? How do molecular assemblies cooperate to form organized or living structures? ('Molecular Organization and Cybernetics').

— Chemical laboratories and industries have created more molecules than there have been found in Nature, particularly pharmaceutics and polymers. Even such physical properties as superconductivity or ferromagnetism are no longer limited to classical metallic materials, but may also be found in molecular materials ('Molecular Synthesis and Engineering').

— Biological specificity and immunity are understood today basically as molecular phenomena related to the DNA and protein structures. Tiny structural modifications in these macromolecules may lead to metabolic deficiencies or other functional disorders ('Molecular Pathology').

— Communication within and between cells and organs in a living organism, as well as between individuals (particularly in sexual activity) in a species, and between species in an ecosystem, occurs very often through molecular interactions ('Molecular Communication').

Most of these and other related questions were dealt with in the Symposium, the Proceedings of which are published in this Series. Future volumes in the Series are expected to develop specific topics related to these questions.

The Symposium was sponsored by various bodies and companies, which are listed in Table 2. They are all gratefully acknowledged for their (material or moral) help, which made possible this gathering. The international honorary committee,

TABLE 2

SPONSORS

Ministère de l'Education Nationale
Ministère des Relations Extérieures
Ville de Paris

Centre National de la Recherche Scientifique
Commissariat à l'Energie Atomique
Institut National de la Santé et de la Recherche Médicale
Institut National de Recherche Pédagogique

Université Paris VI
Université Paris VII
Ecole Supérieure de Physique et Chimie Industrielles

World Association of Theoretical Organic Chemists
Fondation Louis de Broglie

Rhône-Poulenc
Moët-Hennessy
Amstrad France
Alain-Vaneck Promotion

COMMITTEES
Centre de Mécanique Ondulatoire Appliquée

and

International Honorary Committee	*Local Organizing Committee*
Sir D. Barton (*U.K.*)	R. Acher (*Biological Chemistry*)
J-P. Changeux (*France*)	D. Blangy (*Molecular Biology*)
M. Eigen (*F.R.G.*)	C. Bonnelle (*Physical Chemistry*)
J. I. Fernàndez-Alonso (*Spain*)	P. Caro (*Inorganic Chemistry*)
K. Fukui (*Japan*)	P. Claverie[†] (*Theoretical Chemistry*)
G. Herzberg (*Canada*)	I. G. Csizmadia (*Organic Chemistry*)
F. Jacob (*France*)	J-E. Dubois (*Molecular Systemics*)
W. N. Lipscomb (*U.S.A.*)	A. Laforgue (*Theoretical Chemistry*)
P. O. Löwdin (*Sweden*)	R. Lefebvre (*Molecular Photophysics*)
H. M. McConnell (*U.S.A.*)	J-M. Lehn (*Supramolecular Chemistry*)
C. A. McDowell (*Canada*)	G. Lochak (*Quantum Mechanics*)
Sir G. Porter (*U.K.*)	P. MacLeod (*Molecular Neurobiology*)
I. Prigogine (*Belgium*)	J. Maruani (*Molecular Physics*)
B. Pullman (*France*)	P. Rigny (*Physical Chemistry*)
M. Simonetta[†] (*Italy*)	J. Serre (*Theoretical Chemistry*)
[†] Deceased in 1986.	[†] Deceased in 1988.

also given in Table 2, involved fifteen distinguished scientists from ten different countries, including eight Nobel Laureates. May I express my gratitude to all of them, especially to those who managed to participate actively in the Symposium. The local organizing committee involved mostly French scientists belonging to different fields (Table 2), reflecting the interdisciplinarity of the meeting. They are all most gratefully thanked for their help and encouragement. Special thanks go to Prof. I. G. Csizmadia, who helped enormously in the early stages of the organization of the meeting, and to Dr P. Claverie, recently deceased, who helped in the late stages of the organization and also in the selection of the papers for these volumes. Finally my thanks go to Bernard and Isabelle Decuypère, who prepared the indexes, and to the Staff of Kluwer Academic Publishers, for their pleasant and efficient cooperation.

I hope these books will prove to be of as much interest to the reader as the meeting was to the participants.

JEAN MARUANI

Preface to Volume 4:
Molecular Phenomena in Biological Sciences

The importance of molecules in biological sciences was recognized when it became clear that they are directly responsible not only for the structure, but also for the most specific functions of living organisms.

Protein molecules were the first to be identified as "molecules of life" [1], due to their enzymatic properties. Since they could control, by means of their catalytic effects, the setting off of biochemical reactions, thus orienting the metabolism of a cell or an organism through specific channel resulting in specific structural or physiological properties, they were viewed as responsible for the creation and maintenance of the kind of order observed in living beings, and were compared to actual Maxwell demons [2].

At the same time, they kept the secret of their synthesis since, until the 1960s, they were among the main organic molecules which could not be synthesized *in vitro*. Indeed, this secret was an important one since its unveiling turned out to be linked to the major breakthrough of biology in the 20th century, the discovery of the genetic code.

The transmission of genetic material from one generation to another and its expression in hereditary characters — structural and functional — could be directly understood from the structural properties of a set of macromolecules. In addition to enzymatic proteins, the double-helix DNAs were identified as the gene molecules. RNAs of different kinds — messenger, transfer, ribosomal — were found to be responsible for specific protein syntheses, i.e. the translation of gene structures into functional enzymes. This remarkable breakthrough brought about a new era in biology, that of molecular biology: biological *functions* most charac-teristic of living organisms were directly related to specific macromolecular *structures*. As a result, macromolecular physical chemistry acquired a new dimen-sion, since functional properties within a complex organism could be attributed to individual molecules. For a physicist, a molecule is a unit in a thermodynamic statistical ensemble, devoid of any individual specific function and for which birth and death are meaningless concepts. For a molecular or cellular biologist, a molecule may be an individual entity, endowed with a very specific function appearing at some time when it is synthesized and disappearing when it is destroyed or 'denatured' (i.e. when its functional structure is modified).

This new dimension required a new theoretical framework. It so happened that the computer and information sciences, developed at the same time, provided the concepts, or at least the words, needed to describe these phenomena without having to resort to magic teleological properties. These peculiar molecules of life

Jean Maruani (ed.), Molecules in Physics, Chemistry, and Biology, Vol. IV, xxi—xxvi.
© 1989 *by Kluwer Academic Publishers.*

were viewed as information-carrying molecules. Teleology, i.e. apparently pur-
poseful activity, was transformed into teleonomy, i.e. 'non-purposeful end-seeking
processes' (Pittendrigh [3]). The matching of the linear structures of the gene DNA
molecules and of the enzymatic and structural proteins was found to be the same
in all living species and was termed a 'code': the universal genetic code. Finally, the
development of a complex living organism from an egg, following a specific order
in time and space, was viewed as the execution of a 'program': the genetic program.
It is not certain that these metaphors, borrowed from the computer and information sciences, were absolutely necessary to describe the facts of molecular biology
and to unify them in a theoretical framework. It is clear that the same facts,
regarding the mechanisms of genetic determination and protein synthesis, would
have been discovered and given different names if computer and information
sciences had not existed at that time; but these were convenient and had then — as
they still do — a very powerful heuristic value.

However, these metaphors should not be taken too literally. Rather, they must
be extended and modified to account for the basic difference between the object of
computer sciences and that of biology: the first one deals with artificial machines
built by man to achieve well-defined goals, which have been purposefully set up
from the beginning; the other one deals with natural machines, created by nature
with no obvious purpose in mind. This difference can be seen clearly when one
ponders the question of meaning in biological information. If meaning is function
— and one can rightfully say that the meaning of genetic information is the
enzymatic function that it codes for — the meaning of a program is set up from the
beginning, namely to perform the task explicitly defined *a priori*, for which it has
been written. This is not the case in a program without a programmer, or a self-
programming program, where the meaning must be found *a posteriori*, as an
emerging property of the building of the program. Already in 1962, A. Lwoff [4]
warned against abusive use of information-theory concepts in biology, where
information is defined on the basis of structural properties of macromolecules,
without taking account of their biological meaning.

The understanding of biological functions could not be limited to molecular
structures: cooperating interactions between different molecules or classes of
molecules had to be considered. Only by considering different kinds of such
interactions (cooperativity, positive and negative feedback loops, etc.) could the
problem of regulation of enzyme activity, and more particularly of gene expression, be approached effectively. The lactose operon model of Monod and Jacob has
set up a paradigm for the molecular mechanisms of gene expression regulation.

Since then, information-carrying molecules have become the key to understanding other basic biological phenomena, such as hormone action, intercellular
communication, antigen—antibody interactions, and so on. New molecules, which
transmit signals modifying states of neural activity (neurotransmitters), of immunological activity (interleukines), or of growth and differentiation (growth factors),
are constantly being discovered.

As is obvious from the reports which appear in this volume, the study of molecular phenomena in biological sciences is achieved on two parallel levels. One is the molecular structure itself, usually studied by physical methods which enable one to better understand the three-dimensional structure of proteins and nucleic acids and their modifications under different conditions — *in vitro* and *in vivo*. The other is the correlation between given molecular structures and biological functions, as they are observed at a supramolecular level — cellular or organismic.

The chemical physics of macromolecules, which constitutes the first category of works, continues to provide a wealth of information on the spatial properties and intramolecular dynamics responsible for the biological activity of individual molecules: identification of subunits and active domains in cytoplasmic and nucleic proteins and in membrane receptors is in line with the pioneering discovery of the DNA double-helix structure by X-ray crystallography and the demonstration of its role in the semi-conservative nature of gene replication. Energy transfer mechanisms within macromolecules such as chlorophyll, rhodopsin, cytochromes, membrane channels and active transporters have helped us to understand the molecular nature of complex phenomena such as photosynthesis, retinal light stimulation, respiration and transport of ions and nutrients across cell membranes respectively.

The second category of works, more closely related to supramolecular integrated phenomena, has led to unexpected discoveries and new problems, when the techniques of molecular biology, originally developed in microorganisms, have been applied to the study of eukaryotic cells and multicellular organisms. The 'central dogma' of molecular biology, which relates the molecular structure of a gene to the biological character of the organism in the one-to-one, oriented correspondence: 'one gene—one enzyme—one characteristic', had to be modified. More than one gene contributes to the expression of one characteristic and one gene contributes to the expression of several characteristics. Different combinations of genes are expressed together by means of a transcriptional control at the level of the messenger RNA synthesis. Transposon units in some organisms have been found to be responsible for an unexpected degree of gene mobility within the genome. Short repetitive DNA sequences, called homeoboxes, coding for some relatively small regulatory peptides, have been shown to be responsible together for turning on or off batteries of structural genes disseminated in the genome. As 'master genes' responsible for large-scale gene regulation during embryonic development, they seem to strengthen the metaphor of the genetic program, since they operate in a manner similar to program instructions calling for 'subroutines'. However, this is not the case since regulatory genes themselves are structural genes as well and, as stated by Gehring, who discovered these master genes: "even if the homeobox-containing genes regulate many other genes, they must be regulated themselves. Discovering how the regulators are regulated will be another significant accomplishment, and one that could lead to the identification of the factors in the egg cytoplasm that provide the positional information" [5].

In other words, we are back to the cytoplasm with its regulatory proteins and its network of coupled transports and reactions. This amounts to 'delocalizing' the program and extending it to the whole cellular machinery . . . that it is assumed to program!

Recently [6] the role of long-lived DNA-methylating enzymes, transmitted during cell division, has been stressed in what is called an epigenetic inheritance, i.e. a non-direct genetic mechanism for gene regulation.

In shaping modern biological theories, the impact of molecular biology went beyond providing a substrate for the metaphorical, still operational, concept of genetic programming. It provided the required material substrates for concepts employed in the Darwinian theory of evolution: genetic random mutations are the source of the genetic variability upon which natural selection is assumed to operate. However, measurements of evolution rates between species at the molecular level led to far-reaching modifications of the theory, by the introduction of the new concepts of neutral mutations and genetic drift as additional mechanisms for evolution, not driven by natural selection. Also in this field, cooperative network interactions are being used to model complex genotype—phenotype relationships. Such models are necessary to account for the observation of rates of evolution which differ at different levels, and for evolutionary theories based on steep non-linearities such as Eldredge—Gould punctuated equilibria [7]. This trend seems to provide an alternative to, and therefore seems to attenuate, what had previously seemed an inescapable conclusion of molecular biology in its early stages: adaptive properties and goals in living systems, assumed — among other things — to drive the evolutionary processes, had to be transposed from the organism to the genes, from the functional phenotypic properties to the DNA molecule replication ability.

In its most provocative form, this conclusion was stated as a theory of 'selfish genes' or 'selfish DNA' [8]. Additional support for this theory came from the discovery of large segments of repetitive DNA in eukaryotic cells, both coding and non-coding. Since no evidence has been provided for the obvious functional utility of these non-coding repetitive sequences, they can be seen as kinds of parasites making use of the organism for their own 'purpose' of replication. On the other hand, they can be seen as a reservoir of redundancy to generate future diversity in evolution under the effects of the 'noise' produced by random mutations. The selfish gene idea is a molecular traduction of Samuel Butler's statement on "a hen which is nothing but the efficient way a hen's egg has found to produce another hen's egg". However, it is aimed at eliminating all traces of anthropomorphic purpose still contained in the ideas of functional phenotypic adaptation of genetic 'programs', where natural selection would operate as a programmer who has a goal in mind when writing the program. As such it had to be extended to a point where the idea of phenotype itself had to be modified. The 'extended phenotype' [9] is no longer limited to the organism, but is extended to the set of organisms a population where a gene or a 'bloc of genes' (a replication unit) is present. This unit is also assumed to be a selection unit, in the sense that natural selection operates at that level in such a way that the individual organisms seem to be used merely as means for the selfish 'replicators' to replicate.

As mentioned above, this way of thinking is not necessarily the only one compatible with the purely mechanistic, causal descriptions that molecular biology has imposed on biology in eliminating vitalist finalistic theories. Consideration of hierarchical organization, with different levels of integration and networks of interactions at each level, enables one to describe self-organizing phenomena, with the emergence of specific properties at a given level. As stated by S. J. Gould: "a new and general evolutionary theory will embody this notion of hierarchy, and stress a variety of themes either ignored or explicitly rejected by the modern synthesis: punctuational changes at all levels, important non-adaptive changes at all levels, control of evolution not only by selection, but equally by constraints of history, development and architecture — thus restoring to evolutionary theory a concept of organism" [9]. This achieved without having to resort to anything other than a mixture of deterministic mechanistic interactions and of some randomness or noise at some stage (in setting up initial conditions, in orienting bifurcations, through perturbations by a temperature of some sort, etc.).

This last picture, previously suggested by the cybernetics of self-organization [11, 12] (what has been termed the 'second cybernetics'), has benefited from recent works in the physics of complex systems such as spin glasses and chaotic turbulent structures [13, 14]. As an alternative to the simplistic and unrealistic idea of a computer program written in the nucleotide sequences of DNA molecules, it should deprive the recently suggested idea of sequencing the human genome of its fascinating appeal. It is clear that such a list of nucleotides cannot be interpreted as the listing of a computer program — what has been called the Book of Man (!) [15, 16, 17] — since it does not take into account the interactions between the genome and the organism where biological meaning is generated. If sequencing a whole genome were to be of any interest, it would be that of an experimental animal which would allow for experiments aimed at understanding the laws of such interacting networks.

The network of interacting elements is probably only the latest in a series of models borrowed from physics to describe the organization of living systems ever since the discovery of the molecular structure of matter. The ordered, crystallized state of a homogeneous, statistical thermodynamic ensemble at equilibrium was followed by information-carrying molecules working like Maxwell demons. Then, self-organizing properties of coupled fluxes far from equilibrium seemed more suited to the conditions of thermodynamically open systems prevailing most often in living organisms. Most recently, hierarchical ordered—disordered metastable states in interacting discrete networks, like spin glasses, may provide a desired association of chemical energy-producing homogeneity and signal-like information-producing singularities.

HENRI ATLAN

References

1. The Molecules of Life, *Scientific American*, special issue, Oct. 1985.
2. J. Monod: *Le Hasard et la Nécessité*, Paris, Seuil, 1970.
3. C. S. Pittendrigh: Adaptation, Natural Selection and Behavior, in *Behavior and Evolution*, A. Roe and G. G. Simpson, eds., New Haven, Conn., Yale Univ. Press, 1958, pp. 390—416.
4. A. Lwoff: *Biological Order*, Cambridge, Mass, MIT Press, 1962.
5. S. Gehring: The Molecular Basis of Development, *Scientific American*, Oct. 1985, pp. 153—162.
6. R. Holliday: The Inheritance of Epigenetic Defects, *Science* **238**, 1987, 163—170.
7. S. J. Gould & N. Eldredge: Punctuated Equilibria, the Tempo and Mode of Evolution Reconsidered, *Paleobiology* **2**, 1977, 115—151.
8. L. E. Orgel & F. H. C. Crick: Selfish DNA: the Ultimate Parasite, *Nature* **284**, 1980, 604—607.
9. R. Dawkins: Replicator Selection and the Extended Phenotype, *Z. Tierpsycol.* **47**, 1978, 61—76.
10. S. J. Gould: Is a New and General Theory of Evolution Emerging? *Paleobiology* **6**, 1980, 119—130.
11. H. Atlan: *L'Organisation Biologique et la Théorie de l'Information*, Paris, Hermann, 1972.
12. H. Atlan: Self Creation of Meaning, *Physica Scripta* **36**, 1987, 563—576.
13. W. Anderson: Suggested Model for Prebiotic Evolution: The Use of Chaos, *Proc. Natl. Acad. Sci. USA* **80**, 1983, 3386—3390.
14. D. J. Amit: Neural Networks, Achievements, Prospects, Difficulties, in *The Physics of Structure Formation*, W. Guttinger & G. Dangelmayr, eds., Heidelberg, Springer-Verlag, 1987, pp. 2—21.
15. A. Wada: Automated High-Speed DNA Sequencing, *Nature* **325**, 1987, 771—772.
16. J. Palca: Human Genome, *Nature* **326**, 1987, 429.
17. L. Philipson & J. Tooze: The Human Genome Project, *Biofutur* (Paris) **58**, 1987, 94—101.

Biomolecular Evolution

Randomness and Natural Selection in Genome Evolution

GIORGIO BERNARDI and GIACOMO BERNARDI
Laboratoire de Génétique Moléculaire, Institut Jacques-Monod, 2, Place Jussieu, 75005 Paris, France.

1. Introduction

The evolution of living organisms is caused primarily by mutations that may subsequently be eliminated or become fixed in the genome. While it is generally agreed that elimination affects deleterious mutations and occurs by negative selection, fixation has been visualized as due either (i) to positive Darwinian selection acting on advantageous mutations or (ii) to random genetic drift acting on selectively neutral (i.e. selectively equivalent) mutations. Since both advantageous and neutral mutations definitely can be fixed in evolution, the issue is of a quantitative and not of a qualitative nature and concerns the predominance of deterministic or stochastic events in genome evolution. That the issue is a difficult one is proved by the fact that this debate has gone on, in the form just stated, for almost twenty years.

We will summarize here a new approach to the problem (see [1—3] for recent reports). The question we have asked basically concerns the degree of freedom in the fixation of mutations. According to the neutral theory [4—7], the fixation of mutations is free of constraints except for 'functional constraints', like the requirement of given aminoacids in certain positions of polypeptide chains. As a consequence, fixation of mutations in third codon positions and in non-coding sequences was considered to be essentially random. Our starting observation was that this certainly was not the case in the genome of warm-blooded vertebrates [8].

2. The Compositional Compartmentalization of Genomes

The genomes of warm-blooded vertebrates are highly compartmentalized, in that they mainly consist of a mosaic of very long ($\gg 300$ kilobases) DNA segments, the *isochores*, which (i) belong to a small number of classes characterized by different GC* levels and by fairly homogeneous base compositions (at least in the 3—300 kbase range); and (ii) seem to correspond to the DNA segments present in Giemsa and Reverse chromosomal bands. Since the families of DNA molecules derived from the isochore classes (or genome compartments) can be separated, it is possible to study the genome distribution of any sequence that can be detected with an appropriate probe. This approach revealed (i) that the distribution of

---

* GC = mole % of deoxyguanosine + deoxycytidine.

Jean Maruani (ed.), Molecules in Physics, Chemistry, and Biology, Vol. IV, 3—12.
© 1989 *by Kluwer Academic Publishers.*

genes is highly biased towards the GC-rich isochores (which are either absent or poorly represented in cold-blooded vertebrates) and tends to be conserved within birds and within mammals; and (ii) that the GC levels of both coding and non-coding sequences (e.g. introns, families of interspersed repeats), as well as of codon third positions, show a linear dependence upon the GC levels of the genome compartments harboring the sequences. While the strongly biased gene distribution largely came as a surprise, the GC relationships indicated the existence of compositional constraints. (It should be noted that a correlation between GC levels in third positions and GC of flanking sequences was independently reported by Ikemura [9] for several vertebrate genes; moreover, a very recent paper [10] has confirmed and extended some of the points made above).

3. Compositional Constraints Affect both Coding and Non-Coding Sequences

The compositional constraints first detected in the genomes of warm-blooded vertebrates have been investigated in over 60 genomes of prokaryotes, viruses and vertebrates. The results of Figure 1 indicate that the GC levels of each codon position of the genes from a given genome (or genome compartment) are linearly related to those of the corresponding coding sequences; slopes and intercepts for each position are very close for all genomes explored (except for a higher slope in the case of second positions of viruses). In the case of vertebrates, plots against the GC levels of genome (or genome compartments) are very close to those against GC levels of coding sequences, in spite of the fact that non-coding sequences represent over 90% of the genome. These relationships indicate the existence of compositional constraints acting on coding sequences (where they also affect the levels of individual bases; not shown), as well as non-coding sequences.

These findings raise two problems, that of their consequences at the RNA and protein levels, and that of their origin.

4. GC Increases in Coding Sequences Affect mRNA and Protein Stability

All GC changes in second codon positions entail changes in the amino acid composition of proteins; so also do most first position changes, and two third position changes. An analysis of the amino acid replacements which accompany the GC increases in codon positions has revealed that they comprise those [11] that lead to thermodynamically more stable proteins (see Table I). Indeed, the amino acids (alanine and arginine) that are most frequently acquired in thermophiles and that most contribute to an increased stability *increase*, whereas those (serine and lysine) that are correspondingly lost and that diminish stability *decrease* with increasing exon GC (not shown). In the case of compartmentalized genomes, these changes may take place within the same genome.

In conclusion, the compositional changes that make DNA thermodynamically

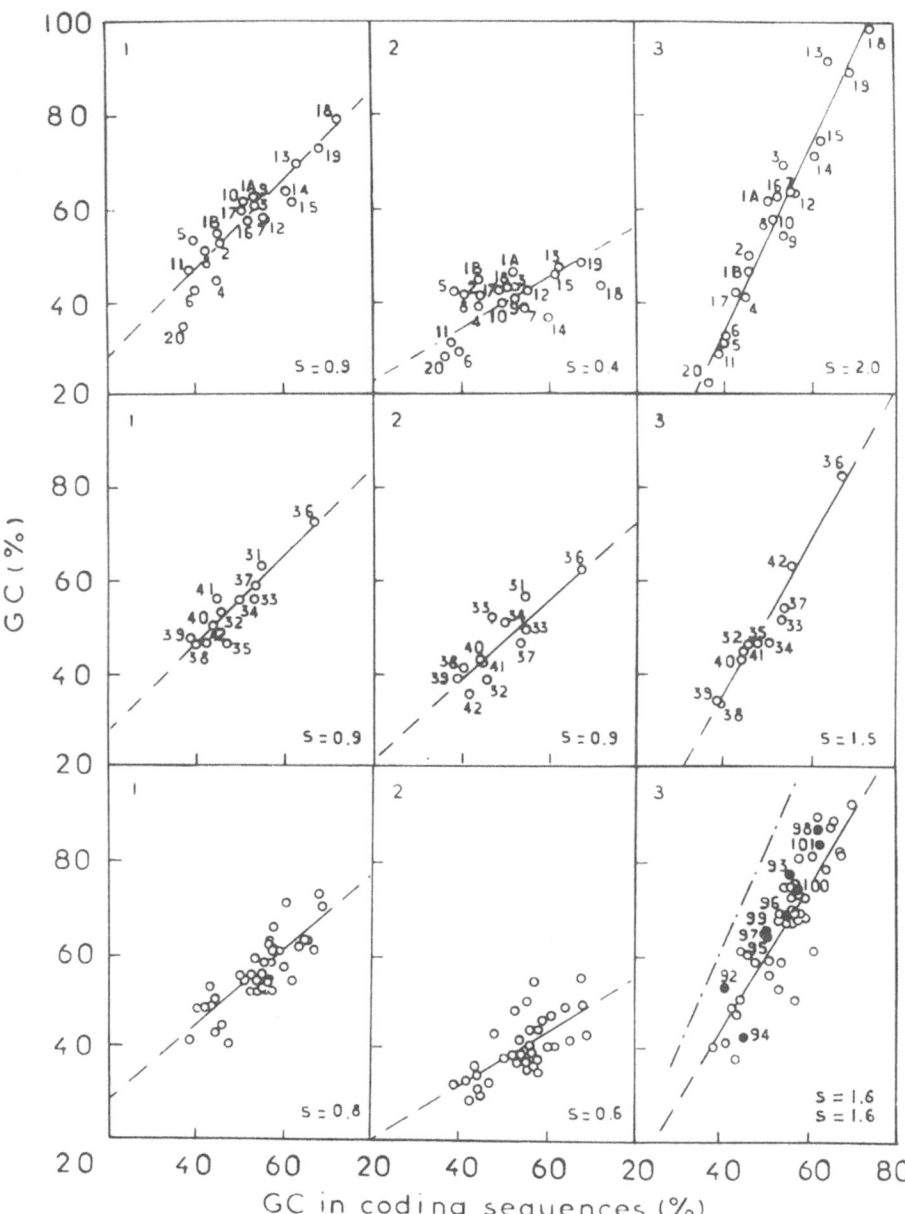

Fig. 1. (Top and middle frames) GC levels of the three codon positions (1, 2, 3) of prokaryotic (top frames) and viral (middle frames) genes are plotted against the GC levels of the corresponding genomes (the list of genomes is given in [1]). The scatter of the points belonging to different genes from the same genome was small and average values, weighted for gene size, were therefore used. Numbers refer to the genomes listed in Table I. Lines were drawn using the least-squares method; the slopes are indicated; correlation coefficients were 0.91, 0.58, and 0.97 for prokaryotic genomes, and 0.91, 0.88 and 0.95 for viral genomes, respectively. (Bottom frames) (O) GC levels of codon positions of individual genes from vertebrates are plotted against the GC levels of the corresponding exons; (●) average values for third position GC of genes belonging to the same compartment of a given genome are plotted against the GC levels of coding sequences of the genome compartment. Correlation coefficients were 0.81, 0.67 and 0.88, respectively. (Dash-and-point line) the third position plot against GC of genome compartments.

Table I. Amino acid exchanges observed in thermophiles and the accompanying GC changes in their codons.[a]

Exchanges			Codon
Mesophiles		Thermophiles	GC
*Gly	→	Ala	0
*Ser	→	Ala	+
Ser	→	Thr	0
Lys	→	Arg	+
Asp	→	Glu	±
Ser	→	Gly	+
*Lys	→	Ala	+

[a] The observed exchanges are given in order of decreasing frequency, asterisked changes corresponding to the largest expected increase in stability [11]. Only the Lys → Ala exchange requires more than one base change per codon. The right column indicates increases (+), decreases (−), and no changes (0) in codon GC levels.

more stable, also increase the thermodynamical stability of the encoded proteins. The same changes obviously also lead to higher GC levels in mRNAs, a factor known [12] to increase their base pairing and stability.

5. Compositional Constraints are Due to Environmental Pressures

In the genome of warm-blooded vertebrates, different compositional constraints are associated with different genome compartments. One way to understand the origin of compositional constraints is, therefore, to investigate the causes for the formation of the GC-rich compartments of warm-blooded vertebrates; (as already mentioned, these compartments do not exist or are poorly represented in cold-blooded vertebrates). We know that the formation of GC-rich isochores is due to regional increases in GC and we attributed such increases to the requirements of chromosome structure and function at the temperatures prevailing in warm-blooded vertebrates [8]. This suggestion has now been tested (Giacomo Bernardi and Giorgio Bernardi, paper in preparation) by comparing the genomes of fishes trapped by geological events in hot springs, streams or lakes, with those of closely related species living in colder environments. The species analyzed so far comprise several *Cyprinodontidae* from the Death Valley basin, California, using the related *Fundulus heteroclitus* as a reference species, and *Tilapia grahami* from Lake Magadi, Kenya, using three closely related *Tilapia* species as 'controls'. In both series of comparisons, the genomes of fishes living at 37°−40° showed GC-rich components that were absent in the reference species living at 20°−25° (Figure 2).

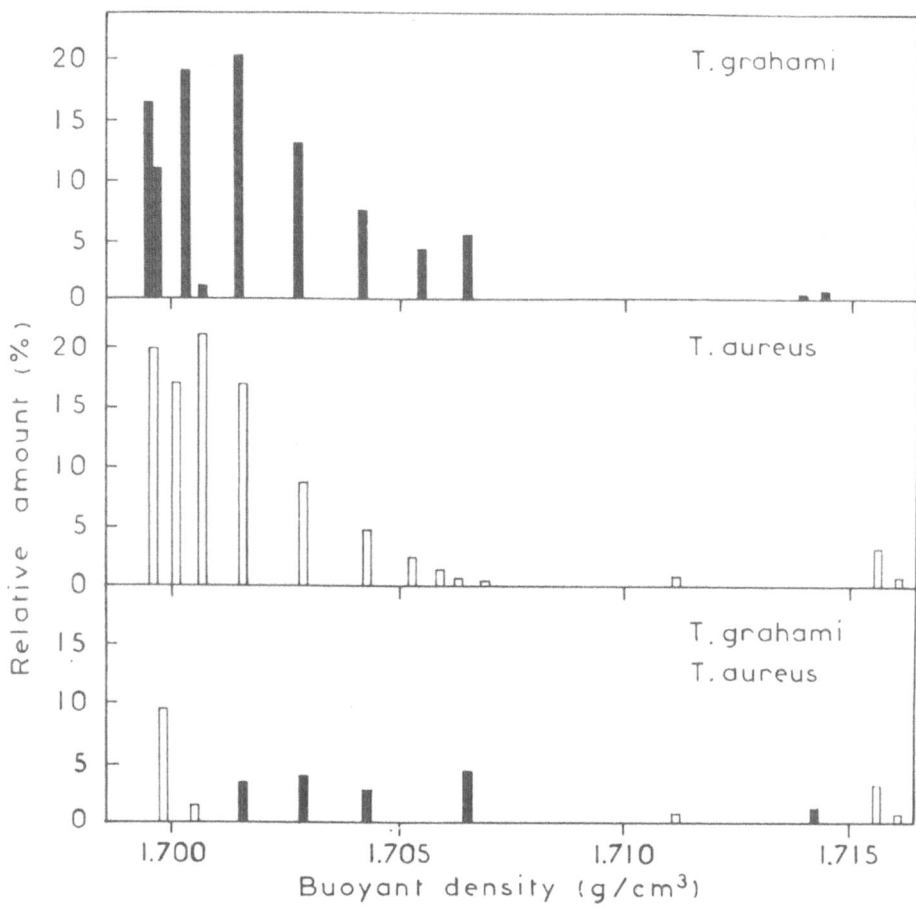

Fig. 2. Histograms showing the relative amounts and buoyant densities in CsCl of DNA fractions obtained by preparative Cs_2SO_4/bis(acetatomercury)dioxane density gradient centrifugation from *Tilapia grahami* and *Tilapia aureus*. The bottom panel shows the difference histogram.

These findings provide precise examples in which an environmental factor, temperature, appears to be responsible for novel compositional constraints in the genome. (The extremely high rates and the underlying molecular mechanisms of such changes and their relevance for the problems of the constancy of mutation rate and gradualism will be discussed elsewhere).

The results presented so far have a direct bearing on two important issues in molecular evolution, namely codon usage and the fixation of mutations.

6. Codon Usage is Largely Determined by Compositional Constraints

Since non-randomness of codon usage was first discovered, several, not mutually exclusive, explanations were provided for this phenomenon. These comprise: (i) the optimization of codon-anticodon interaction energy [13] and the consequent optimization of translation efficiency in highly expressed genes [14, 15]; (ii) the fulfillment of requirements for mRNA secondary structure and stability [12]; (iii) an adaptation of codons to the actual populations of isoaccepting tRNAs [9, 16].

These explanations essentially rested on intraspecific differences in the usage of all synonymous codons. In contrast, our results concern interspecific and inter-compartmental differences in the usage of synonymous codons characterized by different GC levels in third positions; this sub-set of codons corresponds to 2/3 of all synonymous codons. Our results lead to the following conclusions.

(i) Interspecific and intercompartmental differences in codon usage largely depend upon the compositional constraints affecting the genome, or the genome compartments. This provides, to a large extent, a rationale for the 'genome strategy of codon usage' [17] which comprises several 'compartmental strategies' in compartmentalized genomes. It should be noted that a dependence of codon usage upon genome GC was already noticed in some cases of unusual GC content [18, 19].

(ii) The proposals that mRNA structure [12] or the abundance of synonymous tRNAs [9] are the causes and not the effects of codon usage should be reversed. Since third codon positions are under essentially the same compositional constraints as non-coding sequences (Figure 1), the primary phenomenon is at the DNA level and the effects are at the mRNA or tRNA levels. The latter point was already well demonstrated by the changes in tRNA distributions that occur in the silk gland of *B. mori* in connection with the expression of the fibroin and sericin genes [20]. As already mentioned, our results do not bear on the intraspecific differences in codon usage which have been shown in unicellular organisms, like *E. coli* and *S. cerevisiae* [9].

7. Mutations are Mainly Fixed through Natural Selection

The isochore pattern of mammals is very similar, in that roughly the relative amounts of different isochore classes are quite close in different species. Unless one attributes this situation to convergent evolution, this should be seen as the result of the common descent of mammals from a warm-blooded tetrapod endowed with a genome similar to those of present day mammals. Since this common ancestor dates back to over 100 million years ago, one has to reach the conclusion that the base changes which occurred since, preferentially kept low and high GC levels in different codon positions of genes located in GC-poor and GC-rich compartments, respectively. In other words, on the average, changes were conservative as far as base composition is concerned. By analogy with the 'genome strategy of codon usage' [17] one should, therefore, take into consideration a more general 'compositional strategy of coding sequences', which also concerns non-silent changes. This strategy comprises several compartmental strategies in compartmentalized genomes. (b) Changes in non-coding sequences of eukaryotes conform to the same general rules as changes in coding sequences. In eukaryotes, the 'compositional strategy of coding sequences' is therefore part of a 'general compositional strategy', that also affects non-coding sequences. Again this strategy may consist of several compartmental strategies. (c) The CpG level (and the level of potential methylation sites) in both coding and non-coding sequences of vertebrates also appears to be subject to the same compositional constraints as the base changes just discussed; indeed, the CpG shortage is different in different genome compartments and is correlated with their GC levels.

To sum up, intragenomic GC changes clearly indicate that most mutations are fixed, in both coding and non-coding sequences, not at random, but under the influence of compositional constraints, in compliance with a 'general compositional strategy' involving, in all likelihood, both negative and positive selection. Random fixation of neutral mutations [4—7] certainly also occurs, but only to an extent such that the 'general compositional strategy' and the relationships of Figure 1 are not blurred.

As far as intergenomic changes are concerned, two points should be made. (a) GC changes in the three codon positions do not proceed in parallel: second position changes lag behind first position changes which, in turn, lag behind third position changes. Such decreasing extents of change appear to be correlated with the corresponding increasing impacts on amino acid composition of proteins. In other words, the different slopes of Figure 1 are correlated with the different fixation rates that have been detected in different codon positions of a number of genes [4] and indicate the existence of constraints other than the compositional ones. The higher slope of second position GC of viral genomes is likely to reflect lower amino acid constraints in viral proteins. (b) A clear directionality is shown by the amino acid substitutions, the silent base changes, the changes in non-coding sequences and the CpG changes which accompanied the transition from cold-blooded to warm-blooded vertebrates, to lead [8] to the formation of GC-rich genes and GC-rich isochores in the genomes of the latter. These directional changes can only be explained by a positive Darwinian selection acting on mutations that confer selective advantages in relationship with environmental pressures. These advantages have been identified as far as the transition from cold-blooded to warm-blooded vertebrates is concerned. Indeed, silent changes led to an optimization of structure and function at the level of both DNA and RNA, non-silent changes leading, in addition, to an optimization of structure and function at the protein level.

Obviously, our conclusions reverse the proposals of the 'neutral-mutation-random-drift hypothesis' (i) 'that the great majority of evolutionary changes at the molecular level are caused not by Darwinian selection acting on advantageous mutations, but by random fixation of selectively neutral or nearly neutral mutants' and (ii) 'that only a minute fraction of DNA changes are adaptive in nature' [6]. Both proposals rest, in fact, on the classical concept that the phenotype of an organism only corresponds to its 'gene products'; as a logical consequence, 'silent' mutations and changes in non-coding sequences were visualized as having no evolutionary impact. Moreover, random fixation of mutations was perfectly compatible with the limited sequence data analyzed at the time it was proposed.

8. Conclusions

In conclusion, compositional constraints indicate that most mutations are not fixed at random, but in relationship to a 'general compositional strategy' of the genome. This appears to be largely the result of natural selection including positive selection of mutations, that are advantageous as far as environmental pressures are concerned. Neutral mutations no doubt also exist, but their fixation occurs at a level low enough so as not to distort the 'general compositional strategy'. These conclusions lead to two general ideas: (i) that genome evolution depends more on natural slection than on random events; and (ii) that the environment can mould the genome through selection. The latter point has been illustrated here by the effects of temperature on the composition and on the compartmentalization of the vertebrate genome; other environmental factors certainly also play a role and may affect not base composition, but the frequencies of di- and oligo-nucleotides (for instance, ultraviolet light affects the level of pyrimidine doublets in bacterial genomes). Indeed, compositional constraints should in fact be visualized as a sub-set of the 'sequence constraints' acting on the genome [21] and influencing DNA structure [22].

In eukaryotes, both coding and non-coding sequences appear to be under essentially the same compositional constraints, and therefore under the same selection pressures. This finding stresses, first of all, the fundamental unity of the genome, already suggested by the genome strategy of codon usage [17], and contradicts what has been called [23] the 'bean bag' view of the genes within the genome. Second, it confirms the idea [17] that the genome is the unit upon which natural selection acts. Third, it does not support the view that non-coding sequences can be equated with functionless 'junk DNA' [24]. In contrast, it rather suggests that non-coding sequences do play a physiological role, which may have to do with the modulation of basic genome functions. This suggestion, although not a new one [25—28] does not rest anymore on 'adaptive stories', which can be rightly criticized [29—31], but on the newly demonstrated compositional constraints. Interestingly, the same conclusions have been reached on the basis of different evidence for the non-coding sequences of the mitochondrial genome of yeast [32—37].

Finally, compositional constraints identify a new component in the organismal phenotype, which may be called the 'genome phenotype'. Indeed, compositional constraints largely affect the structure and stability of the genome (at its different DNA, chromatin, and chromosome levels), of the transcripts and even of proteins (as exemplified by the stability changes accompanying GC increases in the genome), as well as codon usage. At the same time, they also conceivably touch on a number of basic functions, such as replication, recombination, transcription and translation, that are sensitive to the compositional/structural features just mentioned. This component adds on the other classical component of the phenotype, which is formed by the 'gene products', and is defined by non-silent mutations in the genes and by mutations in regulatory signals.

Acknowledgements

The senior author thanks the Fogarty International Center for Advanced Study in the Health Sciences, National Institutes of Health, Bethesda, 20205, U.S.A. for a scholarship during which this work was initiated. Sequence data treatments were performed using computer facilities at CITI2 in Paris on a PDP8 computer, with the help of the French Ministère de la Recherche et la Technologie (Programme Mobilisateur 'Essor des Biotechnologies').

References

1. G. Bernardi and G. Bernardi: *J. Mol. Evol.* **24**, 1—11 (1986).
2. J. Salinas, M. Zerial, J. Filipski, and G. Bernardi: *Eur. J. Biochem.* **160**, 469—478 (1986).
3. M. Zerial, J. Salinas, J. Filipski, and G. Bernardi: *Eur. J. Biochem.* **160**, 479—485 (1986).
4. M. Kimura: *Nature* **217**, 624—626 (1968).
5. M. Kimura: *The Neutral Theory of Molecular Evolution.* Cambridge University Press, Cambridge, England (1983).
6. M. Kimura: *Phil. Trans. R. Soc. Lond. (Biol.)* **312**, 343—354 (1986).
7. J. L. King and T. H. Jukes: *Science* **164**, 788—798 (1969).
8. G. Bernardi, B. Olofsson, J. Filipski, M. Zerial, J. Salinas, G. Cuny, M. Meunier-Rotival, and F. Rodier: *Science* **228**, 953—958 (1985).
9. T. Ikemura: *Mol. Biol. Evol.* **2**, 13—34 (1985).
10. S. I. Aota and T. Ikemura: *Nuclei Acids Res.* **14**, 6345—6355 (1986).
11. P. Argos, M. G. Rossmann, U. M. Grau, A. Zuber, G. Franck, and J. D. Tratschin: *Biochemistry* **18**, 5698—5703 (1979).
12. M. Hasegawa, T. Yasumaga, and T. Miyata: *Nucleic Acids Res.* **7**, 2073—2079 (1979).
13. H. Grosjean, D. Sankoff, W. Min Jou, W. Fiers, and R. J. Cedergren: *J. Mol. Evol.* **12**, 113—119 (1978).
14. R. Grantham, C. Gautier, M. Gouy, M. Jacobzone, and R. Mercier: *Nucleic Acids Res.* **9**, r43—r74 (1981).
15. J. L. Bennetzen and B. D. Hall: *J. Biol. Chem.* **257**, 3026—3031 (1982).
16. L. E. Post, G. D. Strycharz, M. Nomura, H. Lewis, and P. P. Dennis: *Proc. Natl. Acad. Sci. USA* **76**, 1697 (1979).
17. R. Grantham, C. Gautier, M. Gouy, R. Mercier, and A. Paré: *Nucleic Acids Res.* **8**, r49—r62 (1980).
18. B. P. Nichols, M. Blumenberg, and C. Yanofsky: *Nucleic Acids Res.* **9**, 1743—1755 (1981).
19. Y. Kagawa, H. Nojima, N. Nukima, M. Ishizuka, T. Nakajima, T. Yasuhara, T. Tanaka, and T. Oshima: *J. Biol. Chem.* **259**, 2956—2960 (1984).
20. A. Chevallier and D. R. Garel: *Biochimie* **61**, 245—262 (1979).
21. G. Bernardi, S. D. Ehrlich, and J. P. Thiéry: *Nature* **246**, 36—40 (1973).
22. A. Wada and A. Suyama: *Mol. Biol.* **47**, 113—157 (1986).
23. E. Mayr: *Evolution and the Diversity of Life.* Harvard University Press, Cambridge, Massachusetts (1976).
24. S. Ohno: *J. Hum. Evol.* **1**, 651—662 (1972).

25. R. J. Britten and E. H. Davidson: *Science* **165**, 349—357 (1969).
26. E. Zuckerkandl: *J. Mol. Evol.* **7**, 269—311 (1976).
27. E. Zuckerkandl: *J. Mol. Evol.* **24**, 12—27 (1986).
28. E. H. Davidson and R. J. Britten: *Science* **204**, 1052—1059 (1979).
29. S. J. Gould and R. C. Lewontin: *Proc. R. Soc. Lond.* (*Biol.*) **205**, 581—598 (1979).
30. W. F. Doolittle and C. Sapienza: *Nature* **284**, 601—603 (1980).
31. L. E. Orgel and F. H. C. Crick: *Nature* **284**, 604—607 (1980).
32. G. Bernardi: in: G. Attardi, P. Borst, and P. P. Slonimski (eds.): *Mitochondrial Genes.* Cold Spring Harbor Laboratory, Cold Spring Harbor, New York, pp. 269—278 (1982).
33. G. Bernardi: *Folia Biol.* **29**, 82—92 (1983).
34. M. de Zamaroczy and G. Bernardi: *Gene* **37**, 1—17 (1985).
35. M. de Zamaroczy and G. Bernardi: *Gene* **41**, 1—22 (1986a).
36. M. de Zamaroczy and G. Bernardi: *Gene* **47**, 155—177 (1986b).
37. M. de Zamaroczy and G. Bernardi: *Gene* **54**, 1—22 (1987).

New Insights into the Evolution of Proteins

ROGER ACHER
Laboratory of Biological Chemistry, University of Paris VI, 96, Boulevard Raspail, 75006 Paris, France.

1. The Gene Determinism of the Protein Morphology

Despite the fact that the name 'proteins' means primordial molecules, they have apparently been supplanted as directors of life by genes, the nucleic acids being the vectors of heredity. Any change to become hereditary must, by some means or other, enter the genome, in other words must modify the long nucleotide sequence determining the developing program. Information is inscribed in genes as a one-dimensional reading system, and the deoxyribonucleic acids that constitute the genes are therefore long linear molecules, like cassette tape ribbons. But the living order is three-dimensional and the translation or transferance from a nucleotide sequence (nucleic acid) to an amino acid sequence (protein) is in fact a passage from one-dimensional to three-dimensional system, and therefore essentially a morphogenic mechanism. In contrast to nucleotide sequences, amino acid sequences spontaneously generate specific shapes depending upon the order of amino acids.

1.1. THE EXON—DOMAIN RELATIONSHIP

A single chain can build several conformationally distinct regions or domains. These latter are functionally both autonomous and cooperative such as the organs of a body.

The composite construction of proteins raises the problem of a possible composite evolution with relatively independent evolutions for the constituent domains. In a selective conception of evolution, each domain must co-evolve with a ligand molecule having itself a complementary conformation so that specific interactions can occur.

If we turn now to the manufacture of the proteins, it has been shown that the directing gene, the so-called 'structural gene' is, in eukaryotes, also composite since only some interspersed pieces of the long nucleotide sequence, the exons, are finally translated into amino acid sequence.

The old dogma 'one gene — one protein' could be changed into 'one exon — one domain'. However, data accumulated up to now do not show an exact coincidence between exons and domains, a given domain being sometimes encoded by several exons. Thus, on the biosynthetic point of view, a protein

Jean Maruani (ed.), Molecules in Physics, Chemistry, and Biology, Vol. IV, 13—30.
© 1989 *by Kluwer Academic Publishers.*

resembles a Bunraku Japanese puppet with several manipulators for distinct parts of the body; the coordination between them needs a regulator. Quantitative regulatory mechanisms are involved in gene expression through their regulatory sequences and interacting molecules; on the other hand qualitative regulatory mechanisms can be implied at the level of the ribonucleic acid messenger through alternative splicing, a tissue-specific choice being made among the exons.

1.2. AMPLIFICATION OF THE MORPHOGENIC MESSAGE

Another essential feature of the gene—protein relationship is amplification of the morphogenic message, one gene giving birth to many protein copies. Because of that about half of the dry weight of an organism is made by proteins, the most abundant component of the living matter. The number of protein copies for each structural gene varies following cell differentiation. Assemblies of proteins are made by specific auto-associations and a particular cytoskeleton is built in each type of cell. So not only the shape of each protein is determined but also the number of molecules for each. The architecture program must involve a coordination of many genes but it is difficult at the present time to suggest a mechanism. The polypeptide growth factors play a key role, especially during the development, and steroid hormones are also important gene-activating molecules.

2. The Composite Organization of Proteins

Proteins are long chains of amino acids often fold as strings of globular domains, each having its own function (specific interactions) and keeping its native conformation when excised from the chain. Because very similar domains have been recognized in proteins differing in other parts of their structure, the possibility exists that domains have independently evolved as modules, lately fused by an exon-shuffling [1].

2.1. DIVERSITY OF THE DOMAINS IN A PROTEIN

Several types of domain conformations, deduced from amino acid sequences, have been identified in proteases of blood coagulation and fibrinolysis such as so-called vitamin K-dependent calcium binding module, kringle module, growth factor module and finger module [2] (Figure 1). A given individual module may be repeated in series in the polypeptide chain, for example the kringle module repeated five times in plasminogen [3] (Figure 1) or the globular module also repeated five times in the heavy chain of the immunoglobulin M.

The evolutionary building of a protein therefore involves at the genome level both duplication and fusion mechanisms in order to lead, by several steps, to a structurally and functionally composite molecule. The cooperativity between the domains is generally not clearly understood.

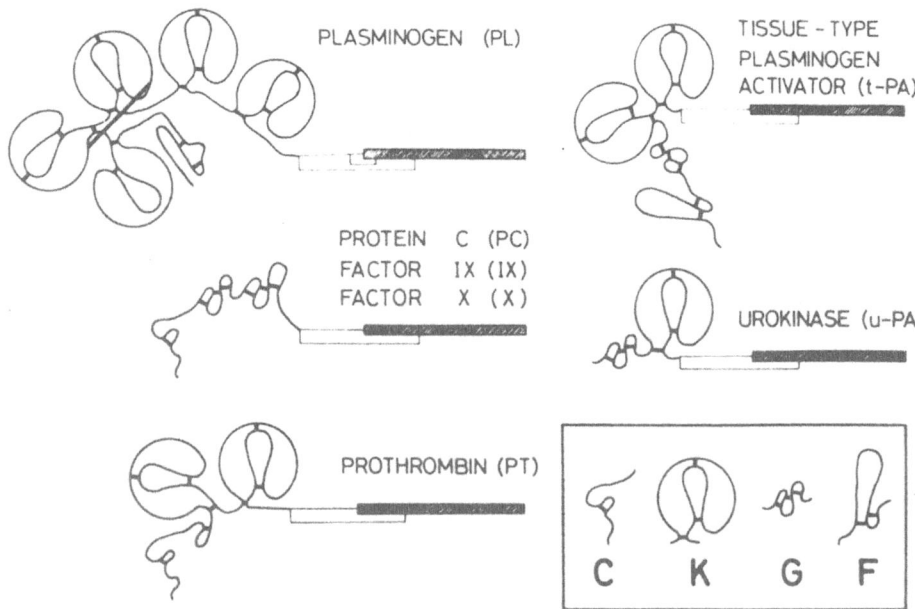

Fig. 1. Structures of plasminogen, Factor IX, Factor X, prothrombin, urokinase, and tissue-type plasminogen activator. The crosshatched bars represent the protease regions homologous to trypsin. The inset shows the different modules of the non protease regions. (C) vitamin K-dependent calcium-binding module; (K) kringle module; (G) growth factor module, and (F) finger module (taken from [2]).

In the case of regulatory serine proteases that act through activation cascades, various 'anchorage' domains allowing multiple ligand bindings on one hand and a catalytic domain on the other are found and cooperation is necessary to reach the full biological specificity. The catalytic domain may exist as free molecules such as the small proteases of digestion. Here, in contrast to the single precisely located cleavage involved in the successive activations of regulatory proteases, neither anchorage nor limited proteolysis are required.

Prokaryotic trypsin (*Streptomyces griseus*), invertebrate trypsin (crayfish) and bovine trypsin show approximately 40 percent identity and a homology of 41—48 percent is found when compared with the respective catalytic regions of human factors IX, X and prothrombin involved in blood-clotting [4]. This could suggest that the catalytic domain was shaped before the prokaryotes—eukaryotes divergence and used later for building the trypsin-like serine proteases through exon duplication and transposition.

2.2. MULTI-DOMAIN PROTEINS AS MULTI-PRECURSORS: THE POLYPROTEINS

Identification of hormone precursor proteins or pro-hormones has shown that virtually all the hormonal or neurotransmitter small peptides are manufactured as longer proteins that are processed into several pieces by maturation enzymes. The fragments may be structurally different and display distinct biological functions or may be nearly identical and have the same biological activity.

2.2.1. *The Precursors of Neurohypophysial Hormones and Neurophysins*

The neurohypophysis of vertebrates secretes in the blood peptide neurohormones whose biological actions have been known since the beginning of the century [5]. Characterization of common precursors for the 9-residue neurohypophysial hormones and the 93/95-residue neurophysins (Figure 2) either directly [6, 7] or deduced from complementary deoxyribonucleic acids [8, 9] has raised the question of the evolutionary relationships between the different domains of the precursors. A stoichiometric and reversible complex between neurohypophysial hormones and neurophysins has been isolated from the mammalian neurohypophysis suggesting a biological significance of the association [10]. On the other hand neurohormones and neurophysins have been identified in isolated neurohypophysial granules indicating that processing of the precursors occurs early in the hypothalamic cell bodies and that processing enzymes are likely also components of the granules.

Isolation and characterization of neurohypophysial hormones have been carried out from species belonging to all vertebrate classes except cyclostomes. Two general conclusions can be drawn, on one hand that, each species having two peptides, an oxytocic one and a pressor one, two evolutionary molecular lineages could be traced in vertebrates, on the other hand that, all peptides processing the same nonapeptide pattern with five invariant residues, they should be encoded by similar genes deriving by duplication from a common ancestral gene.

Four groups can be distinguished: mammals having oxytocin and vasopressin, non-mammalian tetrapods having mesotocin and vasotocin, bony fishes with isotocin and vasotocin and cartilaginous fishes having glumitocin (rays) or valitocin and aspargtocin (sharks) and vasotocin [11] (Figure 3). The lineage isotocin-mesotocin-oxytocin involves two amino acid substitutions in two steps; assuming a single gene, two point mutations are necessary for changing in the first step serine (Ser) into glutamine (Gln) because their codons differ by two nucleotides. In contrast, the other substitutions need only one point mutation.

It has initially been assumed that the change from mesotocin to oxytocin was associated to the transition reptile—mammal and related to the appearance of lactation. However Australian marsupials have mesotocin and not oxytocin whereas the North-American opossum possesses both mesotocin and oxytocin [12]. It is possible that a duplication of the mesotocin gene occurred during the non-mammalian—mammalian transition and one of the copies mutated into oxytocin gene. The second gene copy might have become silent or disappeared in placental mammals.

The vasopressin lineage is simpler because all non-mammalian vertebrates have vasotocin ([Ile^3]-vasopressin), whereas mammals have no vasotocin and at least a vasopressin. The essential change of isoleucin (Ile) to phenylalanine (Phe) in

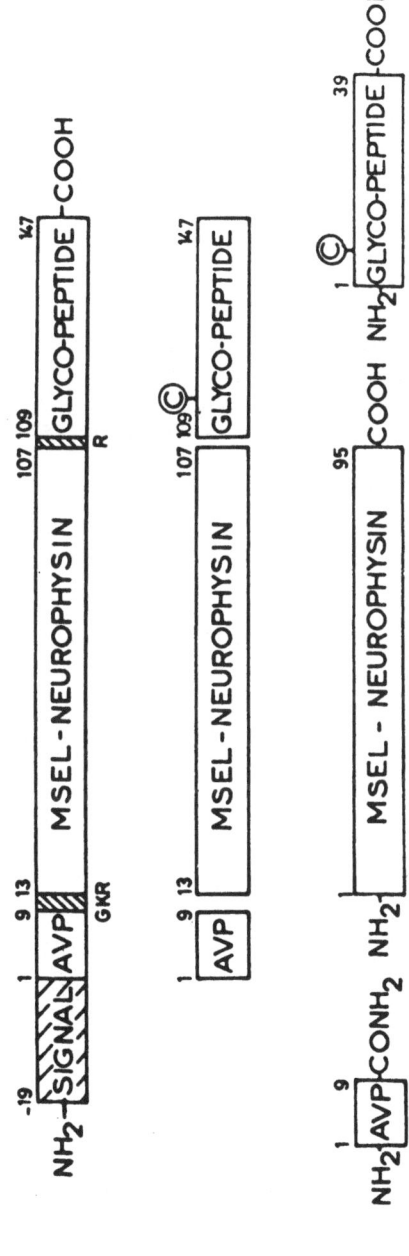

Fig. 2. The bovine vasopressin multiprecursor and its processing. Signal, signal peptide; AVP, arginine vasopressin; G, glycine; K, lysine; R, arginine. The common precursor of vasopressin, MSEL-neurophysin and copeptin (a glycopeptide) is split into three fragments by two types of enzymatic cleavages, one involving a glycyl-lysyl-arginine sequence and leaving a carboxamide group at the C-terminal end of vasopressin, the other involving a single arginine residue and separating MSEL-neurophysin from the glycopeptide (redrawn from [7]).

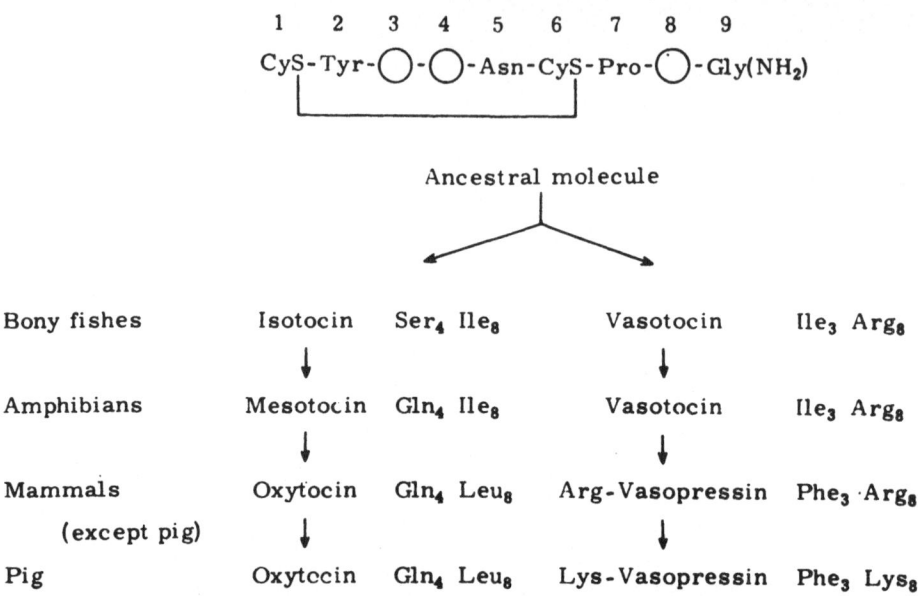

Fig. 3. Hypothetical scheme of the evolution of neurohypophysial hormones. One-gene duplication and a series of subsequent single substitutions in positions 3, 4 or 8 produce two molecular lines. The substituted amino acids and their positions in a hormone are listed to the right of each hormone.

position 3 (Figure 3) needs a single nucleotide mutation. Again marsupials were original: if current vasopressin was found in three Australian families, in the fourth, Macropodidae (which includes kangaroos and wallabies), two pressor peptides, lysipressin ([Lys⁸]-vasopressin) and phenypressin ([Phe²]-vasopressin) have been found in each species. In North- and South-American marsupials, both lysipressin and vasopressin have been identified in each species. It seems therefore that the initial vasopressin gene duplicated early in the marsupial stem and both copies remained often expressed, sometimes under mutated forms. Figure 4 recapitulates biochemical and paleontological data on vertebrate neurohypophysial hormone evolution.

Neurophysins are small proteins (93/95 residues) which constitute the excised second domain of the common precursors (Figure 2). The self-association between neurohypophysial hormones and neurophysins suggests that the two domains should have precise interactions within the precursor itself. Two types of neurophysins, differing essentially by the N- and C-terminal sequences, have been characterized in each mammal investigated. They are distinguished by residues in positions 2, 3, 6 and 7 so that there is a MSEL-neurophysin associated with vasopressin and a VLDV-neurophysin associated with oxytocin. The peculiar feature is the near-identity between the central parts (residues 10 to 76) of the two neurophysins of a given species and the nearly-invariance of this region from species to the other [11, 13]. It is known that oxytocin and vasopressin genes have

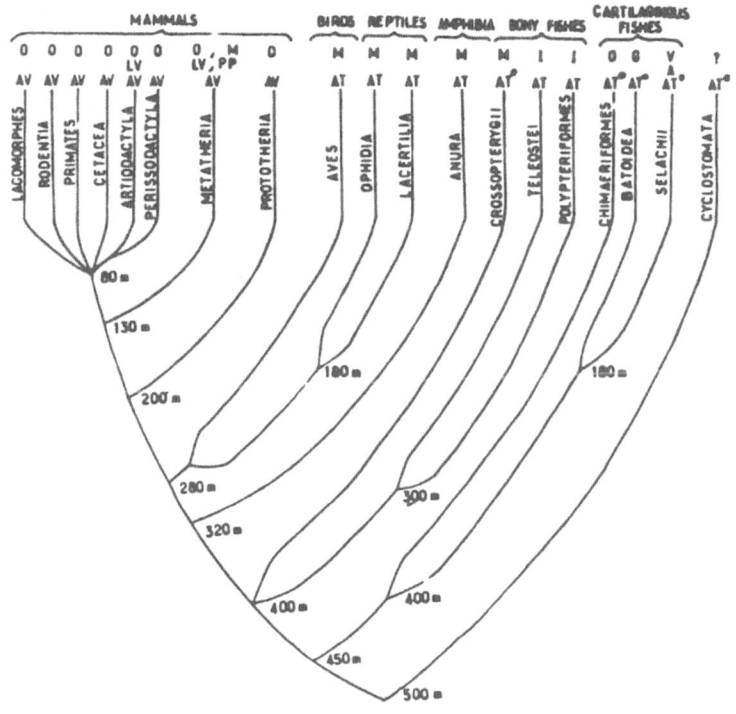

EVOLUTION OF NEUROHYPOPHYSEAL HORMONES

Fig. 4. The two main lineages of neurohypophysial hormones traced in vertebrates. Letters indicate hormones identified in modern representative of groups: AV, arginine vasopressin; LV, lysine vasopressin; PP, phenypressin; AT, vasotocin; O, oxytocin; M, mesotocin; I, isotocin; A, aspargtocin; V, valitocin; G, glumitocin. Numbers give time in millions of years since the divergence.

a similar three-exon organization [14—16] (Figure 5). The two human genes are on the same chromosome linked with 12 kilobases intervening and transcribed from opposite DNA strands [16]. The second exon B of each gene encodes for the neurophysin central part and a recent gene conversion (in gene conversion, gene I interacts with gene II in such a way that the nucleotide sequence of part or all of gene I becomes identical to that of gene II [17]) has been proposed for explaining the nearly-identity of exon B nucleotide sequences in the two genes. This 'similarizing' mechanism, which seems random, would act against the divergence mechanism occurring by random point mutations in the two copies after a gene duplication and could lead to a miscalculation of the time elapsed since duplication when this time is estimated by the number of nucleotide or amino acid substitutions.

2.2.2. *The Precursor of the Epidermal Growth Factor and EGF-Like Peptides*

Epidermal growth factor (EGF) is a 53-amino acid polypeptide that has been isolated from the submaxillary gland of the male mouse and from human urine.

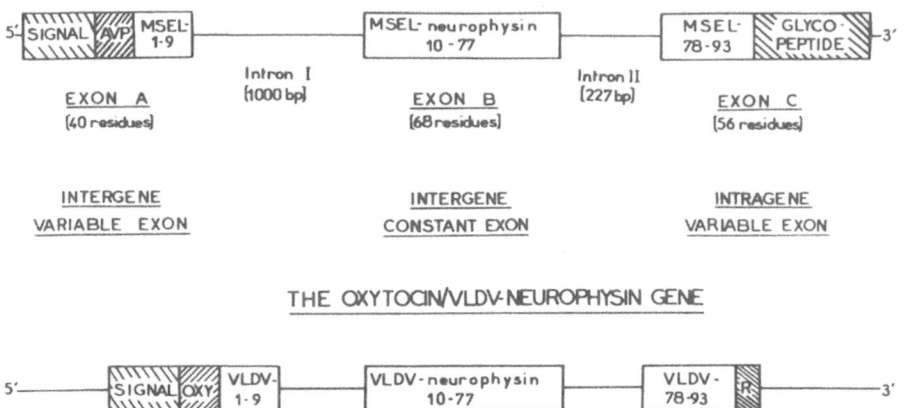

Fig. 5. Rat vasopressin and oxytocin genes and their exons (boxes). AVP or OXY, nucleotide sequences encoding arginine vasopressin or oxytocin, respectively. MSEL- or VLDV-1—9, 10—77, 78—93, nucleotide sequences encoding the corresponding residues of MSEL- or VLDV-neurophysins, respectively. Glycopeptide, nucleotide sequences encoding a 39-residue glycopeptide. R: codon for an arginine residue (redrawn from [12]).

It stimulates the proliferation and differentiation of cells of ectodermal and mesodermal origin. As EGF exerts a number of effects on prenatal and neonatal tissue growth, including precocious eye-opening and incisor eruption, it may play a role in early development. Moreover, because EGF receptors are present in various adult tissues, EGF is presumably involved in the control of growth and function of cells throughout the life.

The amino acid sequence of EGF precursor has been predicted from the nucleotide sequence of deoxyribonucleic acid complementary to the messenger ribonucleic acid. PreproEGF is a protein of 1217 amino acids. The EGF moiety is flanked by polypeptide segments of 976 and 188 amino acids at its amino and carboxy termini, respectively. The amino terminal segment of the precursor contains seven peptides with sequences that are similar but not identical to EGF [18]. The release of EGF from the precursor requires proteolytic processing at both ends of the molecule, in other words excision. Because EGF can be isolated in association with an arginine-specific peptidase, the EGF-binding protein, and has a carboxyl terminal arginine, it has been suggested that the associated peptidase might be involved in processing of the precursor. This concept is supported by the sequence since there is an arginine preceding the first residue of EGF. The boundaries of each EGF-like peptide can be defined by a basic amino acid and thus EGF-like peptides could be released by the same peptidase [18]. This proEGF may be processed to yield a number of different peptides. Six of these EGF-like sequences are tandemly arranged in two groups of three members each. It can be assumed that successive duplications of an EGF-nucleotide sequence occurred within the primordial gene and therefore generated a poly-precursor protein. The seven 'copies' might have changed in number and sequence of amino acids through mutations in the course of evolution. The possibility exists that each peptide has a distinct biological role and a peculiar receptor.

2.3. COOPERATIVITY BETWEEN DOMAINS OF A PROTEIN: THE ENZYME-RECEPTORS

Intercellular messengers such as hormones, neurotransmitters, growth factors, etc. are recognized and bound by target cells. These cells possess specific receptors inserted in their plasma membrane. Receptors may be built with several proteins including a receptor sub-unit that links the messenger and undergoes a conforma-tional change, a transducer sub-unit that transmits the conformational change to a third sub-unit, an enzyme, which in turn is activated. Receptors can also be enzymes themselves directly actived by the messenger. In this case the protein has usually an extracellular ligand binding domain, a transmembrane domain and a cytoplasmic enzyme domain. Among receptors belonging to this type are the insulin receptor and the EGF receptor.

The insulin receptor is a dimeric glycoprotein comprising two identical parts bound by disulfide bridges. Each part includes two chains α and β, bound by disulfide bridges and resulting from cleavage of a single polypeptide chain so that a single gene controls the whole structure. The α-chain (719 residues) builds the extracellular insulin binding domain and contains fifteen consensus sequences for asparagine-linked glycosylation and an unusually large number [37] of cysteine residues.

The β-chain (620 residues) contains three domains: an extracellular amino-terminal domain (194 residues), a transmembrane domain (23 residues) and a cytoplasmic domain (403 residues), which has a potential tyrosine kinase activity [19] (Figure 6).

A similar overall organization has been found for EGF receptor except that a single polypeptide chain is used for building all domains. Furthermore a striking sequence homology is found: the first 444 residues of the mature insulin receptor α-sub-unit are 27% homologous to the first 461 residues of the EGF receptor, 20% of these homologous residues being cysteines and part of the cystein-rich domains of both receptors (Figure 6). Both the insulin receptor and the EGF receptor span the membrane once with a hydrophobic 23-amino acid sequence. Both receptors have an ATP-binding site located precisely 49 residues from the membrane. Sequence homologies are also found in the tyrosine kinase domains, particularly in the sequence surrounding the major phosphorylated tyrosine of the EGF receptor [19].

The cytoplasmic enzymic domains of both receptors display homology with portions of the *src* family, oncogene virus proteins having tyrosine-specific protein

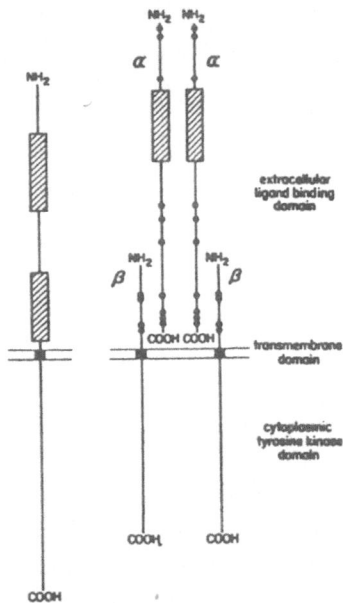

Fig. 6. Schematic comparison of insulin and EGF receptors. Regions of high Cys-residue concentration are shown in hatched boxes, transmembrane domains as black boxes and single cysteine residues, possibly involved in formation of the $\alpha_2-\beta_2$ insulin receptor complex, as black circles (taken from [19]).

kinase activity. The insulin receptor is related equally to each of the *src* family members, which suggests either that it differs from the EGF receptor and is not a cellular proto-oncogene, or that its oncogene counterpart has not been identified. At the gene level, on one hand transforming oncogene virus could have derived from receptor genes, on the other hand duplication of an ancestral receptor protogene could have led to a diversity of specific receptors involved in the cell-cell messages.

3. The Repetitive Propensity within Proteins

3.1. THE REPEAT UNITS OF PLASMODIUM SURFACE PROTEINS

Many large proteins appear built with repetitive sequences of amino acids. Immunity to *Plasmodium falciparum*, the protozoan causing the most severe form of human malaria, is acquired only after extensive exposure over a number of years. Many *P. falciparum* surface proteins are natural immunogens in man. The process of immune evasion might be explained by antigenic diversity among different strains but also by repetition of interspersed epitopes within each protein, causing collectively an overload on the immune system. Seven surface proteins of various *Plasmodium* species have been identified through the complementary deoxynucleic acid technology. All display a *falciparum* interspersed repeat antigen (FIRA), the length of the repeat ranging from six to twelve amino acids and the number of the repeats ranging from 12 to 100. A dominant immunogen protein of *P. falciparum* reveals a structural unit composed of 13 hexapeptide repeats flanked by a highly charged region containing both acidic and basic amino acids [20]. This structural unit is itself repeated, so that blocks of repeats and charged units are interspersed along the molecule [20]. The sequences within the repeats vary much more extensively than those in the charged units. The repetitive building of proteins expresses the duplicative propensity within the genes.

3.2. THE REPEAT 'KRINGLES' IN HUMAN PLASMINOGEN

Internal homology in proteases involved in blood clotting and fibrinolysis was first discovered in prothrombin, whose active form, thrombin, converts soluble fibrinogen into an insoluble fibrin aggregate, the clot. In two regions of 83 residues each, 31 positions have identical amino acid residues, the identities including three disulfide bridges. The bridges in both homologous areas connect half-cystines 1 to 6, 2 to 4 and 3 to 5, forming a 'Kringle-like' structure, referring to a classical shape of Scandinavian cake. The same type of structure has been found later in plasminogen, whose active form, plasmin, digests the fibrin clot. In plasminogen five successive repeat Kringles have been recognized [3] (Figure 1). When the five Kringles of plasminogen and the two Kringles of prothrombin are compared, 20 positions (25%) appear identical. The Kringle structure seems involved in anchorage function, that means specific association with other functional proteins. In particular the plasminogen Kringles are implied in fixation of the zymogen onto fibrin fibers that will be cleaved by plasmin after activation of plasminogen by a plasminogen activator such as urokinase.

Characterization of the prothrombin has revealed that the two prothrombin Kringles are encoded by separate exons [21] and therefore that the repeat units in the protein are determined by internal duplication within the gene. On the other hand the activation site of the zymogen is also encoded by a separate exon and the question of the evolutionary autonomy of the exons arises. The mechanism of exon duplication may not imply the maintenance of the copies in the parent gene and some kind of exon-shuffling might fuse several domains, each evolutionarily shaped for a given interaction [1].

3.3. THE REPEAT MODULES IN FIBRONECTIN

Fibronectin is a multi-domain glycoprotein with a molecular weight of about 450 000 daltons and consisting of two identical polypeptide chains held together by two disulfide bridges near the C-terminus. Fibronectin is a major protein in blood plasma. It is also present on the surface of many cell types and as insoluble multimers in the extracellular matrix of connective tissue. Fibronectin is involved in numerous functions such as cell adhesion, cell spreading and motility, would healing and the maintenance of normal cell morphology. These functions reflect the ability of fibronectin to bind a number of different biological molecules such as collagen, fibrin, heparin, hyaluronic acid, actin, gangliosides, complement component C_3 and C_{1q}, DNA and both vertebrate and bacterial cell surfaces [22]. Fibronectin contains discrete domains of rigid tertiary structure connected by polypeptide strands particularly susceptible to proteinases. About 93% (2100 residues) of the bovine plasma fibronectin monomer have been sequenced [22]. This fibronectin monomer consists of 12 modules of type I homology, two modules of type II homology and 15 modules of type III homology (Figure 7). The module type I (finger module) allows contacts with fibrin, actin and heparin, the module type II (growth factor module) determines interactions with collagen, and the module type III is involved in associations with DNA, cell surfaces and heparin [22] (Figures 1 and 7). The complete sequence of bovine plasma fibronectin (2265 residues) has now been determined [32].

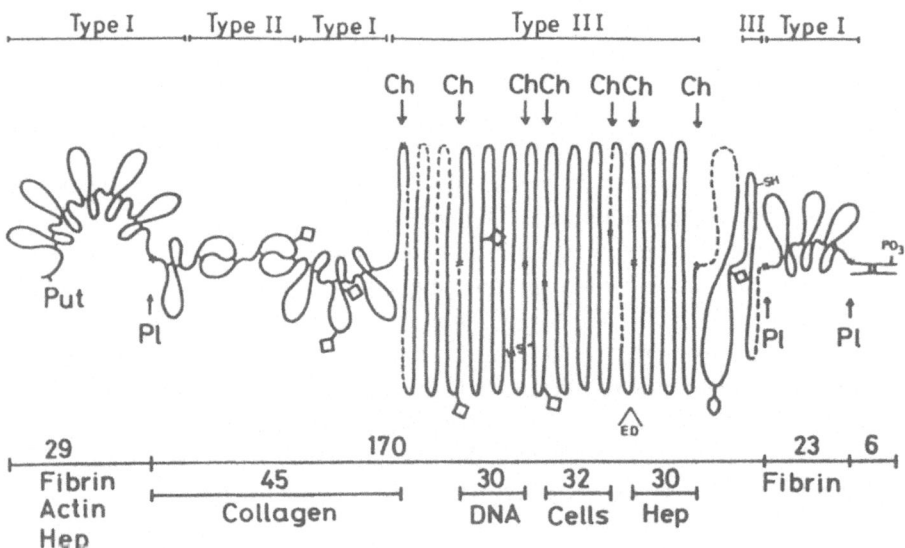

Fig. 7. Model of the bovine plasma fibronectin monomer. Solid lines indicated areas sequenced (approximately 2100 residues, about 93% of the total sequence). Dashed lines indicate areas not yet sequenced (approximately 160 residues). The four fragments indicated in the top line (29 kDa, 170-kDa, 23-kDa and 6-kDa) were obtained by digestion with plasmin (PL). Each of the halves of the 6-kDa fragment belongs to one of the two chains of fibronectin. The fibronectin monomer consists of 12 modules of type I homology, two modules of type II homology and 15 modules of type III homology. The locations of these are indicated at the top of the figure (taken from [22]).

Although there is a single gene coding for fibronectin, there are at least 12 different fibronectins as judged by the messenger ribonucleic acids. Differentiated cells synthesize different cell surface fibronectins and plasma fibronectin is also different from cellular fibronectins. The gene contains over 50 exons but some of these exons can be eliminated at the level of the primary transcript through an alternative splicing. For instance plasma fibronectin, which does not give multimers, is encoded by a messenger ribonucleic acid that has lost the exon coding for a polymerization amino acid sequence. It is assumed that the present-day single gene derived from three primordial minigenes corresponding to the three modules I, II and III, each one having duplicated one or several times on the course of evolution before to be fused. The copies would have evolved independently although keeping common features. Fibronectin, which is also present in the simplest invertebrates, appeared likely before the emergence of pluricellular organisms, an adhesion protein being necessary for joining up cells in ancestral metazoa.

4. Protein Polymorphism and the Multigene Family

For each biological activity, such as proteolytic activity, it has been observed that two or more sequence-related active proteins with the same specificity could often be identified in a given species. These functionally equivalent 'isoproteins' are usually encoded by separate sequence-related genes. The members of the so-called gene family can be regarded as deriving from a common ancestral gene by successive duplications and mutations. This appears likely when the genes are clustered on the same chromosome. The protein products of the multigene family can remain individual molecules acting independently or join up for making a functional multi-subunit assembly.

4.1. DEVELOPMENTALLY REGULATED MULTIGENE FAMILY: THE HEMOGLOBINS

One of the best studied multigene families is the set of genes that code for vertebrate oxygen-carriers, the hemoglobins. In addition to adult hemoglobins, mammals produce embryonic and in some case foetal specific globins. All of these hemoglobins are tetramers of two α-like and two β-like globin polypeptides and amino acid sequence analysis has shown that all of these globins share similar sequence homology and are therefore most likely the products of globin gene duplications that have occurred during evolution. A hierarchy of gene duplications starting with a primordial $\alpha-\beta$ duplication followed by expansion and diversification of both the α- and β-globin gene sub-families has been postulated [23, 24].

In man the α-globin gene cluster includes ζ_2, ζ_1, α_2, α_1 genes sequentially expressed during development in the order they occupy in the chromosome 16, and the β-globin gene cluster comprises ε, Gγ, Aγ, δ and β genes also successively expressed in the order they have in the chromosome 11. A number of tetramer combinations, each involving a pair of α-type chain and a pair of β-type chain, succeed one another in the early development from the embryonic $(\zeta\varepsilon)_2$ until the main adult $(\alpha\beta)_2$ hemoglobin. Independently to the subunit substitution determined perhaps by a functional adaptation, the important new feature that appeared early during vertebrate evolution is the building of contact amino acid sequences in each chain for giving the tetramer assembly [25]. Because functional homotetramers have not been found in present-day vertebrates, it is likely that these contact sequences were made up after the primordial duplication $\alpha-\beta$. Tetramer assembly is essential for the cooperative properties in the oxygen-binding displayed by vertebrate hemoglobins in contrast to invertebrate hemoglobins that are monomeric. Gene evolution built in nucleic acids, aside coding sequences, regulatory sequences such as promoters, enhancers, etc., controlling gene expression. Regulatory molecules operating on these sequences were also generated. The developmentally regulated multigene families constitute integrated systems themselves under control of co-ordinating genes. On the other hand evolutions of both the α- and β-globin gene sub-families must have been in some way co-ordinated despite they are located in separate chromosomes.

4.2. FUNCTIONAL ASSEMBLY OF ISOPROTEINS: THE ACETYLCHOLINE RECEPTOR

The proteins expressed by a gene family may associate themselves for making a functional assembly. The nicotinic acetylcholine receptor (AChR) is a post-synaptic membrane protein complex that alters the ionic permeability of the membrane as a consequence of binding acetylcholine. The AChR from the electroplax of the rays *Torpedo marmorata* or *Torpedo californica* is composed of five subunits present in molar stoichiometry of $\alpha_2\beta\gamma\delta$ and contains both the binding site for the neurotransmitter and the cation gating unit. The complete primary structures of the four polypeptides have been deduced from the nucleotide sequence of the cloned complementary deoxyribonucleic acids [26, 27]. Comparison of the amino acid sequences of the four subunits revealed marked homology among them (54% of the positions with three identical residues have a chemically similar residue to the fourth one). The close resemblance among the hydrophilicity profiles and predicted secondary structures suggests that these polypeptides are oriented in a pseudosymmetric fashion across the membrane. Each subunit contains four putative transmembrane hydrophobic segments of 19—27 residues that may be involved in the cation channel [27] (Figure 8). These segments, designated M_1, M_2, M_3 and M_4, are bounded by charge residues. It is assumed, by analogy with the transmembrane segments of bacteriorhodopsin, that the segments M_1-M_4 form α-helical structures. Studies on selective proteolysis on the AChR subunits have indicated that the amino-terminal two thirds are resistant to proteolysis and are glycosylated. The amino-terminal half of the mature subunits preceding the first transmembrane segment M_1 is most probably exposed on the extracellular surface of the membrane forming a rigid protein core due to secondary, tertiary and quaternary structures. The regions between segments M_3 and M_4 (about one quarter of the mature subunit) is located on the cytoplasmic side of the membrane (Figure 8). It is tempting to suppose that the transmembrane segments are involved in the cation channel of the receptor.

The four proteins of the AChR display self-association and the receptor can be reconstituted in a foreign cell. The cloned complementary deoxyribonuclei acids encoding the four subunits of the *Torpedo californica* AChR, each carried by a simian virus 40 vector, were introduced into COS monkey cells by transfection and the produced α-, β-, γ- and δ-subunit specific messenger ribonucleic acids were recovered, combined in the proper ratio and introduced into *Xenopus* oocytes by microinjection [28]. The injected oocytes, after incubation for 3 days, were electrophysiologically responsive to acetylcholine. All four subunits were required to elicit a normal response. The absence of one subunit may hinder the

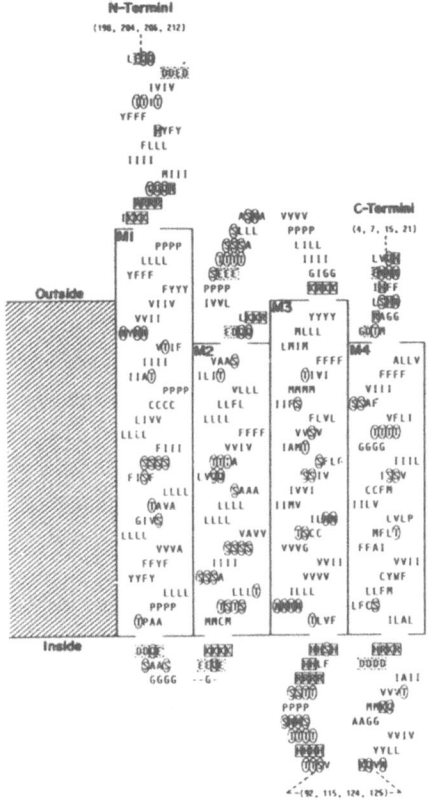

Fig. 8. Proposed orientation of the AChR subunit polypeptides across the membrane. The amino acid sequences of the four putative transmembrane segments (designated as M1—M4 in the direction from amino acid terminus to carboxyl terminus and boxed) and the adjacent regions are shown by the one-letter notation. Set of four letters indicate the residues in the aligned α-, β-, γ- and δ-subunits sequences, respectively, from left to right; gaps (−) have been removed, except at position 283, to permit ready identification of the side chains extending towards one side of an α-helix. Sets of four numbers in parentheses indicate the sum of residues in the mature α-, β-, γ- and δ-subunits sequences, respectively (from left to right), that either extend beyond the ends of the displayed sequences or are inserted between them (taken from [27]).

assembly of the remaining subunits or lead to the formation of an aberrant complex, thus reducing the stability of the constituent polypeptides [28].

Any subunit of the *Torpedo* AChR can be replaced by the corresponding subunit of the calf AChR to form functional channels [29]. These observations are consistent with the high degree of amino acid sequence homology between corresponding subunits of the *Torpedo* and calf AChRs. In contrast to the uniform conductance of the *Torpedo*, calf and hybrid AChR channels (~ 40 pS), their gating behaviour differ widely, the average duration of elementary currents of the calf AChR channel being more than 10-fold that of the *Torpedo* AChR channel. Substitution of the calf δ-subunit in the *Torpedo* AChR alters drastically the gating behaviour of the channel, making it similar to the calf AChR channel [29]. Apparently regions coding in the four genes for assembly amino acid sequences in the polypeptide chains have been preserved during evolution from cartilaginous fishes to mammals, the conductance of the channel remaining virtually the same. However an evolutionary modification in the δ-subunit has increased the duration of elementary currents. This current duration seems determined by at least three reaction steps: channel opening, channel closing, dissociation of the agonist, and the δ-subunit could determine the channel-closing step [29].

5. Conclusion

A very small part of the genomic deoxyribonucleic acid in higher eukaryotes is used for coding proteins. The function of the remaining genome, including introns or intervening sequences within genes, flanking sequences and gene spacers, is obscure. Evolution of regulatory nucleotide sequences modulating gene expression is not perceptible at the protein level. On the other hand transposable sequences that modify protein individuality may have been key agents in eukaryotic evolution. Direct molecular analysis of genomic deoxyribonucleic acid will certainly reveals new mechanisms involved in gene rearrangements.

However a fundamental question remains: what is the driving force in molecular evolution? Neo-darwinists hold that genetic events occur at random and natural selection only directs phylogenic lineages. In contrast the neutral theory claims that the great majority of evolutionary changes at the molecular level are caused not by Darwinian selection but by random fixation of selectively neutral or nearly neutral mutants [30]. The majority of protein and deoxyribonucleic acid polymorphisms would be also selectively neutral.

Another concept is that deoxyribonucleic acids are changing under their own physico-chemical laws and that 'selfish' genes are operating for themselves [31]. It is however unlikely that proteins are only servants, prisoners in organisms.

References

1. W. Gilbert: 'Why gene in pieces?' *Nature* **271**, 501 (1978).

2. L. Patthy: 'Evolution of the proteases of blood coagulation and fibrinolysis by assembly from modules', *Cell* **4**, 657—663 (1985).
3. L. Sottrup Jensen, H. Claeys, M. Zajdel, T. E. Petersen, and S. Magnusson: 'The primary structure of human plasminogen: isolation of two lysine-binding fragments and one "mini"-plasminogen (MW 38.000) by elastase-catalyzed-specific limited proteolysis'. In *Progress in Chemical Fibrinolysis and Thrombolysis*, Vol. 3 (J. F. Davidson, R. M. Rowan, M. M. Sama, and P. C. Desnoyers, Eds.), Raven Press, New York, pp. 191—209 (1978).
4. H. Neurath: 'Evolution of proteolytic enzymes', *Science* **224**, 350—357 (1984).
5. R. Acher: 'Chemistry of the neurohypophysial hormones: an example of molecular evolution'. In *Handbook of Physiology*, Section 7, Endocrinology, Vol IV, Part 1 (E. Knobil and W. H. Sawyer, Eds.), pp. 119—130 (1974).
6. M. J. Brownstein, J. T. Russell, and H. Gainer: 'Synthesis, transport and release of posterior pituitary hormones', *Science* **207**, 373—378 (1980).
7. M. T. Chauvet, J. Chauvet, R. Acher, D. Dunde, and A. N. Thorn: 'Structure of a guinea pig common precursor to a MSEL-type neurophysin and copeptin', *Mol. Cell. Endocrinol.* **44**, 243—249 (1986).
8. H. Land, G. Schütz, H. Schmale, and D. Richter: 'Nucleotide sequence of a cloned cDNA encoding bovine arginine vasopressin neurophysin II precursor', *Nature* **295**, 299—303 (1982).
9. H. Land, M. Grez, S. Ruppert, H. Schmale, M. Rehbein, D. Richter, and G. Schutz: 'Deduced amino acid sequence from the bovine oxytocin-neurophysin I precursor cDNA', *Nature* **302**, 343—344 (1983).
10. R. Acher: 'Neurophysin and neurohypophysial hormones', *Proc. Roy. Soc. B* **170**, 7—16 (1968).
11. R. Acher: 'Molecular evolution of biologically active polypeptides', *Proc. Roy. Soc. B* **210**, 21—43 (1980).
12. M. T. Chauvet, F. Hurpet, G. Michel, J. Chauvet, and R. Acher: 'Two multigene families for marsupial neurohypophysial hormones? Identification of oxytocin, mesotocin, lysipressin and arginine vasopressin in the North American opossum (*Didelphis virginiana*)', *Biochem. Biophys. Res. Commun.* **123**, 306—311 (1984).
13. R. Acher: 'Neurophysins: molecular and cellular aspects', *Angew. Chem. Int. Ed. Engl.* **18**, 846—860 (1979).
14. R. Ivell and D. Richter: 'Structure and comparison of the oxytocin and vasopressin genes from rat', *Proc. Natl. Acad. Sci.* **81**, 2006—2010 (1984).
15. S. Ruppert, G. Scherer, and G. Schutz: 'Recent gene conversion involving bovine vasopressin and oxytocin precursor genes suggested by nucleotide sequence', *Nature* **308**, 554—557 (1984).
16. E. Sausville, D. Carney, and J. Battey: 'The human vasopressin gene is linked to the oxytocin gene and is selectively expressed in a cultured lung cancer line', *J. Biol. Chem.* **260**, 10236—10241 (1985).
17. D. Baltimore: 'Gene conversion: some implications for immunoglobulin gene', *Cell* **24**, 592—594 (1981).
18. J. Scott, M. Urdea, M. Quiroga, R. Sanchez-Pescador, N. Fong, M. Selby, W. J. Rutter, and G. I. Bell: 'Structure of a mouse submaxillary messenger RNA encoding Epidermal Growth Factor and seven related proteins', *Science* **221**, 236—240 (1983).
19. A. Ullrich, J. R. Bell, E. Y. Chen, R. Herrera, L. M. Petruzzelli, T. J. Dull, A. Gray, L. Coussens, Y. C. Liao, M. Tsubokawa, A. Mason, P. H. Seeburg, C. Grunfeld, O. M. Rosen, and J. Ramachandran: 'Human insulin receptor and its relationship to the tyrosine kinase family of oncogenes', *Nature* **313**, 756—761 (1985).
20. H.-D. Stahl, P. E. Crewther, R. F. Anders, G. V. Brown, R. L. Coppel, A. E. Bianco, G. F. Mitchell, and D. J. Kemp: 'Interspersed blocks of repetitive and charged amino acids in a dominant immunogen of *Plasmodium falciparum*', *Proc. Natl. Acad. Sci. USA* **82**, 543—547 (1985).
21. S. J. Friezner Degen, R. T. A. Macgillivray, and E. W. Davie: 'Characterization of the complementary deoxyribonucleic acid and gene coding for human prothrombin', *Biochemistry* **22**, 2087—2097 (1983).

22. K. Skørstengaard, M. S. Jensen, T. E. Petersen, and S. Magnusson: 'Purification and complete primary structures of the heparin-, cell-, and DNA-binding domains of bovine plasma fibronectin', *Eur. J. Biochem.* **154**, 15—29 (1986).

23. V. M. Ingram: *The Hemoglobins in Genetics and Evolution.* Number XXII of the Columbia Biological Series, Columbia University Press, New York and London (1963).

24. A. J. Jeffreys, S. Harris, P. A. Barrie, D. Wood, A. Blanchetot, and S. M. Adams: 'Evolution of gene families: the globin genes'. In *Evolution from Molecules to Men* (D. S. Bendall, Ed.), Darwin College, Cambridge University Press, pp. 174—208 (1983).

25. M. F. Perutz: 'Regulation of oxygen affinity of hemoglobin: influence of structure of the globin on the heme iron', *Ann. Rev. Biochem.* **48**, 327—386 (1979).

26. J. Giraudat, A. Devillers-Thiéry, C. Auffray, F. Rougeon, and J. P. Changeux: 'Identification of a cDNA clone coding for the acetylcholine binding subunit of *Torpedo marmorata* acetylcholine receptor', *EMBO J.* **1**, 713 (1982).

27. M. Noda, H. Takahashi, T. Tanabe, M. Toyosato, S. Kikyotani, Y. Furutani, T. Hirose, H. Takashima, S. Inayama, T. Miyata, and S. Numa: 'Structural homology of *Torpedo californica* acetylcholine receptor subunits', *Nature* **302**, 528—532 (1983).

28. M. Mishina, T. Kurosaki, T. Tobimatsu, Y. Morimoto, M. Noda, T. Yamamoto, M. Terao, J. Lindstrom, T. Takahashi, M. Kuno, and S. Numa: 'Expression of functional acetylcholine receptor from cloned cDNAs', *Nature* **307**, 604—608 (1984).

29. B. Sakmann, C. Methfessel, M. Mishina, T. Takahashi, T. Takai, M. Kurasaki, K. Kufuda, and S. Numa: 'Role of acetylcholine receptor subunits in gating of the channel', *Nature* **318**, 538—543 (1985).

30. M. Kimura: 'DNA and the neutral theory', *Phil. Trans. R. Soc. Lond. B* **312**, 343—354 (1986).

31. W. F. Doolittle and C. Sapienza: 'Selfish genes, the phenotype paradigm and genome evolution', *Nature* **284**, 601—603 (1980).

32. K. Skørtengaard, M. S. Jensen, P. Sahl, T. E. Pedersen, and S. Magnusson: 'Complete primary structure of bovine plasma fibronectin', *Eur. J. Biochem.* **161**, 441—453 (1986).

Biomolecular Chirality

Origin of the L-Homochirality of Amino-Acids in the Proteins of Living Organisms

A. JULG
Laboratoire de Chimie Théorique, Université de Provence, Marseille, France.

La dissymétrie crée le phénomène — P. Curie

1. Introduction

Symmetry is equilibrium, appeasement, and, in the limit, death. By contrast, the breaking of the symmetry generates motion, animates forms, sprouts Life. Among the many examples which can be given to illustrate this law of Nature, the most typical is certainly that of natural substances. As early as the last century, the attention of chemists had been drawn to the fact that most of the substances from plants possess a rotatory power (oil of turpentine, solutions of sugar, of camphor, etc.). Around 1900, Fischer showed that the majority of the natural sugars belong to the same stereochemical series (D-series). His student, Freudenberg (1924) was one of the first to realize that the amino acids of proteins belong to L-series (Figure 1). Moreover, certain D-amino acids have subsequently been found in some organisms (bacteria, annelids, insects, octopus) as constituents of specific molecules (e.g. luciferine) but without being susceptible to incorporation in proteins [1]. More recently, at last, the discovery that, after death, amino acids of

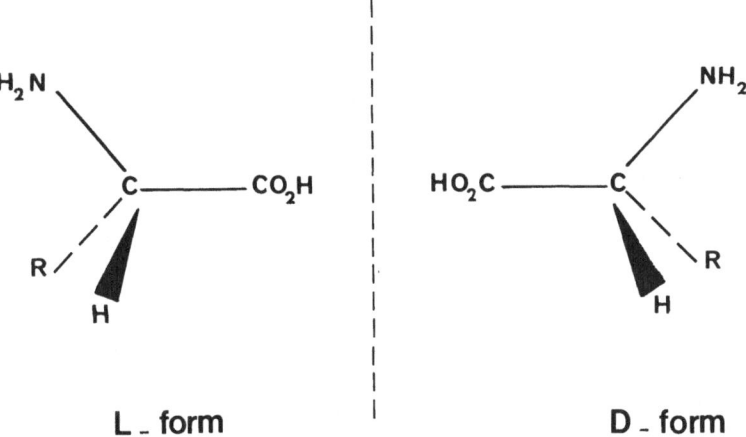

Fig. 1. L- and D-conformations of amino acids. N.B.: with few exceptions, L-forms are dextro-rotatory, and D-forms levo-rotatory.

Jean Maruani (ed.), Molecules in Physics, Chemistry, and Biology, Vol. IV, 33—52.
© 1989 *by Kluwer Academic Publishers.*

the collagen of bones progressively racemize [2, 3] has reinforced the certainty of the strong connection which exists between chiral dissymmetry and Life.

Since the discovery of this dissymmetry in the natural substances, physicists and chemists, of course, attempted to find its cause. But, up to these recent years, no phenomenon based on classical physics has been found to solve the problem, as we shall see. The recent discovery of the parity violation arising from weak interactions [4] threw a new light on the question. But, if the solution has to be searched in this way, all the difficulties are not solved. The aim of this paper is to review the present situation and the chief ways in which the researches are continued.

2. Stability of L-Forms

In fact, the dominance of L-amino acids in living organisms does result from two phenomena which we must well distinguish: on the one hand, the choice of one form (L) rather than the other (D), and, on the other, the stability of the L-population.

The study of amino acids of proteins from fossil bones, indeed, allows us to assert that, for 100 million years at least, L-forms have been, as today, the exclusive components of proteins [3]. Moreover, given the continuity observed for the paleospecies since the appearance of living organisms, it is very probable that the first organized cells were built up on L-amino acids.

The importance of the stability of the L-population does not seem to have been understood in the last century. This explains why the discussions often turned on physical mechanisms which would work at the very time of the synthesis of the molecule under consideration, without reference to the structure of the cell in which the synthesis is performed.

At the present time, we know that the syntheses of the molecules which are necessary to Life occur on substrates whose geometrical structure insures the replicating of the fundamental molecules with great accuracy. This is the 'key and lock' mechanism, proposed by Fischer [5] at the end of the last century, which allows the transmission of the initial structures from generation to generation by means of essentially chemical processes. Besides, it is to notice that the synthesis of proteins in the living kingdom occurs step by step. The amino acids are essentially synthesized and polycondensed into proteins by plants. These proteins are absorbed by herbivores which hydrolyze them to set the amino acids free, from which new proteins are synthesized. Carnivores devour herbivores, using the amino acids of the proteins of the latter to synthesize the proteins which are necessary to them. Omnivores, such as man, utilize the amino acids both from vegetables and animals. But the most important point is that, during these successive syntheses, the amino acids are transferred from one organism to another without their configuration being changed. The construction of proteins of high molecular mass, for steric reasons, requires the exclusive using of amino acids belonging to the same configuration (L or D). The incorporation of a D-amino acid in an α-helix of L-amino acid destabilizes the system significantly [6], so that, if accidentally the configuration of an amino acid is inverted — by a chemical reaction, for instance — this molecule becomes unserviceable. In this connection, let us recall that D-amino acids arising from too long a cooking of foodstuffs, which provokes the racemization, are not assimilated and produce wastes which have to be eliminated.

The situation becomes completely different after the death of the organism. The water-soluble proteins are easily attacked by microorganisms and quickly destroyed with formation of CO_2, NH_3, N_2, H_2O or cut by chemical reagents (H^+ in particular) into peptides and even into free amino acids which are more or less quickly racemized. By contrast, certain fibrous proteins, collagen of bones in particular, can be kept sheltered from any chemical or bacterial degradation. Nevertheless, their amino acid components do not eternally keep the L-configuration. Just like all the optically active molecules, independently of any chemical reaction, amino acids racemize. The swiftness of the phenomenon depends essentially on the height of the potential barrier which separates the two enantiomers. The racemization is very quick for amines (10^{-12} s) [7]. For phosphines and for arsines, the necessary times are equal to a few months and one year, respectively [7]. For amino acids of rigid structures, such as collagen, complete racemization is reached after a few hundred thousands of years only. The phenomenon is used to date fossil bones [2, 3, 8]. Given its slowness, this spontaneous racemization does not practically affect the amino acids in the tissues during the lifetime of the organism. Anyhow, D-amino acids which would appear would be eliminated, as we have seen. A very particular case where the racemization occurs during the life of the organism is that of dentine from mammal teeth. The racemization, indeed, starts as early as the formation of the tooth [9]. But it is well known that teeth are dead tissues, contrarily to bones: broken bones knit again, but teeth do not.

To summarize, the present dominance of L-amino acids of proteins arises essentially from the replicating of an initial model with a continuous elimination of D-forms incompatible with Life. Consequently, the cause of the homochirality we observe in living organisms has to be searched for at the very origin of Life. But why L-forms and not D-forms? And how has the selection been reached? *A priori*, indeed, an L- and D-coexistence might have existed. Nature does offer a curious example of such coexistences. The quasi-totality of gasteropods, snails in particular, are right-handed. The left-handed shells are extremely rare ($< 10^{-6}$), except for certain genera, as *Amphidromus*, living in Java, which include as many right- as left-handed forms (Figure 2).

3. In Quest of Factors Able to Explain the L/D Discrimination

Even before the question of the origin of the dominance of L-amino acids in proteins had been set, many experiments were performed to discover factors able

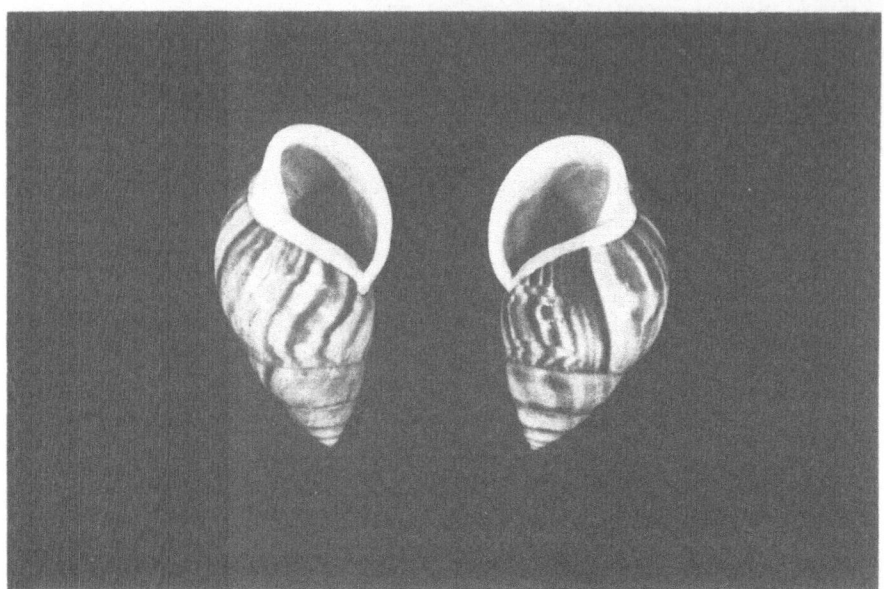

Fig. 2. Snails (*Amphidromus*) from Java (5 cm, author's collection): an example of genus which does offer right- and left-handed forms in equal proportions. With very rare exceptions, the shells of the other gasteropods are right-handed.

to modify the ratio of the two enantiomeric forms in organic and inorganic optically active substances, with a view to explaining the chirality observed in the natural compounds.

As early as Pasteur started his research on this topic, he realized that only causes capable of breaking the local symmetry of space can produce any discrimination between two enantiomeric forms. He thought, in particular, of the effect of polar fields. In 1846, Faraday had discovered that magnetic fields induce an optical rotation in isotropic media. This led Pasteur to grow crystals in magnetic fields [10]. No discrimination did occur. Then, he attempted to create the dissymmetry by rotation fields. He grew plants on a clockwork mechanism to invert the effect of the Earth's rotation [10]. He also used a heliostat in order to give the appearance to the plants that Sun was rising in the west and setting in the east [10]. All these attempts were negative, so that, in his later years, Pasteur concluded that only a still unknown cosmic phenomenon could explain the dissymmetry observed in the natural substances, arguing that the solar system itself presents a dissymmetry [11]. At the time of Pasteur the general belief was that planets and their satellites are rotating on their own axes and around the Sun in the same sense. At present, we know several exceptions to this rule.

Le Bel (1874) proposed the use of left- and right-handed circularly polarized radiation to induce a dissymmetry in photochemical reactions where racemic mixtures appear [12]. After various fruitless attempts [13], the effect has been observed only in 1930 by Kuhn *et al.* [14] who obtained a low dissymmetry ratio, employing circular ultraviolet radiation. Moreover, a stereoselective photodegradation under the effect of elliptical radiations has also been observed for alanine, glutamic acid, and tartaric acid [15]. But such a phenomenon does not seem to have played a role in the L/D discrimination, because the dextro-circularly polarized light, which would predominate at the Earth's surface, remains, in fact, a simple conjecture without evidence [16].

In 1894, Curie suggested combining a magnetic field and a parallel or antiparallel electric field [17]. In fact, as de Gennes [18] demonstrated a long time later (1970), if the fields are uniform and constant, their superposed application does not affect the equilibrium enantiomeric population. A very weak discrimination might nevertheless occur if the transition state of the reaction involves a change of the magnetic moment [19].

The centrifugal force with the spin axis parallel or antiparallel to the gravitation field of the Earth has also been used. The effect of collinear electric and magnetic fields in a centrifuge has also been tested [20, 21]. Positive outcomes have been claimed for the oxidation of isophorone by hydrogen peroxide [22], but the order of magnitude of the enantioselection does not correspond to the expected value [20].

Many other experiments to test the effect of rotation of the medium have been performed [23]. All of them led to negative or not significant conclusions.

At last, experiments in which a more or less complete separation of the two isomers occurs under the effect of rotation and of electric and magnetic fields, have been realized with success [24]. But they do not explain the dominance of one form with respect to the other. For this reason, they do not interest us.

To conclude, no classical chiral field seems to be able to explain the L/D discrimination observed for amino acids in proteins. Even if, sometimes, a weak discrimination between two enantiomers of certain substances is effectively obtained, the conditions in which these experiments have been performed (in centrifuge or within very strong electromagnetic fields) are too different from the natural conditions in which amino acids and proteins are synthesized for these mechanisms to be envisaged.

In a parallel way to this work on the physical causes, chemists attempted to realize asymmetric syntheses, i.e. to produce two enantiomers in different proportions. Many syntheses were effectively achieved with success. But such a discrimination can be obtained only if one reagent is either optically pure or a mixture of two enantiomers in different proportions. Fischer, indeed, showed that, if a molecule is asymmetric, its extension also proceeds in an asymmetric sense [25].

The question of the dominance of one enantiomeric form was not solved, but simply shifted, so that certain chemists, Japp in particular [26], thought that only a directive vital force which would have worked at the moment of organic creation could have explained the phenomenon. For others, the particular choice of L-amino acids is a matter of chance. It is, indeed, well known that supersaturated racemic solutions spontaneously give only *d*-crystals or *l*-crystals, the selection being due to a chance selection of micro-environment [27]. Another alternative which had its supporters, such as Kelvin and Arrhenius [28], consists of admitting that Life arises from specific enantiomers come to Earth from outer space, carried by radiation pressure. Such a hypothesis, nevertheless, is not very tenable at the present time.

Another approach to the problem is the synthesis from achiral molecules on chiral substrates. The effect of substrates was suspected for a long time. Pasteur, for instance, ascribed the crystallization of a solution of sodium-ammonium racemate into an optically active isomer to some dissymmetric influence in the glass of the crystallizing vessel [29]. In fact, it would rather seem that this discrimination arose from the presence, on the walls of the vessel, of optically active germs. The atmosphere of a laboratory and the vessels are, indeed, always polluted by such traces arising from previous experiments. Every chemist knows that, if a substance which has persistently refused to crystallize for a long time, eventually does crystallize, then its crystallization subsequently becomes very easy, as if by magic.

However that may be, various experiments have been performed to verify the effect of chiral substrates. Let us quote the dehydrogenation and dehydratation of 2-butanol on quartz [30], either pure or metal (Pt, Pd, Ni ...) covered, the hydrogenation of tiglic acid CH_3—CH=$C(CH_3)$—CO_2H [31], that of α-pinene [32]. Chromatography on quartz has also been utilized to try to separate various enantiomers [33], in particular of butanol, at room temperature and at dry-ice temperature [34]. But the results were always negative or not conclusive, so that certain authors thought that, contrary to a contemporary opinion and theoretical forecasts [35], chiral substrates cannot affect the population of enantiomeric forms [34]. Only recently, the effect has been proved by using more accurate measurement techniques. The adsorption of radioactive D- and L-alanine hydrochloride (Cl^-, NH_3^+—$CHCH_3$—CO_2H) in 10^{-5} molar dimethylformamide solution on *d*- and *l*-α-quartz shows that *d*-quartz preferentially adsorbs D-alanine and, conversely, *l*-quartz adsorbs L-alanine [36]. The extent of asymmetric preferential adsorption is about 1.0 to 1.8 percent, at the 99.9 percent confidence level. The values obtained for the asymmetric ratio allow the estimation of the difference in the adsorption energy of L- and D-forms on *l*-quartz as being equal to 0.03—0.04 kcal mol^{-1}. Two very recent theoretical calculations performed by means of two completely different methods corroborate this conclusion [37, 38]. L-forms of positive ion (H_3N^+—$CHCH_3$—CO_2H) [37] and of zwitterion (H_2N^+—$CHCH_3$—CO_2^-) [37, 38] are preferentially adsorbed on the $(10\bar{1}0)$ lateral faces of *l*-α-quartz. For the positive ion, the energy difference is found equal to 0.07 kcal mol^{-1} [37]. The order of magnitude agrees with the experimental value. Finally, a calculation concerning the adsorption of alanine on kaolinite gives similar results, as we shall see further. Consequently, asymmetric adsorption on chiral substrates is a factor which can be held as a working hypothesis to explain the L/D discrimination of amino acids in the first living organisms.

The discovery of the parity-violation from the weak interactions cast a new light on the problem [4]. Electrons emitted by radionucleotides, indeed, are longitudinally left-handed polarized in proportion to the velocity of the particles to that of light. From this the idea arose to irradiate racemic mixtures or inactive substances whose cleavage can give optically active molecules, to obtain an asymmetric decay or synthesis. In spite of a few outcomes claimed as being positive [39], at the present time, it seems that no significant conclusion has been obtained.

On the contrary, the unification of the electromagnetic and the weak interactions through the weak neutral current mediated by neutral bosons Z^0, implies a difference between the electronic binding energy of two enantiomeric molecules. The corresponding operator is the following [40]

$$H' = - \frac{G_F}{4\sqrt{2}mc} \sum_a \sum_i N_a \{p_i \cdot \sigma_i \cdot \delta^3(r_i - r_a)\}_+$$

where G_F = Fermi weak coupling constant; m = electron mass; c = speed of light; N_a = neutron number of the atom a; p_i = momentum of the electron i; σ_i = Pauli spin matrix operator of the electron i; $\delta^3(r_i - r_a)$ = three-dimensional Dirac function representing the charge density of the electron i at the atomic nucleus a; $\{\ldots\}_+$ denotes an anticommutator. The differences in energy are very weak but significant. For instance, L-alanine is found more stable than D-alanine by ca. 10^{-20} eV $\sim 3 \times 10^{-19}$ kcal mol^{-1} [24, 41]. In aqueous solution the stabilization is greater: 7×10^{-19} eV $\sim 2 \times 10^{-17}$ kcal mol^{-1} [41]. Similar values are obtained for L-serine and L-aspartic acid in aqueous solution: 5×10^{-19} eV $\sim 10^{-17}$ kcal mol^{-1}, and 8×10^{-19} eV $\sim 10^{-17}$ kcal mol^{-1} respectively [42]. More generally, the stabilization per peptide unit in polypeptides in the α-helix or the β-sheet conformation is approximately estimated as being equal to 10^{-19} eV, L-forms being always the more stable [43]. Another interesting example is that of D-glyceraldehyde which is found more stable than L-conformation by ca. 4×10^{-19} eV $\sim 10^{-17}$ kcal mol^{-1} [44].

Consequently, not only the weak interactions do provoke a discrimination between two enantiomeric molecules, but also the sign of the discrimination does agree with experimental facts: L-amino acids are found more stable than D-forms, and D-glyceraldehyde from which natural sugars derive, more stable than its L-form. Nevertheless, for this latter, we must remark that D-series are chemically connected with L-amino acids, so that the argument is not entirely convincing. Be that as it may, the energy differences are extremely weak, so that it may be asked whether such energy differences are able to provoke the complete discrimination we observe in Nature. Besides, although the values obtained from the asymmetric adsorption are much larger than those from weak interactions, they are sufficiently weak for the question of the homochirality to be posed. Simple thermodynamical considerations cannot explain the phenomenon.

4. The Amplification Mechanism

Independently of the actual processes which have been orienting the natural substances towards homochirality, various mathematical models [45, 46] have been built, showing that from an extremely weak difference in the behavior of two enantiomers, the complete discrimination can be obtained.

These mechanisms are based upon the concepts of catastrophe theory. A classical schema is the following [46, 47]:

i. Production of L- and D-enantiomers from achiral or racemic reactants:

$$A \underset{K_{-1}}{\overset{K_1}{\rightleftarrows}} L \quad \text{and} \quad A \underset{K'_{-1}}{\overset{K'_1}{\rightleftarrows}} D$$

ii. Autocatalytic production of L and D:

$$L + A \underset{K_{-2}}{\overset{K_2}{\rightleftarrows}} 2L \quad \text{and} \quad D + A \underset{K'_{-2}}{\overset{K'_2}{\rightleftarrows}} 2D$$

iii. Mutual inhibition of L and D through an irreversible reaction:

$$L + D \overset{K}{\rightarrow} B$$

$K_1, K_{-1}, \ldots$ being the corresponding rate constants. In addition, one assumes an input of A maintaining the concentration of the initial reactants at a constant or very slowly variable level.

In the absence of any behavior difference between L- and D-isomers ($K_1 = K'_1$; $K_2 = K'_2, \ldots$) the overall process remains racemic if the input remains small. On the contrary, if the concentration of the initial reactants A is sufficiently large, the system is metastable, i.e. under the effect of the slightest perturbation it evolves towards a homochirality (L or D). But this selection is entirely random, so that thermodynamic fluctuations lead to a racemic mixture (L + D) (Figure 3).

The situation is completely different if the rate constants corresponding to the two enantiomers are not rigorously equal ($K_{\pm 1} \neq K'_{\pm 1}$ and $K_{\pm 2} \neq K'_{\pm 2}$). In the absence of exterior perturbation, for sufficiently large concentration in A, the system does evolve to a chiral production of one of the two enantiomers, the choice between the two ways being determined by the values of the ratios K/K'.

In accordance with the Arrhenius law, the rate constants are proportional to $\exp(-E^{\neq}/kT)$, $E^{\neq}$ being the corresponding activation energy. It follows that, for a given reaction (i or ii), K/K' is equal to $\exp(\Delta E^{\neq}/kT)$, $\Delta E^{\neq}$ being the difference in activation energy between the enantiomers. Moreover, the activation energies of both (i) and (ii) reactions are of the same order of magnitude, so that, practically,

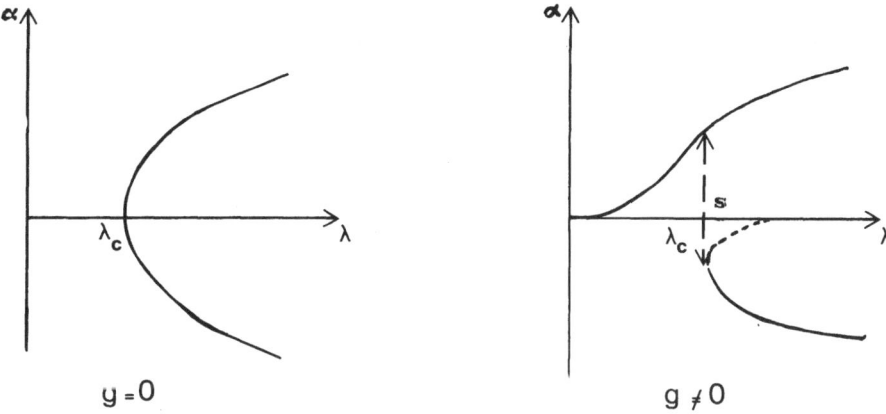

Fig. 3. Schematic representation of a metastable system governed by the equation $A\alpha^3 + B(\lambda - \lambda_c)\alpha + g = 0$. $\alpha = $ (L-D) limit concentration; $A, B, g = $ parameters depending on the rate constants. $g = 0$ if L- and D-forms possess the same behavior with respect to the other reactants. $\lambda = $ parameter depending on the initial concentrations; $\lambda_c = $ critical value of λ. $g = 0$: $\lambda < \lambda_c$: [L] = [D]. $\lambda > \lambda_c$: the system is metastable. It evolves to homochirality under the effect of fluctuations. $g \neq 0$: homochirality if $\lambda > \lambda_c$ and fluctuations $< s$.

the behavior of the system is governed by one parameter only, namely the value of $\Delta E^{\neq}/kT$ corresponding to the reaction (i).

In fact, thermal fluctuations tend to cancel the trend towards the L- or D-homochirality. In order that the difference of the properties of the two enantiomers prevails over thermal fluctuations, it is necessary for the ratio $\Delta E^{\neq}/kT$ to be larger than a certain critical value, determined by the values of the rate constants of the system under consideration (Figure 3). From usual values for the rate constants, this critical threshold can be estimated as being equal to 10^{-15}–10^{-17}. Given that $kT \sim 1$ kcal, we deduce that $\Delta E^{\neq}$ must be larger than 10^{-15} to 10^{-17} kcal mol^{-1} [46].

In the case of the weak interactions, we can put $\Delta E^{\neq}$ equal to the energy difference between the two enantiomers ΔE, so that the L- or D-homochirality will appear if $\Delta E > 10^{-15}$ to 10^{-17} kcal mol^{-1}, which does correspond to the order of magnitude of the calculated energy difference. On the contrary, the chiral interaction energies due to the combination of electronic, magnetic, gravitational and centrifugal fields, even if these fields are extremely strong, are always less than 10^{-19} [47]. This explains the negative outcomes of the experiments, and, consequently, excludes their role in the L-homochirality.

As an illustration of the model, the time necessary to obtain complete homochirality in a lake of 1 km $\times$ 1 km $\times$ 4 m has been calculated using plausible rate constants. Values located in the range of 10 000 to 15 000 years were obtained [47]. These values, certainly, do not correspond to reality, since they arise from a model without reference to the conditions in which Life appeared. They show, nevertheless, that an extremely weak difference in the behavior of the enantiomers can lead to homochirality. In order to make the model more precise, it is necessary to examine more carefully how Life appeared on Earth.

5. A Possible Schema for the Appearance of Life

The problem of the origin of Life on Earth is certainly the oldest question which preoccupies the mind of men. Our purpose is not to dissert on the many myths which have been proposed to explain the appearance of Life. We just want to emphasize that, starting from the fact that Life cannot emerge *ex nihilo*, but is necessarily transmitted by living things, conscious of its chtonian origin, men have been led to call out a Being which insufflated the life to clay or to some wooden statuette. Spontaneous generation, having its origin in an atheistic rationalism, did not hold long.

At the present time, although the problem is far from being solved, we begin to have some ideas on the mechanisms which might have occurred [48, 49, 50, 51, 52].

According to the traces found in precambrian rocks, Life appeared ca. 3 billion years ago, that is approximately 1.5 billion years after the birth of Earth. At that epoch, the atmosphere did not contain oxygen, but a large proportion of CO_2. Temperature was higher than today, but water, nevertheless, was liquid ($t <$ 100 °C) [53]. Abundant precipitations washed the granite basements, under the conjugated effects of CO_2 and water. Feldspar of granites were decomposed and transformed into clays (kaolinite chiefly, given the drastic washing conditions) which deposited in lagoons or on the periphery of the continents. The decomposition of micas gives illites, and subsequently montmorillonite. Quartz remains unaltered, forming sand or sandstone.

On the other hand, the first prebiotic molecules (amino acids in particular) were synthesized in free waters (oceans, lakes or lagoons) under the effect of the solar radiations from simpler components: CO, NH_3, CH_4 in all likelihood (in the absence of oxygen, indeed, the medium was reducing). From the present quantity of organic matter, one can estimate that the average concentration of amino acids, assumed uniformly dispersed in the oceans, would have been equal to 10^{-15}– 10^{-30} mole per liter [48]. In such a concentration, the polycondensation into peptides and proteins would have been impossible. Therefore, we must guess that amino acids have been either concentrated at privileged points before polycondensation, or more probably synthesized on suitable substrates and polycondensed on the spot. Given the well-known adsorbent properties of clays, it is logical to think that these minerals were the substrates on which amino acids have been synthesized and polycondensed.

Many experiments corroborate this hypothesis, clearly showing the catalytic role of clays in such reactions. Let us quote the chief ones among them: the synthesis of amino acids by means of ultraviolet irradiation of solutions of formaldehyde and ammonium salts in the presence of clays [54]; the synthesis of various amino acids and formation of purines and pyrimidines by reacting CO, H_2 and NH_3 on montmorillonite [55]; the synthesis of polypeptides of high molecular mass ($> 10^4$) by condensing amino acids on kaolinite [56]; the formation of polypeptides from amino acids adenylates on montmorillonite [57]; the formation of carbohydrates and lipids from paraformaldehyde on kaolinite [58].

Whatever the detail of the mechanisms may be, the polycondensation of amino acids allowed the organization of these polymers into cellular systems able to reproduce themselves. Very probably the first polycellular living organisms were bacteria (*Isua sphaera* found in Greenland), then blue-green algae (-2×10^9 years) which progressively fixed CO_2 of the atmosphere, freeing oxygen. The first animals appeared much later on (-1.4×10^9 years). In fact, it seems that the process was not successful at the first attempt. Indeed, one observes an alternance of green and brown layers in very old rocks corresponding to a reducing and oxidizing atmosphere, respectively. In the absence of oxygen, rocks contain Fe^{2+} ions, whereas Fe^{3+} ions are preponderant in the presence of oxygen [53].

6. The Asymmetric Adsorption on Kaolinite

If clays seem to have played a determinant role in the polycondensation of amino acids into proteins, it is necessary to make sure that these minerals can effectively induce any asymmetry, the polycondensation experiments we have quoted, indeed, were performed from optically active amino acids.

In fact, only adsorption on kaolinite has to be considered. Indeed, as we have said, this mineral was the most abundant clay at the time of the appearance of Life. On the other hand, the other clay minerals (montmorillonite, illites, . . .) do not present any asymmetry [59].

Kaolinite, $Si_4Al_4O_{10}(OH)_8$, occurs as minute hexagonal plates ($\sim 1 \mu$) constituted by layers stacked in a triclinic arrangement ($a = 5.14$; $b = 8.93$; $c = 7.37$ Å; $\alpha = 91.8°$; $\beta = 104.5°$; $\gamma = 90°$) [59]. Each layer is constituted by a sheet of regular tetrahedra about Si atoms, topped by octahedra about Al atoms (Figure 4).

The triclinic arrangement is the cause of the asymmetry. Kaolinite does possess two enantiomeric forms according as the (**a**, **b**, **c**) trihedron is direct or indirect. Unfortunately, we do not know which form is dextrorotatory and which is levorotatory. Moreover, we do not know whether one form is predominant with respect to the other. This latter condition, however, is fundamental for a discrimination to occur. On the analogy of α-quartz whose *l*-form is slightly more abundant than *d*-form (ca. 1%) [60], we can guess that the phenomenon is general, so that the *l*- and *d*-populations of kaolinite must be slightly different. Moreover, it seems that L-amino acids are polymerized on kaolinite at an appreciably higher rate than the corresponding D-forms [61]. This would require, on the one hand, that *d*- and *l*-populations of kaolinite are effectively different, and, on the other, that D- and L-forms of amino acids are adsorbed on a given (*d* or *l*) structure in

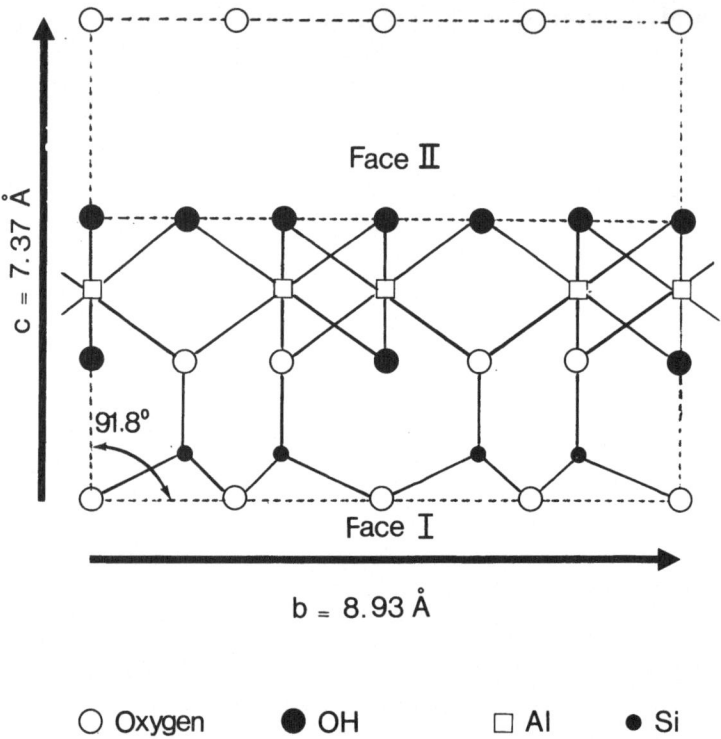

○ Oxygen   ● OH   □ Al   • Si

Fig. 4. Structure of kaolinite [59]: projection on a plane perpendicular to the a-axis.

different proportions. In the absence of direct experimental data concerning this latter point, we have performed a theoretical calculation [62].

This calculation necessitates some preliminary explanations. In the general case, the study of adsorption is a difficult problem. The phenomenon, indeed, is governed by the Gibbs energy of the whole system (substrate + molecule + solvent). Nevertheless, in the case we consider, the terms corresponding to the solvent-substrate and solvent-molecule interactions are the same for the two enantiomeric forms, so that they cancel out in the difference of adsorption energies. Likewise for the entropic terms which arise essentially from the vibrations of the molecule. Consequently, the difference in behavior of two enantiomers results from that between the molecule-substrate interaction energies only.

In order to calculate this interaction energy, we will proceed as follows. First, the actual substrate is replaced by a finite number of suitable point charges located at the sites of the various atoms so that the cluster is neutral and possesses the symmetries of the mineral [63]. Then, the interaction energy between the molecule and the cluster can be easily calculated by means of the *ab initio* self-consistent field (SCF) method. Let Δ_{SCF} be the difference between the energy of the molecule in the isolated state and that on the substrate ($\Delta_{SCF} > 0$ corresponds to an attraction). This difference corresponds only to the electrostatic interactions and the polarization of the molecule under the effect of the crystal field. In order to take the finite size and the polarizability of the atoms into account, we must introduce a dispersion-repulsion term, by means of a pair-potential for instance. But, on ionic substances, the dispersion-repulsion contribution which appears is weak with respect to the SCF term (10% in the case of alanine on quartz [37]) and the difference Δ_{DR} between the two enantiomeric forms is typically equal to 10^{-3} times Δ_{SCF} [37], so that, finally, the differential adsorption energy is essentially governed by the difference of the Δ_{SCF} terms.

Given the quick decreasing of the interaction energy versus the molecule-layer distance, we can reduce the actual crystal of kaolinite to two layers only. At last, each layer has been replaced by 115 point charges: 16 (+3.0) corresponding to Si atoms; 20 (+2.1) to Al; 45 (−1.4) to O; and 34 (−0.8) to OH groups. The values of these charges have been chosen, on the one hand, to agree with the values theoretically obtained for various crystals (SiO_2, silicates [64]), and, on the other, to insure the electroneutrality of the cluster.

As we have seen, at the time that the first living organisms appeared, CO_2 was very abundant so that the adsorption reactions occurred in a slightly acidic medium. Now, it is well-known that amino acids do exist in three forms according to pH: Positive ion H_3N^+—CHR—CO_2H in acidic medium; zwitterion H_3N^+—CHR—CO_2^- at the vicinity of the neutrality (6 < pH < 7); negative ion H_2N—CHR—CO_2^- in basic solutions. Given that carbonic acid is weak, we will consider positive ion of alanine (Ala^+) and also zwitterion ($Ala^{\pm}$), neglecting the negative ion.

The results are the following. In both cases (Ala^+ and $Ala^{\pm}$), on the structure corresponding to the *direct* (**a, b, c**) trihedron, L-forms are preferentially adsorbed, so that the uncertainty concerning the pH vanishes. The difference adsorption energies between L- and D-forms are found equal to 0.03 kcal mol^{-1} for Ala^+ and 0.01 for $Ala^{\pm}$ respectively. In both cases (Ala^+ and $Ala^{\pm}$) the preferential site is the center of a hexagon of oxygen atoms belonging to the SiO_4 tetrahedra (Face I). The positive ion is standing in a tripod position, whereas the zwitterion is lying (bridged position) (Figure 5). Moreover, insofar as we can admit that the molecules are adsorbed independently of one another, they can form alignments favorable to the polycondensation, the molecules (Ala^+ or $Ala^{\pm}$ according to pH) being disposed with respect to one another in a suitable orientation to give syntactic polymers, i.e. built up on amino acids belonging to the same configuration (Figure 6).

Preliminary calculations give the same result for cysteine H_2N—$CH(CH_2SH)$—CO_2H, proline

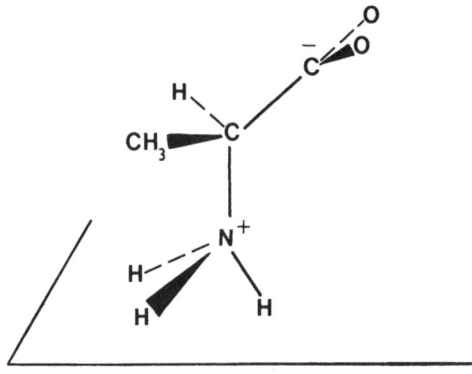

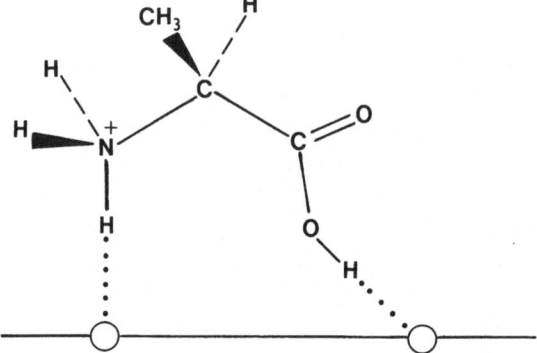

Fig. 5. Tripod position of the zwitterion and bridged position of the positive ion of alanine on the face I (Figure 4) of kaolinite.

and aspartic acid NH_2—$CH(CH_2$—$CO_2H)$—CO_2H, so that the asymmetric adsorption on kaolinite is a good candidate to explain the dominance of L-forms over the D's, under the conditions, of course, that:

i. d- and l-forms of kaolinite occur in slightly different proportions,
ii. the more abundant form corresponds to that on which the calculations indicate that L-forms are preferentially adsorbed $[(\mathbf{a} \times \mathbf{b}) \cdot \mathbf{c} > 0]$.

Here still, according to the amplification mechanisms we have previously seen, the weak difference in adsorption energies does not preclude homochirality from being reached after a sufficiently long time.

But kaolinite can have played another role. Experiment, indeed, shows that on montmorillonite a racemization occurs in acidic medium [65]. Given that montmorillonite is constituted by layers whose two faces are made up of SiO_4 tetrahedra, we can think that the basal face of kaolinite constituted by SiO_4 tetrahedra, is also able to produce racemization. This opinion is supported by the

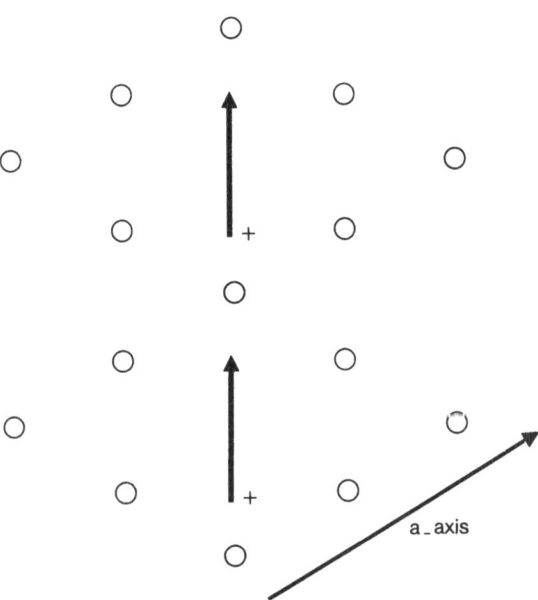

Fig. 6. Preferential adsite of both the positive ion and the zwitterion of alanine on the face I (Figure 4) of kaolinite. The ions (represented by arrows) are all oriented in the same direction so that their polycondensation does give syntactic polymers.

following results we have obtained by means of theoretical calculations. In the isolated state, the net charge of the asymmetric carbon in the alanine positive ion is equal to $+0.04$. On the substrate, this charge becomes equal to -0.03. This sign inversion of the net charge allows the fixation of H^+:

$$-CHR-+H^+ \rightleftarrows -C^+H_2R-.$$

The complex thus obtained is unstable and the removal of one of the two hydrogen atoms gives either the starting isomer or its inverse.

7. Weak Interactions or Asymmetric Adsorption?

In conclusion, it seems that two mechanisms have been retained to explain the chirality in the proteins of living organisms, namely the difference of energy between the enantiomers arising from weak interactions (ΔE_{WI}) and the asymmetric adsorption on kaolinite. Independently of the actual mechanisms which have produced the complete L/D discrimination by means of amplification processes, given the large difference between ΔE_{WI} and the adsorption energies of L- and D-forms (ΔE_A), at first sight, we might think that the asymmetric adsorption played the predominant role. In fact, the problem is much more complex.

As we have said, in a lake, complete homochirality can be obtained in about 15 000 years. Obviously, such a time is too short with respect to what is currently admitted concerning the speed of the passage from mineral to living. For the asymmetric adsorption, given that $\Delta E_A \sim 10^{15}\Delta E_{WI}$, the homochirality would be practically reached instantaneously, which is unacceptable.

This leads us to return to the mechanisms which can have worked. According to what we have seen, it is impossible not to take the role of clays into account. A plausible schema is the following. Amino acids are synthesized on kaolinite according to reactions (i) of Section 3. But the yield of the discrimination is proportional to the very small excess of one form with respect to the other. Moreover, as we have said, a parasitic racemization can appear. So that, finally, everything occurs as if we had reactions (i) only, but with constants $K_{\pm 1}$ considerably reduced. Then, on kaolinite, the polycondensation occurs according to reactions similar to (ii) of Section 3, but with formation of L_2, L_3, L_4, ... Reaction (iii) might correspond to the production of unserviceable LD dimers which stop the growth of the peptidic chains, and which are eliminated from the cycle, cleaved into simpler molecules.

Under these conditions, we understand that the time necessary to obtain complete L-homochirality is, in fact, much longer than that calculated from the difference ΔE_A using rate constants corresponding to standard reactions.

Likewise, the effect of the weak interactions is considerably reduced by the racemization on clays, so that it is possible that the critical threshold cannot be reached, which would exclude the role of weak interactions in the amplification mechanism. In other words, the L-homochirality observed for the amino acids of proteins would arise only from the asymmetric adsorption on kaolinite, thwarted by racemization, under the conditions, of course, that the l- and d-forms of kaolinite are in different proportions.

On the analogy of the parity-violating energy differences obtained for small chiral molecules, we can think that the phenomenon is general and applies to chiral crystals. But for these latter, the energy differences must be considerably larger. As a first approximation, indeed, we can evaluate ΔE_{WI} for such a crystal as the sum of the contributions arising from chiral units. For instance, two enantiomeric quartz crystals of 0.1 mm along each main edge, which is corresponding to 10^{15} unit cells, might present a difference in energy which would produce the weak discrimination by the 1% observed in Nature [66], without any amplification mechanism having to be assumed. The fact that no example of discrimination has been reported in the laboratory might arise simply from too small a difference of d- and l-populations, the case of quartz in Nature [60] being exceptionally favorable. A very small ratio $(1 - d)/(1 + d)$ of 10^{-6}, for instance, which is largely sufficient to induce the homochirality by amplification, cannot be detected for kaolinite with the means we have available at the present time.

Consequently, even if Life appeared thanks to asymmetric adsorption on kaolinite, the *first cause* of its appearance would be the weak interactions, but at the level of minerals and not at that of molecules, as certain authors claim [41—44].

In any case, Life has been able to appear only when the concentration in components, from which amino acids were synthesized and polycondensed, reached a certain value, and temperature was suitable. Afterwards, when the concentration fell below the critical amplification threshold, any spontaneous formation of new prebiotic peptides ceased.

The case of Titan, the biggest satellite of Saturn, is typical concerning the influence of temperature. The atmosphere of this moon, indeed, is very similar to that of the primitive Earth [67]: the major components are N_2, CH_4, H_2, and, in weaker concentrations, Ar, CO, HCN, C_2N_2. Oxygen is absent. But the surface of Titan is covered by a $CH_4 + C_2H_6$ ocean, and the temperature is very low (ca. 100 K), so that although the components which would be able to produce biomolecules were present, Life did not appear because, on the one hand, mineral catalysts were absent, and, on the other, the temperature has remained too low.

8. Is the Universe Chiral?

The dominance of L-amino acids in living organisms on Earth being strongly linked to mineral asymmetry conditions, we may wonder whether the sign of the homochirality we observe is peculiar to our planet or universal.

Unfortunately we have no material for a direct answer. No trace of Life, which would be able to give us any information about this problem, has been detected outside the Earth. The amino acids found in meteorites are completely racemized [68], so that we cannot know whether they are biological or abiotic in origin. Moreover, they come, for a large part, from terrestrial contamination. Their number, indeed, is decreasing from the surface of the meteorite to the inside. The number of minerals collected on the Moon is too small to allow any conclusion. At last, the optical properties of two enantiomers being the same, no information can be obtained concerning the molecules which have been identified in the interstellar space.

In fact, the answer lies in the knowledge of the proportion of matter and anti-matter in the Universe. If all the particles were changed into their corresponding antiparticles (electron → positron, proton → antiproton, . . .) the sign of the parity-violating energy differences would be changed [69], so that if the Universe was built up only on anti-matter, D-amino acids would be predominant in the corresponding living organisms, d-quartz more abundant than l-quartz, and so on.

Various cosmological theories have been proposed concerning the presence and the proportion of anti-matter in the Universe [70]. A recent model is that matter and anti-matter are in equal proportions, separated from one another by a gap [71]. Under these conditions, in the half of the Universe to which Earth belongs, the chirality of amino acids would be L, whereas in the other half, amino acids would be D, so that the average chirality of the whole Universe would be equal to zero, as its electric charge.

Not only Life would arise from a molecular and mineral asymmetry, but also the Universe itself would exist only because matter and anti-matter would be separated from one another, creating a cosmic dissymmetry, which, far from clearing up the mystery of the Universe, would make it still more mysterious. Its history, indeed, appears as a succession of events — from the big bang to Man — which are too well imbricated in one another for us not to think that its evolution was already programmed at the first instants of the Universe.

Note Added in Proof

Recent calculations (A. Julg, *Compt. Rend. Acad. Sc. Paris* **II 305**, 563 (1987) and in press) show that: (i) on the direct form of kaolinite, iminium ions (arising from nitriles in acidic reducing medium) are adsorbed so that the fixation of CN^- (Miller's reaction) affords L-amino acids by hydrolysis; (ii) the direct form of kaolinite is more stable (and consequently slightly more abundant) than the inverse form.

   Both these results corroborate the ideas developed in this paper concerning the role of kaolinite in the appearance of Life on Earth.

References

1. J. J. Corrigan: *Science* **164**, 142 (1969); R. Rogers Yocum, D. J. Waxman, and J. L. Stroninger: *TIBS* (April 1980) 97.
2. J. L. Bada and R. Protsch: *Proc. Nat. Acad. Sci.* **70**, 1331 (1973).
3. W. G. Armstrong, L. B. Halstead, F. B. Reed, and L. Wood: *Phil. Trans. Royal Soc. London* **301**, 301 (1983).
4. S. Weinberg: *Phys. Rev. Lett.* **19**, 1264 (1967); A. Salam: Proc. 8th Nobel Symp. — *Elementary Particle Physics* (ed. by N. Svartholm) Almquist and Wiksell, Stockholm, p. 367 (1968).
5. E. Fischer: *Chem. Ber.* **27**, 3189 (1894).
6. G. Wald: *Ann. New York Acad. Sc.* **69**, 352 (1957); E. R. Blout and M. Idelson: *J. Am. Chem. Soc.* **78**, 3857 (1956).
7. C. C. Costain and G. B. M. Sutherland: *J. Phys. Chem.* **56**, 321 (1952); R. E. Weston: *J. Am. Chem. Soc.* **76**, 2645 (1954).
8. J. L. Bada and P. M. Helfman: *World Archeology* **7**, 160 (1975); R. Lafont, G. Perinet, F. Bazile, and M. Icole: *Compt. Rend. Acad. Sc. Paris*, Série II, **299**, 447 (1984); A. Julg, *L'anthropologie* **91**, 235 (1987).
9. J. L. Bada and S. E. Brown: *TIBS* (Sept. 1980) III.
10. L. Pasteur: in *Oeuvres de Pasteur* I. *Dissymétrie moléculaire* (ed. by Pasteur Valery-Radot) Masson, Paris, 1922.
11. L. Pasteur: *Compt. Rend. Acad. Sc. Paris* **78**, 1515 (1878).
12. J. A. Le Bel: *Bull. Soc. Chim. France* **22**, 337 (1874).
13. A. Cotton: *J. Chim. Phys.* **7**, 81 (1909).
14. W. Kuhn and F. Braun: *Naturwissenschaften* **17**, 227 (1929); W. Kuhn and E. Knopf: *Z. Phys. Chem.* **B7**, 292 (1930).
15. B. Nordén: *Nature* **266**, 567 (1977).
16. P. D. Richie: *Asymmetric Synthesis and Asymmetric Induction*, Oxford Univ. Press, 1933; W. A. Bonner: in *Exobiology* (ed. by C. Ponnamperuma), North Holland, Amsterdam, 1972, p. 170; J. R. P. Angel, R. Illing, and P. G. Martin: *Nature* **238**, 389 (1972).
17. P. Curie: *J. Physique* **3**, 409 (1894).
18. P. G. de Gennes: *Compt. Rend. Acad. Sc. Paris* **B270**, 891 (1970).
19. W. Rhodes and R. C. Dougherty: *J. Am. Chem. Soc.* **100**, 6247 (1978); R. C. Dougherty, *J. Am. Chem. Soc.* **102**, 380 (1980).
20. C. A. Mead and A. Moscowitz: *J. Am. Chem. Soc.* **102**, 7301 (1980).
21. A. Peres: *J. Am. Chem. Soc.* **102**, 7389 (1980).

22. P. Gerike: *Naturwissenschaften* **62**, 38 (1975); D. Edwards, K. Cooper, and R. C. Dougherty: *J. Am. Chem. Soc.* **102**, 381 (1980).
23. C. Honda and H. Hada: *Tetrahedron Lett.* **16**, 177 (1976); B. Nordén: *J. Phys. Chem.* **82**, 744 (1978); R. V. Jones: *Proc. Roy. Soc. London* **A349**, 423 (1976); P. W. Atkins: *Chem. Phys. Lett.* **74**, 358 (1980).
24. S. F. Mason: *Int. Rev. Phys. Chem.* **3**, 217 (1983).
25. E. Fischer: *Chem. Ber.* **27**, 2985 and 3231 (1894).
26. F. R. Japp: *Nature* **58**, 452 (1898).
27. K. Harada: *Naturwissenschaften* **57**, 114 (1970).
28. H. Kamminga: *Vistas in Astronomy* **26**, 67 (1982).
29. L. Pasteur: *Rev. Scientifique* **7**, 2 (1884).
30. G. M. Schwab and L. Rudolf: *Naturwissenschaften* **21**, 363 (1932).
31. G. M. Schwab, F. Rost, and L. Rudolf: *Kolloid Z.* **68**, 157 (1934).
32. A. P. Terentjev, Je. I. Klabunowski, and W. W. Patrikejev: *Dokl. Akad. Nauk. SSSR* **74**, 947 (1950).
33. Jc. I. Klabunowski and W. W. Patrikejev: *Dokl. Acad. Nauk. SSSR* **78**, 415 (1951); G. Karogonnis and G. Goumoulos: *Nature* **142**, 162 (1938); G. K. Schweitzer and C. K. Talbot: *J. Tenn. Acad. Sc.* **25**, 143 (1950); A. Nakahara and R. Tsuchida: *J. Am. Chem. Soc.* **76**, 3103 (1954).
34. A. Amariglio, H. Amariglio, and X. Duval: *Helv. Chim. Acta* **51**, 2110 (1968).
35. D. P. Craig and D. P. Mellor: *Topics Curr. Chem.*, Springer-Verlag, Berlin, Vol. **63**, p. 1 (1976).
36. W. A. Bonner, P. R. Kavasmaneck, F. S. Martin, and J. J. Flores: *Science* **186**, 143 (1974).
37. A. Julg, A. Favier, and Y. Ozias: *Surf. Sc.* **165**, L53 (1986).
38. L. Vega, L. Breton, C. Girardet, and L. Galatry: *J. Chem. Phys.* **84**, 5171 (1986).
39. D. W. Gidley, A. Rich, J. Van House, and P. W. Zitzewitz: *Nature* **297**, 639 (1982); R. A. Hegstrom: *Nature* **297**, 643 (1982).
40. C. C. Bouchiat and M. A. Bouchiat: *J. Phys.* **35**, 899 (1974) and **36**, 493 (1975); R. A. Hegstrom, D. W. Rein, and P. G. H. Sandards: *J. Chem. Phys.* **73**, 2329 (1980).
41. S. F. Mason and G. E. Tranter: *Chem. Phys. Lett.* **94**, 34 (1983); *Mol. Phys.* **53**, 1091 (1984).
42. G. E. Tranter: *Mol. Phys.* **56**, 825 (1985).
43. S. F. Mason and G. E. Tranter: *J. Chem. Soc., Chem. Comm.*, 117 (1983).
44. G. E. Tranter: *J. Chem. Soc., Chem. Comm.*, 60 (1986).
45. F. C. Frank: *Biochem. Biophys. Acta* **11**, 459 (1953); B. Nordén: *J. Mol. Evol.* **11**, 313 (1978); C. Fajszi and J. Czégé: *Origins of Life* **11**, 143 (1981).
46. D. K. Kondepudi and G. W. Nelson: *Phys. Rev. Lett.* **50**, 1023 (1983); **314**, 438 (1985).
47. D. K. Kondepudi and G. W. Nelson: *Physica* **125A**, 465 (1984).
48. J. D. Bernal: *The Physical Basis of Life*, Routledge and Kogan Paul, London (1951).
49. A. I. Oparin: *The Origin of Life on Earth*, Academic Press, New York (1957).
50. M. Calvin: *Chemical Evolution*, Clarendon Press, Oxford (1969).
51. B. K. G. Theng: *The Chemistry of Clay-Organic Reactions*, Adam Hilger, London, p. 274—275 (1974).
52. A. G. Cairns-Smith: *Genetic Takeover and the Mineral Origin of Life*, Cambridge University Press (1982).
53. J. Labeyrie: *L'homme et le climat*, Denoël, Paris (1985).
54. T. E. Pavlovskaya, A. G. Pasinskyi, and A. I. Grebenikova: *Dokl. Akad. Nauk SSSR* **135**, 743 (1960).
55. D. Yoshino, R. Hayatsu, and E. Anders: *Geochim. Cosmochim. Acta* **35**, 927 (1971).
56. J. J. Fripiat, P. Cloos, B. Calicis, and K. Makay: *Proc. Int. Clay Conf. Jerusalem* **1**, 233 (1966); E. T. Degens and J. Mathéja in *Prebiotic and Biochemical Evolution* (ed. by A. P. Kimball and J. Oró) North Holland, Amsterdam, p. 39 (1970).
57. M. Paecht-Horowitz, J. Berger, and A. Katchalsky: *Nature* **228**, 636 (1970).
58. G. R. Harvey, K. Mopper, and E. T. Degens: *Chem. Geol.* **9**, 79 (1972).
59. G. W. Brindley and K. Robinson: *Min. Mag.* **27**, 242 (1946); *Crystal Structures of Clay Minerals and their X-ray Identification*, ed. by G. W. Brindley and G. Brown, Mineral Soc. London (1980).
60. C. Palache, H. Berman, and C. Frondel: in *Dana's System of Mineralogy*, J. Wiley, New York, 7th edn. vol. III, p. 16 (1962).

61. T. A. Jackson: *Chem. Geol.* **7**, 295 (1971).
62. A. Julg: *Compt. Rend. Acad. Sc. Paris* **II 303**, 1773 (1986).
63. A. Julg and D. Létoquart: *Phil. Mag.* **33**, 721 (1976).
64. A. Julg: *Crystals as Giant Molecules*, Lecture Notes in Chemistry, vol. **9**, Springer-Verlag, Berlin (1978); A. Julg, A. Pellegatti, and F. Marinelli: *Nouv. J. Chim.* **6**, 31 (1982).
65. M. Frenkel and L. Heler-Kallat: *Chem. Geol.* **19**, 161 (1977).
66. G. E. Tranter: *Nature* **318**, 172 (1985).
67. D. Gautier: *The Atmospheres of Saturn and Titan*, Proc. Int. Workshop, Alpbach, Austria, ESA SP-241, p. 75 (1988).
68. J. Oró: *Space Life Sciences* **3**, 507 (1972); J. Oró, S. Nakaparksin, H. Lichtenstein, and E. Gil-Av: *Nature* **230**, 107 (1971).
69. L. D. Barron: *Chem. Phys. Lett.* **79**, 392 (1981); *Molec. Phys.* **43**, 1395 (1981).
70. H. Alfvén: *Rev. Mod. Phys.* **37**, 652 (1965).
71. H. H. Fliche, J. M. Souriau, and R. Triay: *A. Astrophys.* **108**, 256 (1982); J. M. Souriau: *Compt. Rend. Acad. Sc.*, Série générale, II, 213 (1985); F. X. Désert and E. Schatzman: *A. Astrophys.* **158**, 135 (1986).

Right-Handed and Left-Handed Molecular Structures

A. LAFORGUE
Laboratoire de Mécanique Ondulatoire Appliquée, Université de Reims, B.P. 347, 51062 Reims Cedex, France.

1. Introduction

Right-handed and left-handed structures exist in crystals, molecules [1] and living matter. Chirality, and especially molecular chirality, has played a major role in the history and philosophy of science.

It is in particular the subject of *asymmetric chemistry*, where work has become increasingly abundant. It is also a meeting point between the sciences of matter and the life sciences. *Curie's principle* is still a topical subject of discussion [2]; the existence of enantiomers has created a fundamental problem for *quantum mechanics*, for which no definitive solution seems to have been found; the origin of the chirality of living matter, linked to the origins and evolution of life, has given rise to nothing more concrete than hypotheses [3, 4].

It was a significant moment in the history of science when Pasteur separated crystals of sodium ammonium tartrate by hand [5]; solutions of the crystals manifested right and left rotatory powers of the same value and of opposite signs; Pasteur had indirectly separated right-handed and left-handed acid molecules which were called optical antipodes or enantiomers, and which were considered like the crystals they form to have a symmetrical mirror image.

More generally speaking, a compound which is identical to its own mirror image has no rotatory power; a compound whose mirror image is not identical may or may not be devoid of rotatory power: if it is optically active, it exists in two forms which have opposite rotatory powers.

In this outline the two aspects of Curie's principle will be recognized: *the symmetry of causes is to be found in effects, but effects can be more symmetric than causes.*

Molecular symmetry is a special case, the general case of course being that of the asymmetric molecule. The dissymmetric motives which usually lead to optical antipodes have however been catalogued: asymmetric atoms, i.e. with a bond direction pointing to the summits of a tetrahedron (C, N, S, Si, etc.) or an octahedron (transition metals), occupied by different groups, allenic bonds, diphenylic systems with steric hindrance, etc.: these are chromophores of optical activity [6].

The *delimitation* of molecules is therefore not indispensable for a discussion of

Jean Maruani (ed.), Molecules in Physics, Chemistry, and Biology, Vol. IV, 53—76.
© 1989 *by Kluwer Academic Publishers.*

the chirality of living matter; it is enough to recognize the chromophores of optical activity.

Two enantiomers can sometimes be distinguished by odour or by taste. They can have different physiological roles, and can in some cases be respectively toxic and non-toxic.

The physical properties of two enantiomers are identical except with regard to polarized light. The right-handed and left-handed components of natural light are unequally absorbed (circular dichroism). As a result [7], the plane of polarization of light which is rectilinearly polarized turns to the right or to the left, thus giving the two enantiomers their identifications. Their chemical properties are identical as far as symmetric reagents are concerned, but not in the case of asymmetric reagents (or in the presence of asymmetric catalysts [8]). The same applies to their properties of adsorption, dialysis, complexing, ...

They can be produced by asymmetric synthesis and used in other asymmetric syntheses. They are synthesized selectively by living matter and exist in living matter in different quantities. In essential biomolecules, only one enantiomer exists for each chemical species.

For example [6], cholic acid — which has 11 chromophores — has about 2000 isomers, of which only one is known to exist in living organisms. However, there are exceptions to 'optical purity' (cf. Section IV).

The fundamental conclusion arrived at by Pasteur was that only living organisms can produce chirality. Chirality could not occur on earth before life. The need for a living organism to exist in order to separate a right-handed substance from a left-handed substance would thus be connected with the impossibility of spontaneous generation, ...

But asymmetric chemistry has led to absolute asymmetric synthesis [9—12], i.e. asymmetric synthesis brought about by the action of physical agents alone. Only one path is in fact absolutely certain: the action of polarized light in the absence [10] or in the presence [11] of a magnetic field; to this may be added the radiolysis of a racemic compound by polarized light [9, 12].

Yields have remained low and none of the production modes used amounts to a preparation. But their existence raises the problem of the relationship of succession and of the causal relationship between the appearance of chirality and the appearance of life.

The theory of molecular optical properties [13, 14] shows that during the action of physical agents of suitable symmetry an achiral molecule assumes chiral properties, in particular in the magneto-optical effect. Craig, Power and Thirunamachandran [15] have calculated that a chiral molecular environment directed an induced circular dichroism for an achiral molecule. Experimental evidence of this is extensive [6].

Does the chiral molecule play the same role in relation to electronic optics as in relation to electromagnetic optics? The unification of the two optics, particularly in terms of their effects on matter, is a necessary condition of de Broglie mechanics;

however, this principle has been only sparingly applied with regard to chiral molecules. Some effects of electronic optics which were expected *a priori* seem never to have been set out, even in the form of hypotheses. Such would seem to be the case of induced circular dichroism. The synthesis of chiral molecules under the effect of circularly polarized electrons has only rarely been considered, and the differential radiolysis of racemic compounds is still disputed (see Section 3.4.1).

Three points raised in this introduction seem to us to be fundamental.

What justification is there for speaking of a chiral molecule?

What is the origin of molecular chirality?

What are the links between life and chirality?

2. The Chiral Molecule

We will examine in turn:

— the conservation of the information bit linked to chirality in relation to the stability of the asymmetric molecule (Section 2.1);
— the requirement that an asymmetric molecule should be optically active, in relation to the status of enantiomers in quantum mechanics (Section 2.2).

2.1. THE REPRODUCTION OF MOLECULAR CHIRALITY AND THE LIFE-TIME OF THE CHIRAL MOLECULE

Left-handed quartz and right-handed quartz occur in unequal abundance on Earth. This inequality is likely to continue to obtain in view of the permanence of quartz.

Molecular chirality also perpetuates itself, but by a different (and in a way opposite) path. Asymmetric organic molecules, which are like all organic molecules out of thermodynamic equilibrium, have a finite life-time; they reproduce naturally in the course of the vital processes, this being globally equivalent to autocatalysis [16]; they can also be manufactured in whatever quantity is needed by asymmetric chemistry.

But at the same time the enantiomers' property of spontaneous inversion comes into play.

A molecule is in fact defined within a domain of stability or pseudostability. It can escape from that domain in various ways, either as a result of the action of a reagent, or in the absence of a reagent by the reactions of isomerization or dissociation. Each of these reactions is controlled by a potential barrier.

Organic molecules are out of equilibrium in the sense that crossing the dissociation barrier can lead to an energy state lower than that of the molecule.

In the case of an asymmetric molecule, one of the possible isomerizations is the inversion of the molecule leading to its optical isomer, which has a domain of stability equal to the previous one (Figure 1). On the figure each of the domains of stability is symbolically represented by a plane area. Corresponding to the

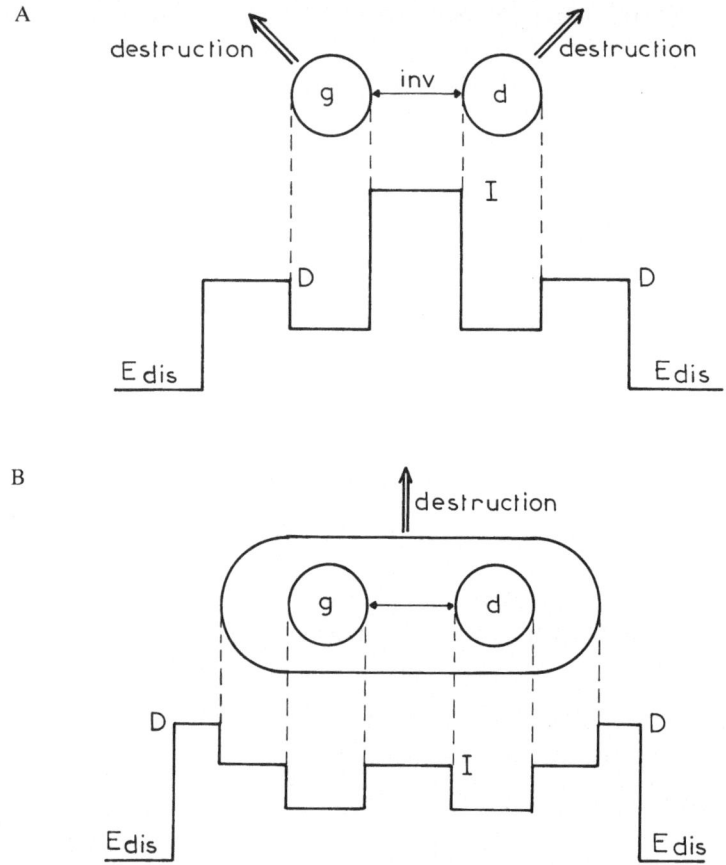

Fig. 1. Domains of stability in asymmetric molecules: A. Non-connected domains for two enantiomers; B. Double well in a connected domain. **g**: domain of left-handed structure (A enantiomer, B geometrical state); **d**: domain of right-handed structure (A enantiomer, B geometrical state); I: easiest inversion barrier; D: easiest destruction barrier; E_{dis}: energy of compound totally dissociated into atoms; inv: path of easiest inversion; destruction: paths of destruction.

operation of symmetry which causes the transition from the left-hand isomer to the right-hand isomer, there is an operation which causes the transition from the right-hand domain of stability to the left-hand domain of stability, which is symbolically represented in the figure as a symmetry. As regards the other combination or dissociation reactions, we will only consider those which:

(1) destroy asymmetry;
(2) result in an energy level lower than that of the asymmetric molecule.

Figure (1A) shows the path of easiest destruction and the path of easiest inversion in schematic form.

The opposite case (Figure 1B) is that of a global domain of stability for both forms, including two subdomains (of higher stability) of the right-hand form and of the left-hand form.

In case A, a life-time can be defined which is the same for each of the enantiomers **g** and **d**, controlled essentially by identical barriers of easiest destruction. In case B, a life-time of the asymmetric molecule can be defined, controlled essentially by the common barrier of easiest dissociation.

Let us consider case A. The life-time of the enantiomer is shorter than the probable time of its spontaneous inversion. Let us now suppose that a chain of biosynthesis enables the reproduction of the molecule in the form **d** or **g** in which it exists. This results in a multiplication of the starting form, and gives for suitable reproduction time and life-time values a population which varies in a regular manner, or for example in a constant manner. The only possible disruption of this process is the eventuality of spontaneous inversion.

The impurities thus introduced into the molecular population will be rarer the smaller the ratio of the life-time of the enantiomer to the probable time of spontaneous inversion. It can indeed be said that although the reproduction rate is determined by the conservation of a stationary molecular population, the quicker the spontaneous destruction of the chiral molecule the more effectively chirality is conserved.

This gives rise to reflections on the meaning and usefulness of death in situations which although far more complex are comparable. (In the case of pluricellular organisms, the death of the individual whose cells threaten to mutate one by one ensures the conservation of the species).

From another viewpoint, the quantity of asymmetric matter which can be synthesized in a given molecular species is unlimited. Such asymmetric synthesis is rather like the mass production of a manufactured object. Seen in this light, spontaneous inversion transitions are comparable to manufacturing errors. The shorter the life-time in relation to the inversion time, the rarer they are.

To conclude, it may be said that the shorter the life-time of the asymmetric chromophore and the longer its inversion time, the better the quality of the chiral information **d** or **g**.

To give a numerical example, in the case of alanine, the simplest of the amino acids, the following values are given in the literature: calculated inversion time $\simeq 10^9$ years [17]; destruction time deduced from experimental considerations $\simeq 10^3$ years [18].

2.2. THE SUFFICIENT CONDITION FOR OPTICAL ACTIVITY IN AN ASYMMETRIC MOLECULE AND THE STATUS OF THE CHIRAL MOLECULE IN QUANTUM MECHANICS

Asymmetric molecules which are optically inactive do exist [19]; ammonia NHDT is one example. The problem of finding the sufficient condition for an asymmetric molecule with rotatory power has only recently been raised by P. Pfeifer [19].

After a detailed study he defines the condition that the inversion barrier must be high enough. We will discuss this conclusion later. But it should first be said that, in order to raise this problem, quantum mechanics needed to resolve two more urgent problems, namely: the existence of enantiomers, and the spontaneous oscillation of the molecule without rotatory power.

The wave function of a stationary state of an asymmetric molecular structure can in fact only be a proper function of the complete Hamiltonian operator. Since the outset this has been a thorny problem for quantum mechanics [17, 20] and it is still being discussed today [19, 21, 22].

Some theoreticians [20] have refused to grant enantiomers the status of a molecule. They argue that only proper states of the complete Hamiltonian can be considered as such, and that the set of two enantiomers should be taken as a whole. As a result, given enough time, there will be a spontaneous transition to the other enantiomer. But this position is difficult to defend in quantitative terms [17], and is even more difficult to defend in qualitative terms. In the case of a non-connected domain of stability, which is as we have seen a real case for enantiomers, the concept of the molecule would no longer be that of the chemist.

Over the last sixty years, some theories have been proposed in which optical isomers have been given back the status of molecules without offending against the relevant principles. They state that the molecule under observation must not be considered in isolation, but rather in interaction with the field [19], or coupled with a photon [24], or subjected to a physical potential originating from the weak interaction [25], or sometimes in its molecular environment manifested by a potential which is dependent on time [17], or finally in the form of phonons [22]. The fundamental symmetric level E_S and the first antisymmetric excited level E_A form a quasi-degeneracy which is practically indistinguishable from perfect degeneracy. Nonetheless most theories of this type have been consistently criticized [19].

Bearing in mind the extreme instability of this quasi-degeneracy, Claverie and Jona-Lasinio [26] make a simpler proposal: that the system has become classical as regards a single degree of freedom.

Other theoreticians warn against the abuse of the principles of quantum mechanics. Schrödinger has developed the paradox of the cat. Some propose restrictions or additions based on experimentation [21, 23]. Jörgensen [23] notes that an enantiomer cannot be described by a proper function of a Hamiltonian independent of time. He draws a comparison between this situation and asymmetry through the Jahn-Teller effect, and puts forward a set of paradoxes in which quantum mechanics can suggest proper states which turn out to be unacceptable, such as the non-living/living state of Schrödinger's cat.

It should however be noted that all the authors mentioned above base their discussion exclusively on the inversion barrier. We demonstrated in the previous section that the dissociation barrier, or any other process leading to the destruction of asymmetry, is also important. This leads us to consider simultaneously the path of easiest destruction and the path of easiest inversion for a couple of left-handed and right-handed structures. The two paths set out from the same situation, and their combination gives rise to a one-dimensional domain, where the wave function of the molecule has to be determined. Strictly speaking, we should consider a surface of potential with numerous parameters, but we know (from the theory of the absolute reaction rate) that barriers with several dimensions do not bring about a qualitatively new result compared with the results obtained with single-dimension barriers.

The combination of the path of easiest inversion and the path of easiest destruction constitutes the *path of easiest destruction and inversion* which will hereafter be referred to as P.E.D.I.

In an earlier paper [27], we considered the case of an asymmetric molecule resembling NIIDT. If we represent its energy E as a function of the geometrical parameter x, the distance from the summit A to a plane defined by three atoms $B_1 B_2 B_3$ (Figure 2), the inversion barrier is represented in conventional form. The extension on the side of the rising x is the concrete expression of an inversion path which — one may suspect by reason of symmetry — is the easiest. The energy as a function of x would then constitute the P.E.D.I. of the molecule.

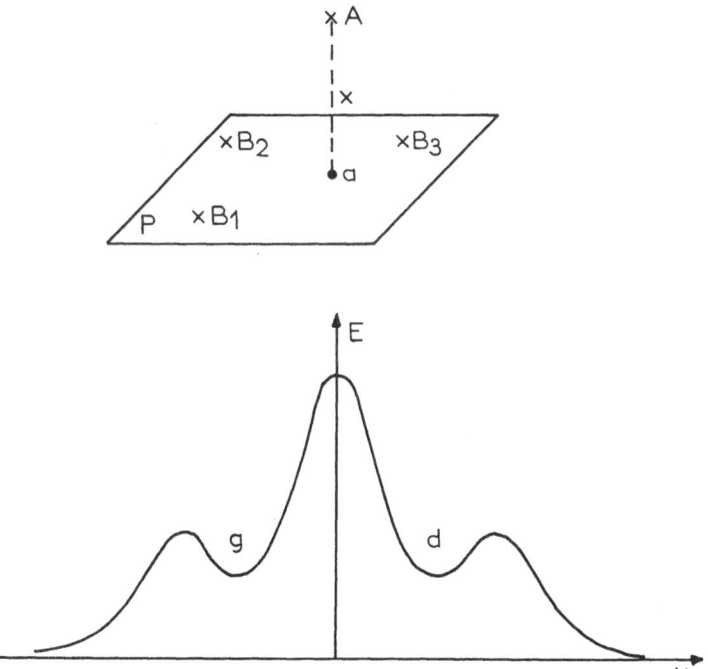

Fig. 2. Example of rectilinear P.E.D.I. (path of easiest destruction and inversion); A B_1 B_2 B_3: atoms or radicals; P: plane of symmetry causing transition from left-handed structure to right-handed structure; E: energy of a configuration; **g**: left-handed configuration; **d**: right-handed configuration; x: distance from atom A to plane of symmetry P.

We will determine the wave function at a single variable from the potential function along the path (Figure 3). As the lowest potential is at infinity, this is a problem which is dependent on time. We will also state an initial condition which is the location of the function according to an amplitude curve close to a normal distribution at the lowest point of the left-hand well. Over an infinite period of time, the wave will escape from the well and will be close to zero everywhere: this signifies only that the probability of spontaneous destruction is not zero. For a time T which is much shorter than the probable destruction time, the organization of the wave will be almost stationary, which signifies only that the molecule is practically stable.

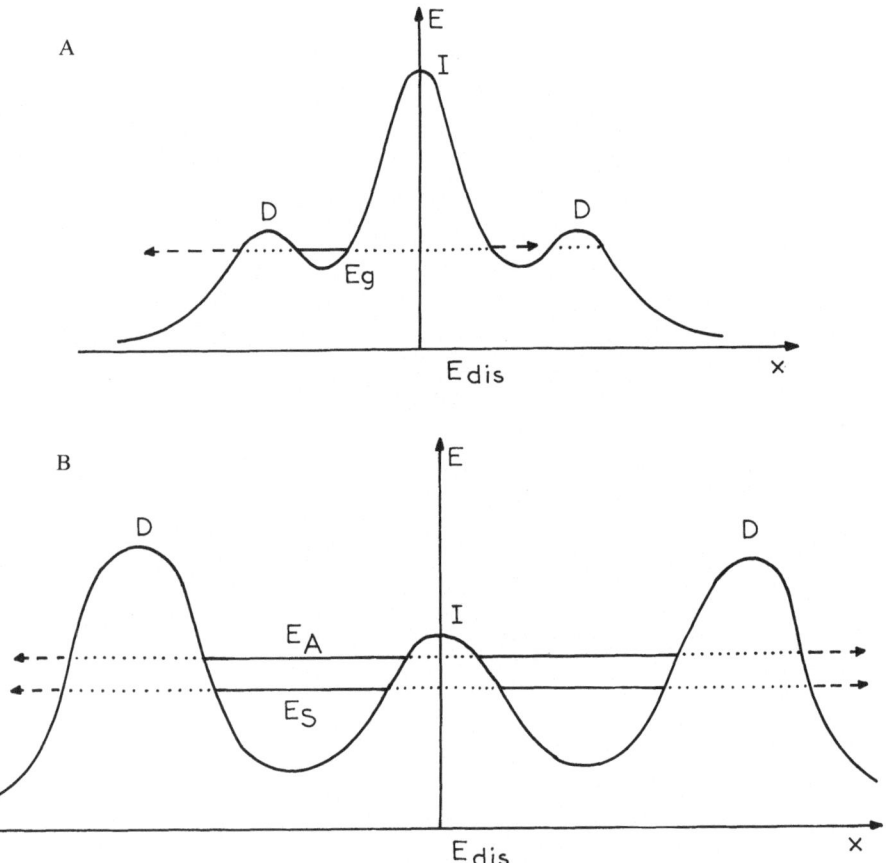

Fig. 3. Potential curve along a P.E.D.I. A. Non-connected domains of stability. B. Double well in a symmetrical domain of stability. x: Curvilinear abscissa along a P.E.D.I.; E: Energy of an x abscissa configuration; E_{dis}: Energy of totally dissociated compound; I: Inversion barrier; D: Dissociation barrier; E_g: Energy of one of the enantiomers (left-handed one shown in figure); E_S: Energy of the symmetric state; E_A: Energy of the antisymmetric state; — Energy level of an almost stationary wave; — — → Energy level of a progressive wave; · · · · · Energy level whose kinetic part is negative.

Let us first suppose that T can be chosen to be far greater than the probable inversion time (this is case B): the wave organizes itself in the double well as in the absence of spontaneous destruction. In an earlier paper [28] we described how a computer simulation could help follow this organization in a double well: the behaviour of the wave is remarkably like a ripple filtering through from the first into the second well. By the end of time T, there has been a return to the classical situation of the infinitely deep double well. If we now consider successive T times the result will only be a crushing of probabilities inside the double well.

Let us now suppose that we are in case A. T is much less than the destruction time, which itself is much smaller than the inversion time. By the end of time T, the wave is organized in the left-hand well. There is only a negligible presence of the wave in the right-hand well. Over successive periods of time, the probability of presence in the left-hand well will have become negligible before the probability of presence in the right-hand well is appreciable. The interference of waves through the median barrier can therefore play no observable role, while structuring into an almost stationary wave in a well takes place in the same conditions as for a symmetric molecule.

At no time did we encounter the difficulty of symmetry, since the problem raised is that of the evolution of the wave function, and not its stationary state.

It is the slow decrease in probability, the edges of the well being more porous than the central compartment, which causes the localization of the enantiomers.

The reason why existing theories are unsatisfactory is that they are based on the supposition of deep double wells and true stationary states, while they neglect the fact that concerned molecules are out of thermodynamic equilibrium, as indeed are metal complexes and organic molecules.

We would also suggest that compounds having comparable inversion barriers and very different destruction barriers would give rise to different experimental observations. Highly stable compounds would only enable the observation of symmetric and antisymmetric states, and compounds with low stability would enable the observation of enantiomers. Clearly there is a need for experimentation, in particular involving independent action on one or other of the barriers by means of suitable catalysts.

Moreover, numerical experiments simulating the evolution of a pseudostationary state on a PEDI should be conducted systematically, with a wide range of variation in inversion and destruction barriers.

3. The Origin of Chirality

The theory we set out in Section 2.2 states that enantiomers can be observed as molecules in a racemic. It does not explain that they can be observed with a probability different from 0.5 before having been artificially separated.

The justification of a probability $P \neq 0.5$ and qualitatively different effects calls into play a causality which is added to the habitual mechanics of molecules.

3.1. POSITION OF THE PROBLEM

A problem which earlier elicited considerable interest — linked to probabilistic notions on the one hand and on the other to the dissymmetry of the universe postulated by Pasteur [5] in 1874 — is now giving rise to fresh interest [4, 6, 16, 27, 29—37] since the dissymmetry was proven [38—40] in 1956. The asymmetry of weak interactions, manifesting itself in the non-conservation of parity, has been extended [41—43] to a series of fundamental laws. This has consequences in nuclear physics [30] and atomic physics [44]. But the problem as to whether they engender chirality at the molecular level remains unresolved.

Some authors [4, 6, 33, 34] have reviewed the various mechanisms which it is thought may be linked to the origin of chirality; Norden [30] has listed and classified some forty of these. He distinguishes between microscopic selection, made on the basis of averages throughout the biosphere, and macroscopic stereoselection applied to the same biosphere, and concludes in favour of the latter.

This approach in fact brings together the two problems of the origin of chirality and the origin of optical purity. If *a priori* we assign decisive importance to an amplification mechanism (either before life, or during biological evolution) the two origins are not separated. Using a more general problematic approach, they must be examined in turn as we propose to do here, and as others have done earlier [44]. This is why we have regrouped these mechanisms — some of which lead directly, and others indirectly to optical purity (and are not therefore in themselves a selection) — according to whether they are the product of chance, given cosmological situations, or a fundamental dissymmetry in the universe.

3.2. THERMODYNAMIC GENERATION

Some hold that the generation of chirality could be explained by the occurrence of an initial fluctuation of sufficient importance in a molecular population, followed by the operation of an amplifying system.

If it is true that Curie's principle [2] is violated by the appearance of the initial fluctuation, and then borne out in its amplification, then these phenomena would be globally equivalent to an autocatalysis [16, 45]. The biological environment may or may not play a role in the appearance of the initial fluctuation or in its amplification [46, 47]. Although autocatalysis plays a significant role in biology, purely physicochemical amplifying systems have been described [48]. This occurs when the combination of the two enantiomers or racemic compound is less stable than the enantiomers themselves [49—52]. In any case, we have known for thirty years how to make optically active crystals appear in laboratory conditions [53, 54], but with no preference for dextrorotatory or levorotatory crystals. Indeed such experiments remain very similar to that of Pasteur. The concept of an initial event at the origin of chirality still remains; if such an event were reproduced, it could have resulted in the inverse of the molecules which are actually given preference. The sign of chirality is not determined either by terrestrial conditions or by the permanent laws of the universe. But no trace of an inverse event has apparently ever been found. (However, there does exist a preferential sign in the crystallization experiments of Thieman, which will be described in Section 3.4.2.)

The idea of a very powerful amplifying mechanism which made what was initially a very local and very small difference into something general and exclusive rapidly won approval. Numerous amplifying mechanisms have been invoked, although proof that any of them has actually been at work is matter of discussion.

It is the random nature of fluctuation which is disputed by the theories of *directed* chirality. These theories derive the origin of molecular chirality from a chiral situation having already arisen on earth, or from the physical laws. Many causes of an asymmetric nature can be described, but according to Curie's principle their effects may or may not be asymmetric. Thus for each hypothesis advanced, proof must be provided that it really leads to the synthesis of chiral molecules or to a differential interaction with a racemic population.

3.3. GEOLOGICAL, GEOGRAPHIC OR COSMOLOGICAL GENERATION

Some theories search for the origin of molecular chirality in space [3, 55—57]. Others again have deduced from the absence of chirality in the amino-acids of a meteorite [47, 58] that optical purity is the result of biological mechanisms. Other theories search for the origin of chirality in the field of geophysics.

We mentioned earlier that right and left quartz occur in unequal abundance over the earth's surface. Could molecular chirality originate from crystalline chirality? Several authors, and in particular A. Julg [59], have demonstrated on the basis of precise quantum-chemistry calculations that the adsorption energy of a right-handed or left-handed alanine molecule is not the same for a right-handed as for a left-handed quartz crystal.

Natural magnets have been considered in some theories [4], as has geomagnetism [60].

More generally it can be said that the superposition of the gravity field, of geographic asymmetry, and of the electromagnetic field creates a chiral situation on the earth which manifests itself in a certain number of ways.

For example, it has been calculated [61] that the quantity of right and left circularly polarized light received by the surface of the earth is not the same, which was verified much later by observing a circular polarization of about 0.1% [62] of sunlight at dawn and at sundown in an IR band (but not in the UV band). Moreover, it has been demonstrated in experiments that decomposition [12] by circularly polarized light is stereoselective and induces optical activity. The same is true of synthesis [9—12]. This theory in fact envisages the natural achievement of the absolute asymmetric synthesis performed in the laboratory. But quantitative estimates [4] are unfavourable, for natural polarization means that light rays travel great distances near the surface of the oceans, which absorbs wavelengths liable to be chemically active.

Finally, the action of natural β-radiation is also a possible geological cause; but not only geological, for the partial polarization of this radiation is the consequence of a permanent and universal law, the asymmetry of the weak interaction.

3.4. GENERATION BY THE ASYMMETRY OF SPACE

Considerable interest has been aroused in the hope that the chirality of living matter can be explained as a consequence of the chirality of physical space.

Numerous authors have pointed to the similarity between the chirality of living matter and that of space.

The asymmetry of fundamental laws (and thus of space) manifests itself at the molecular level [63] essentially in two ways:

(1) two optical antipodes are not exactly symmetrical;
(2) there exists in nature a source of polarized radiation, namely β radioactivity.

Thus the origin of molecular chirality has been sought along these two paths [34—37, 64—70].

3.4.1. *Circular Polarization of Natural β Radiation*

As the β particles emitted by natural radionuclides are polarized circularly and mainly to the left, it was proposed to expose racemic compounds or inactive substances to natural β radiation, and then to examine them for traces of active products, as a decomposition or selective synthesis could have taken place. The same could be true under the action of artificially polarized electrons [34, 69, 70].

These experiments were performed. No synthesis was recorded. Preferential radiolysis of various amino acids was attempted [37, 67—75]. This experimental verification has been published several times, but has never been accepted without criticism [4, 68, 69, 76]. An enrichment in levorotatory amino acid was generally found, while an opposite effect, enrichment in dextrorotatory tryptophan, has also been observed [70].

Moreover, the existence of an autocatalytic effect was demonstrated, the radiolysis rate depending on the number of left-handed and right-handed molecules already decomposed [37].

Another method was proposed [63, 68, 76], challenged [77] and then proposed once more [78]. It is estimated to multiply by a factor of the order of 10^4 the direct observation of radiolysis differences by Gidley, Rich, Van House and Zitzewitz, who bombarded an amino acid powder with a collimated beam of polarized positrons. The positronium thus formed is then detected in its singlet state. The relative difference found for leucine being 4×10^{-2}, the authors deduce that preferential radiolysis would give a relative difference of below 10^{-7}.

Finally, in another type of experiment, Kovacs and Garay [79] crystallized sodium ammonium tartrate (cf. Section I) in the presence and in the absence of a natural radioactive element ^{32}P: the D and L distribution of the crystallization centres was found to be symmetric in the absence of radioactivity and dissymmetric in its presence. The quantity of radioactive substance was related to optical activity. 1 to 5 mg/ml of chiral germ is needed in order to have a significant influence on the optical activity of the crystalline phase, while a thousand times less had the same effect. The global effect is therefore still of the autocatalytic type.

Keszthelyi [34] has classified and discussed the experiments which tend to reproduce the generation of *chirality*. He concluded in 1984 that a connection between the violation of *parity* of weak interactions and chirality of biomolecules is highly improbable. This is the logical conclusion of experimental results: "one can hardly find any experiment which is not refuted by another one", and above all by the result of a comparison between theory and experiment with regard to the relative difference between decomposed molecules in *radiolysis* by *polarized electrons*.

These arguments are in fact not totally irrefutable.

(1) Very different orders of magnitude exist for *paramagnetic* and *diamagnetic* molecules, as we have calculated [27].
(2) The theory predicts the probability of decomposition of the first molecule. The experiment measures the probability of differential decomposition of a population; this process is *autocatalytic* as noted by Garay [37] as early as 1968.

We will develop these two ideas further later, but from a phenomenological viewpoint we would suggest the following conclusions.

(1) Enough experiments have been described in which *optical activity* has appeared to make us consider that *absolute asymmetric synthesis* can be performed by the action of *circularly polarized electrons* on an inactive substance.
(2) An elementary theoretical explanation can be provided by the analogy between electron optics and photon optics.

For a more precise theoretical explanation, other authors [80, 82] have calculated the steric preference in radiolysis (or possibly [81] in the formation of positronium). Unfortunately, molecules are implicitly or explicitly supposed to be diamagnetic. *A non-zero difference of cross section is always shown.* The values calculated by Hegström [81] for the relative difference of cross section between the two forms range from 10^{-10} to 10^{-11} for amino acids irradiated at 100 Kev. They would be 100 times greater in the case of a heavy atom ($Z \approx 100$).

This calculation leads to effects which are lower than experimental results [76]. But, as we have seen, the comparison is not valid. The theoretical calculation concerns the first radiolyzed molecule, not a fraction of the product.

Moreover, circular dichroism in relation to the polarized electrons causes the rotation of the plane of polarization in relation to a beam of rectilinearly polarized electrons [7].

In order to discuss the possible generation of molecular chirality from natural radioactivity, the collision process must be analyzed according to the principles of quantum mechanics. But as the molecules involved are inevitably complicated, this type of processing is difficult.

In 1982 we processed [27] the collision process by the use of a highly simplified model representing an asymmetric atom. The calculation is nonetheless difficult.

Hypotheses which were often very approximate brought us to the following estimates of orders of magnitude:

$$\frac{\Delta\sigma}{\sigma} \sim 10^{-3} \qquad \text{for paramagnetic molecules (with a permanent moment of the order of 5 Bohr magnetons);}$$

$$\frac{\Delta\sigma}{\sigma} < 10^{-7} \qquad \text{for diamagnetic molecules.}$$

σ is the cross section, and $\Delta\sigma$ the difference in cross section, depending on which of the enantiomers is involved in the collision with a polarized electron.

(During the preparation of this paper [27], we were not aware of the previous results, which this paper does not contradict.)

The immediate result of the processing of a transition probability using quantum mechanics is that this probability does not change according to whether the function of the initial wave concerns the asymmetric atom and the symmetric atom, or whether they are taken in reverse order. The first process is preferential radiolysis, and the second asymmetric synthesis.

It does not therefore seem useful to suppose that there are reserves of racemic compound on which natural β radiation would act. It is simpler to suppose that the radiation has caused prebiotic synthesis with chiral preference.

The essential result is the role of the molecule with a permanent magnetic moment. The ratio with the diamagnetic molecule is between 10^4 according to our limit and 10^8 according to the lower limit of Hegström. If preferential radiolysis is possible in the diamagnetic case it should appear very clearly in the paramagnetic case.

These calculations concerning the polarized electron are indirectly confirmed in the calculations of other authors concerning the circularly polarized *neutron*. Harris and Stodolski [25] have shown that a beam of polarized neutrons has a non-zero optical activity if it interacts with the electrons of a closed shell chiral molecule. Gadzy and Ladik [83] point to a difference in cross section for a chiral molecule with a permanent magnetic moment, provided that it has a fixed orientation in relation to the incident beam. The calculation, based on quantum electrodynamics, describes the interaction between the polarized beam and the target molecule by the exchange of circularly polarized virtual photons. Processing could not therefore be directly applied to the electron, the authors note, as it excludes the terms of exchange. This processing is in agreement in qualitative terms with the collision calculation we made for the polarized electron, indicating a difference in order of magnitude between paramagnetic and diamagnetic molecules.

The small size of the effect due to the diamagnetic molecules may explain the disputed success of the experiments. A paramagnetic impurity at a concentration of 10^{-4} can change results in the ratio 10^4. In fact these impurities may originate not only from the products but also [77] from the experimental process.

Moreover, it can be postulated that the induced dichroism demonstrated for electromagnetic optics [15] also exists in electronic optics. Stereospecific radiolysis would then be an autocatalytic phenomenon, the difference in cross section increasing commensurately with the appearance of a disequilibrium in the environment. Simply expressed, this means that the phenomenon starts only with difficulty, but that — if it occurs — it exceeds the results expected by linear extrapolation, an impression which emerges from the published experimental results as a whole.

3.4.2. Dissymmetry Between Two Optical Antipodes

The asymmetry of weak interactions determines a difference of energy between two enantiomers [61 to 66]. To demonstrate this, the weak neutral current linked to parity violation is considered, together with the spin-orbit interaction. Opposite energy terms result for the enantiomers. The total energy difference would seem to be of an order of magnitude lower than 10^{-19} a.u.

Experimental verifications can be no more than indirect [34], and have consisted of experiments of crystallization [84, 85] and polymerization [86]. Comparative preferential dissociation by β and γ radiation [87, 88] could be connected to this energy difference between optical antipodes [34].

In a theoretical discussion [89], Hegström has compared the effect of the energy difference of optical antipodes and differential radiolysis. The paramagnetic hypothesis would cause the balance between the two mechanisms to disappear to the benefit of the polarized electrons. But above all the postulated kinetic is open to dispute for it neglects autocatalysis. This theory does however rightly attract our attention to the combined action of various effects.

3.4.3. Need for a New Model of Generation by Weak Interactions

The reality of absolute asymmetrical synthesis by weak interactions does not in itself prove that it is the cause of the appearance of molecular chirality in nature, since the existing model ('the hypothesis of Vester–Ulbricht') can be disputed [34], and a new model should be constructed.

(1) It is not convincing to show that weak interactions could have selected amino acids in the left-handed form — as some are to be found in right-handed form amongst the antibiotics, in the cell walls of bacteria and blue-green algae, and in snake venoms [90] — that sugars are mostly dextrorotatory (although rhamnose is levorotatory), and that substances — admittedly of incidental importance — such

as terpenes can exist in unequal abundance in both forms. It seems reasonable to think that physical factors have created an instrument of asymmetric synthesis which living organisms have used in accordance with biological laws. The first traces of chirality may have appeared in the form of asymmetric catalysts (the actions of the catalyst can be limited to the excess molecules of the racemic compound).

Such a hypothesis is in agreement with that of Fox [29], according to which *reactions* rather than *substances* have been captured by the reproduction of macromolecules.

(2) According to the hypothesis that chiral generation is produced by natural β radiation, it can be assumed that chirality appeared in paramagnetic molecules, a common example of which is that of organometallics containing a transition atom, which play an important role in redox balances [91]. It has been suggested that the inorganic environment played an important role in the origin of biological evolution and then in molecular evolution.

(3) We should also note that the hypothesis of direct asymmetric synthesis by polarized electrons involves the same probabilities as those [28] of differential radiolysis. Ulbricht suggested as early as 1959 [35] that either a synthesis or a destruction could be envisaged. The first hypothesis would seem to be simpler.

(4) If we wish to constitute a quantitative model, we will have to take into account the autocatalytic character (probably linked to an induced electronic circular dichroism) of asymmetric synthesis reactions.

It would seem that these four hypotheses could form the basis for a qualitative and quantitative theory of chirality generation through weak interactions which could be "refutable and verifiable". Indeed, the alternative models (random or geophysical) leave room, with equal or nearly equal chances, for optical antipodes of the forms observed in nature; it would be enough to find a trace of them to refute the theory. On the other hand, the stages it suggests can be verified by experimentation.

3.5. DISCUSSION

We have just stated only part of the theories proposed to explain the origin of molecular chirality, for the issue of the chirality of weak interactions must be reformulated. It might be expected that we estimate their relative probabilities. But the opposite combination of various factors should also be envisaged. What would be the effect in prebiotic conditions of a cloud of volcanic, radioactive, micro-crystalline, magnetized dust? . . . A major asymmetric fluctuation would be more probable at a point where a chiral situation already existed, etc. . . . Nor is there any objection to a juxtaposition of such factors: the appearance of molecular chirality in many centres under the action of different factors.

In fact generation by the asymmetry of interactions is studied for its epistemo-logical interest. Experimental and above all theoretical results seem significant enough to ensure that its participation in the origin of chirality cannot be dismissed.

4. Life and Chirality

Like many physical phenomena (electricity, principle of equivalence . . .) chirality was discovered as a biological phenomenon. The association between chirality and living matter has continued because of the usefulness of biological concepts in the study of chirality and because of the optical purity of living organisms.

4.1. THE USEFULNESS OF BIOLOGICAL CONCEPTS IN THE STUDY OF CHIRALITY

(1) The asymmetric molecules studied are mainly organic, and are often molecules of natural species. The information bit linked to the chiral molecule is conserved by a process of autocatalysis which simulates vital phenomena and by the death of the molecule as an assemblage of atoms.
(2) The existence of enantiomers appeared for a long time to be a violation of the principles of quantum mechanics, and none of the many theories which sought to explain it was wholly satisfactory.
(3) The optical purity of enantiomers inside living matter appeared initially to be difficult to understand outside the context of a particular biological causality.
(4) Stereospecific toxicity is rather like a very simple form of recognition of the self and the non-self, an essential function in living organisms.
(5) If we consider the hierarchization of organizational systems, it has been noted that one particular form of asymmetry is associated with each of the different levels.

Parity violation is associated with the nuclear level: it may be said [30] that atomic nuclei are intrinsically asymmetric. Chirality is associated with molecules, helicity with macromolecules, and asymmetry of form to organelles, cells, organs and individuals. But the relationship between successive levels of asymmetry is not immediate: the two signs of coiling of a macromolecular chain can exist even if the links have the same chirality. On the other hand the existence of a regular coil is not independent of the optical purity of the links [30] and the proportion of the two forms of helix is not independent of the sign of the optical activity of the links. We have suggested that a no less complex relationship of filiation exists between nuclear asymmetry and the chirality of biomolecules.

A priori molecular chirality is in interaction with that of the upper or lower levels, both with regard to form and stereochemical composition. For example optical purity is maintained in individuals because they assimilate optically pure substances from the biosphere.

(6) In this paper we set out from the commonplace remark that organic molecules

are out of thermodynamic equilibrium. Their domain of pseudostability is limited by destruction barriers. The problem of molecular structure is not to find a stationary wave, but to find a wave which is dependent on time. But the progressive damping of the wave of the molecular system by flow through the destruction barrier has in general no experimental consequence. This is the case if the molecule has no isomers, or has non-equivalent geometric isomers.

It is also the case for an asymmetric molecule if the inversion barrier can be much more easily crossed than the destruction barrier. But there are contrary cases in which there can be no sufficient interference between the waves existing on either side of the inversion barrier, and quasi-stationary organization in each well gives an autonomous molecule.

Molecular chirality, like life, is a phenomenon dependent on time, which cannot offer a true stationary state.

Chirality has attained autonomy as a physical phenomenon, but it remains linked to biology by the nature of the concepts involved: *pseudostationary state, death, reproduction, multiplication, recognition, and hierarchization between the individual and the species.* The unique chirality-related information bit is related to chirality in a manner which can be described by the language of biology. Chirality appears in the context of nature to be a fundamental break.

4.2. THE RELATIONSHIP BETWEEN THE ORIGINS OF LIFE AND THE ORIGINS OF CHIRALITY

Is there a biological causality which cannot be reduced to the causality of physics? Some have argued [92] in favour of non-reductibility that the repetition of a great number of identical experiments is never done in biology. As biological predictions have no constrictive force, the scope left to chance is greater. On the other hand, the impossibility of rapid generation of a living organism at the current state of its evolution forms no obstacle to the progressive generation of the biosphere [29a]. Life would seem to form part of the permanent laws of the universe.

The "origins of life" have become a subject of theoretical and experimental studies, which include the formation of prebiotic molecules. Indeed, the passage of ionizing radiation through a reducing atmosphere [93, 94] containing the elements C, N, O, H gives birth to many organic molecules, in particular asymmetric molecules, from which biomolecules could derive. But in such experiments right-handed and left-handed molecules appear with the same probability. And it has not been demonstrated that this natural synthesis is or is not at the origin of chirality.

Many authors have thus wondered whether this origin took place at the origin of life, or before or after: is it a cause or an effect? does it have some usefulness for life? [6, 30] and above all according to which mechanism [3, 4, 48, 53, 95, 96] did it appear? We have already seen that life is not a necessary condition for absolute asymmetric synthesis, but that it seems unlikely that this process of synthesis could have produced 100% optical purity. What could be the advantage or necessity of chirality for living organisms?

4.3. THE OPTICAL PURITY OF LIVING ORGANISMS

Kizel [6] has reviewed a certain number of advantages which optical purity has for living organisms, such as the solidity of polymer chains [30, 97, 98] the rapidity of biological reactions, and economy of information in the coding and transmission of messages. He has discussed the mechanism of chirodiastaltic forces, which establish a difference between the interaction of the same enantiomer with two optical antipodes and intervene in biomolecular recognition.

Moreover it seems clear that the restriction to a single enantiomer per chemical species increases the repetitivity of biological experiments, and tends to facilitate the learning processes and choices of life.

A more precise idea could be put forward: the vital processes are based on extremely specific reactions [8], and thus on molecular recognitions. The 'lock' which recognizes, which is generally chiral, cannot behave identically in relation to two keys turning in the opposite direction.

Finally, if one enantiomer is stored in the 'self', the other must be ignored or stored in the 'non-self': if one enantiomer is stored in the 'non-self' (and triggers for example an immune reaction) the other must be ignored or stored in the self (and not trigger this reaction). The phenomena would be reversed for the enantiomer lock.

A statistically racemic being, if it can be supposed to exist, would come up against problems of formal logic in the vital functions (in particular immune functions). According to Ulbricht [33, 35], "racemic life" would seem to be a contradictory concept; according to Langenbeck [99] it would seem to be thermodynamically unstable, and the imbalance must be increased by evolution.

Let us imagine the existence of a species with a "perfect immune system", through which any substance would be of necessity classified in the self or the non-self. Any optical impurity would thus be immediately destroyed. If on the contrary a molecule can be ignored by the immune system a third logical value comes into play: we could imagine an evolution from the almost racemic to the almost pure. A being in which right-handed and left-handed forms would coexist in different proportions would evolve towards optical purity excluding the minority form (except in reject substances).

The optical purity of living organisms would thus be guaranteed at two levels: at the molecular level by the life time of the chiral molecules which is lower than their inversion time, as we saw in Section 2; and at the level of overall operation by the set of functions involving molecular recognition.

Morozov [45] has discussed important features in this field: stereoselectivity in the synthesis of biomolecules by both space and time organization; their destruction by stereospecific enzymes which oxydize D-amino-acids [100] (the second method for cumulating the D-form is accumulating it in those structures which do not participate directly in the metabolism); connection between disturbances in the functioning of organisms and chiral impurity [101]. On the other hand, he has proposed a quantitative model for spontaneous symmetry breaking, including a lower limit for the quantity of chiral material in the medium which is capable of breaking the initial symmetry, biological evolution and the hierarchy of organized systems.

5. Conclusion

In accordance with Curie's principle, molecular asymmetry may lead or not lead to chiral properties. Two optical antipodes can be isolated, or rather two levels very close to each other can be distinguished: symmetric and antisymmetric. The domain of stability is connected in the second case, and non-connected in the first.

Chiral information is reproduced by an autocatalytic process; it is lost by the inversion of its molecular support, and statistically conserved by its destruction. The two inversion and destruction barriers govern chiral memory. It may be said that a chiral molecule exists if the destruction time is lower than its inversion time.

We have defined the path of easiest destruction and inversion. To the potential function along this path corresponds — if the molecules are endothermic — a quantum state which is unbound, and in consequence is not subject to the symmetry of the Hamiltonian operator; however its pseudostability does not distinguish it from an achiral molecule of analogous composition: it can live for a long time, even on the geological time-scale. This resolves, without arbitrary postulations, an enigma which has been with us for sixty years.

This explains the distinction between the two enantiomers but not their statistical imbalance. The origin of this imbalance is problematic. In examining the various mechanisms proposed for its resolution, we have concentrated on the interaction of circularly polarized natural β radiation with the molecules.

A quantum mechanical calculation led us to the values of differential interaction of a polarized electron, either left-handed or right-handed, with a carbon atom. This interaction is measured by σ cross section, for a given molecule with a permanent moment of a few Bohr magnetons, $\Delta\sigma/\sigma = 10^{-3}$; for a diamagnetic molecule $\Delta\sigma/\sigma < 10^{-7}$. *The probabilities are the same for the creation or destruction of a chromophore of optical activity.*

Moreover the analogy of the two optics (electronic and photonic) leads us to predict an induced electronic circular dichroism, and the autocatalytic character of radiolysis or preferential synthesis reactions.

These two theoretical predictions can be applied both to the construction of a model of natural absolute asymmetric synthesis and to a discussion of laboratory experiments of differential radiolysis. The autocatalytic characteristic has been recognized from the first experiment and all the others by their very divergence confirm it. A complementary explanation is to be found in the formation of paramagnetic impurities under the action of radiation. Interpreted in this way, the experiments leave no room for doubt about the reality of absolute asymmetric synthesis by polarized electronic radiation. The most satisfactory model of its use by nature remains to be imagined. The direct formation of asymmetric para-magnetic catalysts by polarized electrons would be an intelligible stage.

In the hierarchically organized levels of life, asymmetry appears at the scale of the nucleus, the small molecule, the chain, forms (organelles, cells, organs, individuals) and stereochemical composition (from the macromolecule to the biosphere).

The chiral molecule is in interaction with the other levels. For example the helical coiling of the chains is neither independent of the chirality of the links, nor simply determined by it. It can be postulated that there exists an analogous relationship of determination and autonomy between the nuclear level and the molecular level which would be in agreement with the formation by polarized radiation of asymmetric catalysts. Other indirect interaction paths between levels are not excluded, such as an effect of crystalline chirality perhaps resulting from nuclear asymmetry.

The interaction of the molecular level with the stereophysiology of the individual would seem to be essential, and to be the only explanation of the high optical purities observed in living organisms.

Finally the stereochemical composition of the biosphere results from a set of interactions between levels, of which the two most significant would seem to be nuclear/molecular interaction through β radiation and molecule/individual interaction through the system of recognition of the self and the non-self. But the molecule/macromolecule, individual/species, and organism/biosphere interactions also play an essential role.

Chirality can be seen as a micro-model of life in which all biological information is limited to a single bit, on which the major biological concepts continue to operate: temporality, death, reproduction, selection, recognition, immunity, hierar-chization, and the interaction between hierarchized levels. It could become the key element of a mathematical model or a philosophy, or, simpler, contribute to our understanding of life.

Acknowledgements

I would like to express my gratitude to Professor R. Nataf for numerous discussions over many years on this subject. I would also like to thank Dr C. Gignoux for his observations during a recent discussion.

Note Added in Proof

The interest in this field has produced many recent works, discussed further in [102]. They often confirm our views, reported above. For instance, an important amplification ratio is observed concerning the enantioselective radiolysis [103]. Calculating it — taking care of the interactions of the chiral molecule — is a new challenge for the theory [104]. Our most significant experimental predictions concern chiral breaking.

(1) Very short lifetimes (hence a pseudo-stationary state) could be observed for unusually active molecules.
(2) Very long periods for oscillations showed that the inversion barrier cannot break the symmetry by itself.
(3) Transformations of oscillating into chiral molecules by catalysts or reagents could be obtained.

Starting from this we are trying to develop a general theory of organization of matter by successive spontaneous symmetry breakings [102].

References

1. J. Biot: *Bull. Soc. Philomath.*, 190 (1815).
2. P. Curie: *Oeuvres*, Gauthier-Villars, Paris (1908); P. G. de Gennes: *Symmetries and Biochem. Symmetries in Solid State Physics*, N. Boccara ed., Paris (1981).
3. G. P. Gladyshev and M. M. Khasanov: *J. Theor. Biol.* **90**, 191 (1981); and refs. therein.
4. W. Thieman: *Naturwissenschaften* **61**, 416 (1974); and refs. therein.
5. L. Pasteur: *Ann. Chim. Phys.* **28**, 56 (1850); *C. R. Ac. Sc.* **78**, 1515 (1874).
6. V. A. Kizel: *Soviet Phys. Usp.* **23**, 277 (1980); and refs. therein.
7. A. Fresnel: *Oeuvres complètes* **I**, p. 731, Paris (1866).
8. T. Sugimoto and N. Baba: *Israel J. Chem.* **18**, 214 (1979); and refs. therein.
9. W. Kuhn and E. Braun: *Naturwissenschaften* **17**, 227 (1929); W. Kuhn and E. Knopf: *Z. Phys. Chem.* **7B**, 292 (1930).
10. A. Moradpour *et al.*: *J. Am. Chem. Soc.* **93**, 2553 (1971); H. Kagan *et al.*: *Tetrahedron Lett.* **27**, 2479 (1971); O. Buchardt: *Angew. Chem.* **86**, 222 (1974).
11. D. Radulescu and V. Moga: *Bull. Soc. Chim. Romania* **A1**, 18 (1939); P. Gerike: *Naturwissenschaften* **62**, 38 (1975).
12. W. Kuhn and E. Knopf: *Naturwissenschaften* **18**, 183 (1930).
13. E. U. Condon: *Rev. Mod. Phys.* **9**, 432 (1937).
14. A. Crémieu, P. Smet, and J. Tillieu: *C. R. Ac. Sc.* **257**, 843 (1963); P. Smet and J. Tillieu: *ibid.* **257**, 3123 (1963); *ibid.* **257**, 3319 (1963); *ibid.* **260**, 445 (1965); P. Smet: *ibid.* **258**, 98 (1964); *ibid.* **261**, 2173 (1965); A. S. Garay, J. Czégé, E. Tolvaj, M. Toth, and M. Szabo: *Acta Biotheorica* **22**, 34 (1973).
15. D. P. Craig, E. A. Power, and T. Thirunamachandran: *Proc. Roy. Soc.* **A348**, 19 (1976); and refs. therein.
16. R. Buvet: in *Les origines de la vie*, ARCAM, Paris, p. 13 (1983).
17. F. Hund: *Z. Phys.* **43**, 805 (1927).
18. C. Sagan: *Evolution* **40**, 49 (1957).
19. P. Pfeifer: Dissertation 6551, Swiss Fed. Inst. Techn., Zurich (1980).
20. L. Rosenfeld: *Z. Phys.* **52**, 161 (1929), M. Born and P. Jordan: *Elementar Quantenmechanik*, Springer, Berlin (1930).
21. R. G. Woolley: *Adv. Phys.* **25**, 27 (1976); *J. Am. Chem. Soc.* **100**, 1073 (1978).
22. E. B. Davies: *Ann. Inst. H. Poincaré* **A28**, 91 (1978); *Comm. Math. Phys.* **64**, 151 (1979).
23. C. K. Jorgensen: *Theor. Chim. Acta* **34**, 189.
24. M. Simonius: *Phys. Rev. Lett.* **40**, 1980 (1978).
25. R. A. Harris and L. Stodolsky: *Phys. Lett.* **78B**, 313 (1978).
26. P. Claverie and G. Jona-Lasinio: *Phys. Rev.* **A33**, 2245 (1986), and refs. therein; G. Jona-Lasinio and P. Claverie: paper to be published in Supplement of *Progress in Theoretical Physics* (issue ded. to Y. Nambu); and refs. therein.
27. A. Laforgue: *Actes 2e Sém. Ec. Biol. Théor. C.N.R.S.*, Publ. U. Rouen, p. 107 (1982).
28. A. Laforgue, C. Bruceña-Grimbert, D. Laforgue-Kantzer, G. del Re, and V. Barone: *J. Phys. Chem.* **86**, 4436 (1982).

29. S. W. Fox: *J. Chem. Educ.* **34**, 1972 (1957).
30. B. Norden: *J. Mol. Evol.* **11**, 313 (1978).
31. L. S. Rodberg and W. S. Weisskopf: *Science* **125**, 627 (1957).
32. F. Vester: Seminar at Yale University (7 Feb. 1957).
33. T. L. V. Ulbricht: *Comparative Biochemistry* **4**, ed. M. Florkin and H. S. Mason, Academic Press (1962).
34. L. Keszthelyi: *Origins of Life* **8**, 299 (1977); *ibid.* **11** (1981); *ibid.* **14**, 75 (1983).
35. T. L. V. Ulbricht: *Quat. Rev.* **13**, 48 (1959); T. L. V. Ulbricht and F. Vester: *Tetrahedron* **18**, 629 (1962).
36. F. Vester, T. L. V. Ulbricht, and H. Krauch: *Naturwissenschaften* **46**, 68 (1957); F. Vester: Dissertation, Univ. Saarlandes, Saarbrücken (1967).
37. A. S. Garay: *Nature* **219**, 338 (1968).
38. J. P. Lee and C. N. Yang: *Phys. Rev.* **104**, 254 (1956).
39. C. S. Wu *et al.*: *Phys. Rev.* **105**, 1413 (1957).
40. R. Nataf: *Introduction à la Physique des Particules*, Masson, Paris (1987).
41. J. H. Christensen, J. W. Cronin, J. W. Fitch, and R. Turlay: *Phys. Rev. Lett.* **13**, 562 (1964).
42. P. K. Kabir: *The C, P, Puzzle,* Academic Press (1968).
43. E. M. Henley: *Ann. Rev. Nucl. Sci.* **19**, 367 (1969).
44. C. S. Fajsi and J. Czégé: *Origins of Life* **8**, 277 (1977).
45. L. Morozov: *Origins of Life* **9**, 187 (1979).
46. G. Wald: *Ann. N.Y. Acad. Sci.* **69**, 352 (1957).
47. M. Ageno: *J. Theor. Biol.* **37**, 187 (1972).
48. C. Ponnamperuma: *The Origins of Life*, Duttin, N.Y. (1972).
49. F. F. Seelig: *J. Theor. Biol.* **31**, 375 (1971).
50. P. Decker and A. Speidel: *Z. Naturforschung* **22b**, 257 (1972).
51. I. Prigogine: *Theoretical Physics and Biology*, A. Marois, Amsterdam (1961).
52. R. E. Pincock, R. R. Perkins, A. S. Ma, and K. R. Wilson: *Science* **174**, 1018 (1971); R. E. Pincock, R. P. Bradsaw, and R. R. Perkins: *J. Mol. Evol.* **4**, 67 (1975).
53. M. Calvin: *Chemical Evolution*, Oxford U. Press (1969).
54. E. Havinga: *Biochem. Biophys. Acta* **13**, 171 (1954).
55. D. Buhl and C. Ponnamperuma: *Space Life Science* **3**, 157 (1971).
56. W. Thieman: *Life Science and Space Research* **13**, 63 (1975).
57. J. J. Berzelius: *Ann. Phys. Chem.* **33**, 113 (1834).
58. H. Kvenvolden, J. Lawless, K. Pering, J. Peterson, J. Flores, C. Ponnamperuma, J. R. Kaplan, and C. Moore: *Nature* **228**, 223 (1970).
59. A. Julg: this paper.
60. L. Mortberg: *Nature* **232**, 105 (1971).
61. A. Byk: *Z. Phys. Chem.* **49**, 641 (1904).
62. J. R. P. Angel and R. Illing: *Nature* **238**, 389 (1972).
63. A. S. Garay and P. Hrasko: *J. Mol. Evol.* **6**, 77 (1975).
64. Y. Yamagata: *J. Theor. Biol.* **II**, 495 (1966).
65. D. W. Rein: *J. Mol. Evol.* **4**, 15 (1974).
66. R. A. Hegström, D. W. Rein, P. G. H. Sandars: *J. Chem. Phys.* **73**, 2329 (1980).
67. M. A. Bouchiat and C. Bouchiat: *J. Phys.* **36**, 493 (1975).
68. A. S. Garay, L. Keszthelyi, I. Demeter, and P. Hrasko: *Nature* **250**, 332 (1974).
69. S. V. Starodubstev, M. N. Garski, and A. G. Sizykh: *Dan SSSR* **129**, 907 (1959); V. I. Goldanskii and V. V. Khrapov: *Soviet Phys. JETP* **16**, 582 (1962); W. H. Bernstein, R. M. Lemmon, and M. Calvin: *Molecular Evolution*, eds. D. L. Rohlfing and A. T. Oparin, Plenum Publ. Corp., N.Y., p. 151 (1972).
70. W. Darge, I. Laczo, and W. Thieman: *Nature* **261**, 522 (1976).
71. W. A. Bonner: *J. Mol. Evol.* **4**, 23 (1974); W. A. Bonner, M. A. Van Dort, and M. R. Yearian: *Nature* **258**, 419 (1975).
72. L. Keszthelyi: *Nature* **264**, 197 (1976).
73. L. A. Hodge, F. B. Dunning, G. K. Walters, and G. J. Schroepfer, Jr.: *Nature* **280**, 250 (1979).
74. I. Deszi, D. Horvath, and Z. S. Kajcsos: *Chem. Phys. Lett.* **23**, 549 (1973).
75. W. Brandt and T. Chiba: *Phys. Lett.* **57A**, 395 (1977).

76. A. S. Garay, L. Keszthelyi, I. Demeter, and P. Hrasko: *Chem. Phys. Lett.* **23**, 549 (1973).
77. Yan-ching Jean and H. J. Ache: *J. Phys. Chem.* **81**, 1157 (1977).
78. D. W. Gidley, A. Rich, J. Van House, and P. W. Zitzewitz: *Nature* **297**, 639 (1982).
79. K. L. Kovacs and A. S. Garay: *Nature* **254**, 538 (1975); K. L. Kovacs: *Rad. Effects* **31**, 225 (1977); *J. Mol. Evol.* **10**, 161 (1977); *Origins of Life* **9**, 219 (1979).
80. A. K. Mann and H. Primakoff: *Origins of Life* **11**, 255 (1981).
81. R. A. Hegström: *Nature* **297**, 643 (1982).
82. Ya. B. Zeldovich and D. B. Saakyan: *Soviet Phys. JETP* **51**, 118 (1980).
83. B. Gadzy and J. Ladik: *Chem. Phys. Lett.* **91**, 158 (1982).
84. W. Thieman: *Naturwissenschaften* **61**, 476 (1974); W. Thieman and K. Wagener: *Angew. Chem.* **9**, 740 (1970).
85. W. Thieman: *J. Mol. Evol.* **4**, 85 (1974).
86. W. Thieman and W. Darge: *Origins of Life* **5**, 1963 (1974).
87. M. Akaboshi, K. Kawai, H. Maki, and K. Kawamoto: *Origins of Life: Proc. of the 2nd ISSOL and 5th ICOL*, ed. H. Noda, p. 343.
88. M. Akaboshi, M. Noda, K. Kawai, H. Maki, and K. Kawamoto: *Origins of Life* **9**, 181 (1979).
89. R. Hegström: *Origins of Life* **14**, 405 (1984).
90. R. Bentley: *Molecular Asymmetry in Biology* **1**, p. 222, Academic Press (1969).
91. S. Wherland and H. B. Gray: 'Electron Transfer Mechanisms Employed by Metalloproteins', in *Biol. Aspects of Inorg. Chem.*, The Biorganic Group, Wiley, N.Y. (1977); P. M. Vignais *et al.*: *Current Topics in Bioenergetics* **12**, 115 (1981).
92. Elsasser: *J. Theor. Biol.* **90** (1961).
93. S. L. Miller: *Proc. of the 4th Congress on the Origins of Life*, L. Margulis ed., Springer Verlag, p. 128.
94. J. Duchesne: in *Les origines de la vie*, ARCAM, Paris, p. 86 (1983).
95. A. Amariglio and H. Amariglio: *Molecular Evolution, Evolution and the Origin of Life*, eds. R. Buvet and C. Ponneramperuma, Amsterdam (1971), p. 63; K. Harada, *ibid.*, p. 71.
96. M. G. Rutten: *The Origin of Life by Natural Causes*, Elsevier, Amsterdam (1971).
97. E. R. Blout and M. Idelson: *J. Am. Chem. Soc.* **78**, 497 (1956).
98. K. Harada: *Naturwissenschaften* **57**, 114 (1970).
99. W. Langenbeck: *Die organische Katalysatoren*, J. Springer, Berlin (1935).
100. J. S. Fruton and S. Simmons: *General Biochemistry*, 2nd ed., John Wiley, New York (1959); D. E. Atkinson and S. W. Fox: *Arch. Biochem. Biophys.* **31**, 220 (1952); L. Birkhoher and N. Wetzel: *Z. Physiol. Chem.* **264**, 31 (1940).
101. K. Miescher: *Experimentia* **11**, 417 (1955); W. Kuhn: *Adv. Enzymol.* **20**, 129 (1958); F. Kögl: *Experimentia* **5**, 173 (1949).
102. A. Laforgue: 'The chiral molecule', *Theochem* (in press); 'Chiralité et biologie', *Séminaire de l'Ecole de biologie théorique*, Solignac 1985, éd. Y. Bouligand (in press); 'Les propriétés chirales de la matière vivante et la Physique quantique', *Annales de la Fondation Louis de Broglie* (to be published), and refs. therein.
103. E. Conte, G. Fanfani, M. Pieralice, R. Amerott, and A. d'Addabo: *Origins of Life* **17**, 51 (1986), and refs. therein.
104. A. S. Garay: *Life Science* **II10**, 1393 (1971); A. S. Garay, J. Czege, L. Tilvoy, M. Toth, and M. Szabo: *Acta biotheur.* **22**, 340 (1973).

Molecular Pathology

The Adventure of Molecules in Pathology

GEORGES SCHAPIRA

Institut de Pathologie Moléculaire, 24, rue du Fg St-Jacques, 75014 Paris, France.

1. Introduction

The atomic hypothesis, first formulated around 500 years B.C., was without influence on medicine, when empiricism was at a stage compared to superstition. Hippocrates, the master of Cos Island, and 600 years later Galen, in Rome, were the most famous men in medicine.

In the 17th century, the microscope and the cell were discovered. As a result, in the 19th century, cellular theory and, with the German Virchow, cellular pathology were founded.

At the beginning of our century, in Oxford, Sir Archibald Garrod [1] described some inborn errors in metabolism due to an enzymatic defect. The four diseases he described were uncommon; they all had a hereditary origin, which was confirmed by his friend, the geneticist Bateson, who had heard of the work of the Augustinian monk Gregor Mendel. He suspected that this kind of disease was not rare, and that enzymatic defects would appear more and more frequently. From the four diseases he described we arrive today at 4000!

The notion of molecular disease [2—5] was magnificiently emphasized, in the middle of this century, by the American Linus Pauling, who was not a physician, but a chemical physicist. He obtained two Nobel prizes, one for his work in Chemistry, another for fighting for peace in Vietnam. He could have been awarded a third one for molecular medicine!

2. Mendelian Hereditary Diseases

In the beginning, molecular hereditary diseases were localized in a protein molecule. But soon the genetic aspect was involved, with the help of genetic engineering, developed by several Nobel laureates. The primary cause of a hereditary disease is now localized in DNA, and we can recognize the primary defect on the gene [6]. Two main observations permitted this discovery.

1. The DNA of the chromosome is very long: with restriction endonucleases, one can obtain many short reproducible fragments, which can be separated.
2. The fragments must be amplified in order to be investigated. The recombination of the gene with another DNA, more precisely with a plasmid which is auto-reproducible in bacteria, allows the multiplication of the gene in

Jean Maruani (ed.), Molecules in Physics, Chemistry, and Biology, Vol. IV, 79—86.
© 1989 *by Kluwer Academic Publishers.*

these bacteria. Among the bacteria clones, techniques are available to isolate the gene, especially hybridation by complementarity with either mRNA or deoxyoligonucleotides, synthesized according to the protein formula, using the genetic code.

The main possible defects of a gene are a substitution of a nucleotide base or an addition, a deletion or an inversion of bases, or crossing-over; all can occur in exon, intron or promoter.

A radioactive gene-probe is combined by complementation with the gene in order to be localized, and the defect is directly determined. When the gene probe is not available, one can make use of a special phenomenon: often an altered gene is hereditarily linked to an extra-gene part of the DNA, whose polymorphism is detected by fragments obtained by restriction endonucleases; this polymorphism is a marker of the abnormal gene [7, 8].

When investigating a fetus, one has to perform the examination in the early months of pregnancy, because the consequence, according to a medical point of view, may be an abortion. One may examine fetal DNA either by amniotic punction or by chorionic sampling.

(a) THE SICKLE-CELL DISEASE

Some people, all of African origin, are afflicted with a disease of the red cell, a hemolytic variety of anemia plus thrombosis, called sickle-cell disease. The red cells look normal and are circular at the beginning of the examination; but when oxygen disappears from the medium the red cells lose their circular form and take the form of a sickle: they become rigid and cause obstruction of the capillaries.

As all chemists and biologists, Pauling knew that hemoglobin (Hb) is the carrier of oxygen. He thus made the hypothesis that the disease was not a cellular one, but resided in the molecule of Hb itself. He asked one of his graduate students, Itano, to prove this hypothesis. For two years, Itano tried different physico-chemical methods without success and, desperate, wanted to stop; but Pauling refused.

At that time, in Sweden, Tiselius had described the first electrophoresis method, and one of his first apparatuses arrived at CalTech, in Pauling's laboratory. The technique of measuring protein mobility in the electrical field was then tedious: only one estimation could be made a day; this is a big contrast with the current method used nowadays, which allows more than 100 separations per day.

Itano was immediately successful with this new technique. He found that the migration of Hb from ill people was different from that of Hb from normal people. A father and a mother, heterozygotes for the disease, have two different Hb molecules: one normal and one similar to that of their ill child. This molecular disease is due to a single amino-acid substitution in the protein, and then to a single base change in a nucleotide in DNA.

Applications of this method to other proteins and enzymes have given a new impulse to the study of inherited metabolic diseases of Mendelian type.

What was the difference between the normal and 'sickle' hemoglobin molecules? You must remember that there are two peptide chains α, and two peptide chains β, in the Hb molecule. Vernon Ingram, in England, prepared a tryptic digest of Hb A, the normal molecule, and of Hb S, the 'sick' molecule. Then, by electrophoresis and chromatography, a fingerprint of all peptides was obtained, and one of these showed abnormal mobility. The analysis of this peptide indicated the substitution of one amino-acid, Glu, by another one, Val, at the position 6 of the β chain, entailing a loss of negative charge.

(b) THALASSEMIAS

Another hereditary anemia, which is observed around the Mediterranean Sea and in South-East Asia, is called *thalassemia* (from the Greek thalassa = sea). Whereas sickle-cell anemia is due to a single genetic defect, thalassemia is due to many different mutations [10].

These mutations occur in the gene corresponding to the α and β chains of hemoglobin. If two allelic genes of β chains are involved, the homozygote is ill, whereas if only one is involved, the heterozygote is not ill. The situation for α chains is more complicated, because there are two couples of α genes ($\alpha 1$, $\alpha 2$). When due to the deletion of one or two α gene, α-thalassemia is without consequence. Severe anemia accompanies the mutation of three genes, and the new-born is dead when all four genes are altered.

In β thalassemia (we exclude heterozygotes who are not ill) there are two varieties, depending on the β chains which are stopped or not. In β^0 thalassemia the synthesis of β chains is stopped and no Hb is synthesized, while some is in β^+ thalassemia.

The consequence is a severe hemolytic anemia. The mutation can occur on an exon or an intron, or a promotor. In all α thalassemia, it is usually a deletion. In β^0 thalassemia they are many causes; it may be due to a mutation of the junction intron-exon; it may be due to a mutation of an intron or an exon with a new site or with a suppression of *splicing*. A punctual mutation in the 17th or the 38th position can determine a non-sense mutation: the synthesis of β chain is stopped. A large deletion is observed in cases from India and Pakistan.

The case of β-δ thalassemia is very special. Its cause is a deletion of β gene and δ which are linked to the γ gene: the result is an activation of the γ gene. There may be no illness, if γ combines enough to α. $\alpha^2\gamma^2$ is fetal hemoglobin, which is a good substitute for $\alpha^2\beta^2$, hemoglobin A. Hybrids can occur between α^1 and α^2, or β and δ (Lepore hemoglobin). One has observed in thalassemia more different mutations than in any other hereditary disease.

The prenatal detection is more difficult, but we have to remember that these variations occur in different ethnic groups.

(c) GLUCOSE-6-PHOSPHATE DEHYDROGENASE DEFICIENCY

Generally, hemolytic anemia occurs only in special situations, mainly after the use of certain drugs such as antimalarials, and it caused problems during the Korean war, mainly among blacks. In this case the drug is contra-indicated [12]. A special case can be reported. A young French physician had to undergo a surgical procedure. He gave the list of all the drugs he should not take, which included sulfamides. The surgeon forgot that sulfamides are used not only orally, and at the end of the operation he sprinkled sulfamide powder on the wound. As a result the patient was very ill, but he is now one of the heads of Paris hospitals.

(d) DISEASES OF THE MIND

The body is involved in all these hereditary diseases, but the mind is also sometimes afflicted. Some of the diseases can be prevented, because a treatment is available at birth. Some cannot be treated, and the 'treatment', after antenatal diagnosis, is abortion.

Phenyl ketonuria is the best example [14, 15]. The absence of an enzyme, phenylalanine-oxidase, stops the catabolism of phenylalanine, which accumulates in plasma where its increase can be estimated: this estimation is obligatory in France for all new-borns. The treatment is a diet low in phenylalanine which can prevent idiocy.

The situation is the same for *hypothyroïdism*. If thyroxin is found to be decreased in the blood, or if thyroid-stimulating hormone is increased, thyroxin must be administered for a long time. If not, cretinism will occur [16].

The cost of systematic detection is covered by the reduction in expenses for prolonged treatment. In other hereditary diseases where offsprings are at-risk for idiocy, like galactosemia, neonatal screening is not obligatory.

Galactosemia is a disease which is not always recognized, because there is no systematic detection, and which can be treated. Cow's milk, as well as mother's milk, is toxic for babies afflicted with *hereditary galactosemia* [17]. The deficiency of an enzyme, galacto-phospho-uridyl transferase, prevents transformation of galactose 1 P into glucose 1 P. If milk, theoretically the best food for infants, is suppressed, everything returns to order.

Often there is no treatment for mental deficiency. We must develop preventive medicine. Genetic counseling can recognize cases where husband and wife are heterozygotes for the same disease. This is an indication for an early exploration of the fetus, provided the idea of an abortion is accepted by the couple.

A good example of successful genetic counseling is *Tay-Sachs disease* [18], frequent among the Jews from Eastern Europe, which has been extensively studied, mainly in Baltimore-Washington.

3. Polygenic Diseases

There are many diseases due to a single gene which obey Mendel's laws. Some very frequent diseases seem to be due to more than one gene, plus environmental factors: atherosclerosis, arterial hypertension, diabetes, cancer, mental diseases, ageing, are still not well understood. We will discuss two examples.

(a) ARTERIAL HYPERTENSION [19]

This is due to many causes, overproduction of adrenaline, aldosterone, renine are often the cause of acute hypertension. Permanent arterial hypertension is rarely due overproduction of angiotensin II (the most active hypertensor).

Molecular explanation is now available only for adrenaline, which can serve as a model. Adrenaline binds to a protein receptor of smooth muscle membrane in the wall of peripherical arterioles; it binds to the receptor and this first messenger is relieved by a 2nd messenger, cyclic AMP (adenosine monophosphate) and by ionic K^1 exchange; it determines a change of organisation of myosin and actin and then the contraction of arterioles occurs, followed by arterial hypertension.

(b) MANIC-DEPRESSIVE ILLNESS

Lithium is the first example [20] of the concept of the involvement of a molecule in a mental disease. Testing a hypothesis regarding the etiology of manic-depressive illness presaged the beginning of modern psycho-pharmacology. Cade [20] reasoned: could manic-depressive illness be analogous to thyrotoxicosis and myxedema, mania being a state of intoxication by a normal product of the body circulating in excess, whilst depression is the corresponding deprivative condition? The intoxicating substance in excess may well be excreted in the urine. The best plan seemed to use the crudest form of biological fluid in a preliminary investigation. Guinea pigs were used and fresh concentrated urine was injected intraperitoneally. Some specimens of urine from manic patients were far more toxic than any of the control specimens from normal subjects, schizophrenics and melancholics . . . . All that had been demonstrated so far was that any concentrated urine in sufficient quantity would kill a guinea pig, but that urine from a manic subject often killed more readily. It was not very surprising, to find that urea was perhaps the guilty substance, but, in fact, quantitative estimation showed that urine from manic subjects did not differ significantly from that of non-maniacs or controls in urea content.

The specimens were more toxic than could be explained by the concentration of urea actually present, even if the urea effect was being maximally enhanced by uric acid . . . . It appeared important to estimate more accurately how much uric acid increased the toxicity of urea. The practical difficulty was the insolubility of uric acid in water, so the most soluble urate was chosen, lithium salt. And that is how lithium came into the story and became the first treatment for a psychosis, and more precisely a prophylaxy for mania.

4. Acquired Molecular Diseases

The interpretation of acquired molecular diseases has not benefited from the same energetic approach as hereditary diseases. Whatever the aggression to which an organism is submitted (infectious, parasitic, mechanical, nutritional, toxic aggressions), there is always a primary chemical modification one of the organism constituents, that is not always easy to detect.

(a) PRIMARY BIOCHEMICAL LESIONS [21]

In avitaminosis these precede the appearance of anatomic signs, as proved in Oxford by Peters. Pigeons lacking vitamin B1 or thiamin (a water soluble vitamin) are afflicted with beriberi which is characterized by contractures and convulsions. The cerebral tissue of such pigeons is anatomically and histologically intact, but in contrast to normal pigeons, the cerebral tissue is unable to metabolize pyruvic acid. The pyruvic acid oxidizing system is impaired, pyruvic acid and its hydrogenation product, lactic acid, accumulate, while the consumption of oxygen is decreased.

The administration of vitamin B1 results in the spectacular disappearance of the generalized contractures, retractions of the head, and convulsions. At the same time, the *in vitro* cerebral metabolic disturbances also disappear. The addition of B1 to sections of cerebral tissue from animals lacking vitamin B1 restores oxygen consumption and leads to the disappearance of pyruvic acid resulting from a prolonged diet of washed and polished rice (which removes vitamin B1). Thus the first mechanism of action of a vitamin was discovered. It is the cofactor, the coenzyme of the pyruvic oxidizing system. Its absence results in an enzymatic deficiency which leads to the pathological process.

(b) INTOXICATIONS

These occur when the molecular cause is well defined, but often the disease cannot be explained on a molecular basis. A reaction with a biochemical component has to be recognized.

In this perspective, carbon monoxide intoxication is a good example. Carbon monoxide, due to a defective heating apparatus, combines with blood hemoglobin, oxygen cannot combined any more and tissue respiration stops.

During the Second World War, research concerning the chemical weapon, poisonous gas, led to therapeutic applications. Gas had been used by the Germans in 1916; yperite or mustard gas, or vesicating gas, causes a biochemical lesion characterized by the blockage of the thiols in tissue proteins. 60 millipicograms per m^2 of skin surface could be fatal, generally due to lesions of the internal lung surface, resulting in acute oedema of the lung. It was keratine from the hair of

Oxford students that served as the experimental material, because of its richness in thiol functional groups.

A compound was sought that could restore the thiol group blocked in a disulphide form by the poisonous gas, in the hope of avoiding development of the lesion. An antidote was discovered and therapy was successful if given sufficiently soon after the attack. The product was called 'British anti-Lewisite' or 'B.A.L.' and developed by the English Stocker.

Tobacco smoke [22] is the cause of cardiac and pulmonary disorders (for which nicotine is responsible) and also of many cancers, principally of the lungs. This has been demonstrated by epidemiological investigations and by the careful analysis of tobacco smoke. Among the 2000 compounds, about 50 are carcinogenic.

Exogenous morphine competes as a drug for receptors with endogenous [23] cerebral morphines. A better understanding will perhaps permit a more successful battle against the very dangerous phenomenon of addiction.

Chronic alcoholism [23] is only partly explained by the inhibition of some enzymes: alcohol dehydrase, aldehydrase dehydrase. Paradoxically, it is perhaps the least known of all social poisons.

(c) INFECTIOUS DISEASES

Among the acquired diseases, the infectious diseases, microbial or viral, were recognized, with their specificity, by Louis Pasteur and his school. Although their molecular understanding was still very incomplete, the subsequent molecular basis of immunity has considerably progressed: the structure of the immunoglobulins is well known, the selection theory of preexistant antibodies well established [24].

Bacteriology and immunology have resulted, thanks to vaccination, in an almost complete disappearance of infectious diseases in the industrial countries. The technique of vaccination has regularly improved: the naturally attenuated virus with Jenner; the artificially attenuated microbes or viruses with Pasteur, and non-viral envelope proteins or artificial peptides similar to these proteins are being investigated [25].

All diseases, and not only hereditary diseases, are molecular diseases; many acquired diseases have a hereditary predisposition, an idiosyncrasy.

The summary of this story was given, in the 17th century, by Philippe Aurelien Theophrastus Bombastes von Hohenheim, known today as Paracelsus:

Man is a chemical complex. The cause of diseases is some alteration in his chemical components; we therefore need chemical medicines in order to fight against diseases.

References

1. A. F. Garrod: *Inborn Errors of Metabolism.* Reprinted by Harris, Oxford University Press (1963).

2. J. B. Stanbury, J. B. Wyngaarden, and D. S. Frederikson: *The Metabolic Basis of Inherited Diseases.* McGraw Hill, Toronto (1983).
3. D. J. Weatherhall: *The New Genetics and Clinical Practice.* Oxford University Press (1985).
4. M. F. Perutz: 'Fundamental research on molecular biology relevant medicine'. *Nature* **262**, 449 (1976).
5. G. Schapira and J. C. Dreyfus: *Pathologie moléculaire.* Masson, Paris (1975).
6. F. Gros: *Les secrets du gène.* Ed. O. Jacob, Paris (1986).
7. L. Pauling and H. A. Itano: 'Sickle-cell anemia: a molecular disease'. *Science* **110**, 543—548 (1949).
8. H. Wajcman and D. Labie: 'Aspects actuels de la biologie de la drépanocytose'. *Ann. Med. Int.* **132**, 548—594 (1981).
9. Y. W. Kan and A. M. Dozy: 'Antenatal diagnosis of sickle-cell anemia by DNA analysis'. *Lancet* **I**, 910 (1900).
10. D. Labie, C. Trougos, and H. Wajcman: 'Les bases moléculaires des thalassémies'. *Ann. Méd. Int.* **136**, 923—945 (1985).
11. D. J. Weatherhall and J. B. Clegg: *The Thalassemia Syndromes.* Blackwell Sc. Publ. (1981).
12. A. Kahn: 'G6PD variants'. *Hum. Gen. Sup.* **I**, 37—44 (1978).
13. Y. Y. Hsia: 'Detection and treatment of inborn errors of metabolism associated with mental deficiency'. *Enzymes in Mental Health.* Edited by C. J. Martin and B. Kiseh. J. B. Lippincott, Philadelphia (1966).
14. W. L. Nyhan: 'Asbjorn Folling and phenylketonuria'. *T.I.B.S.*, 71 (1984).
15. S. C. Woo, A. S. Lidsky, F. Guttler, T. Chandrat, and R. Robson: 'Cloned phenylalanine hydroxylase gene allows prenatal diagnosis'. *Nature* **306**, 151—155 (1983).
16. J. H. Dassault: 'Hypothyroïdie congénitale'. *Med. Sci.* **1**, 261—263 (1985).
17. V. E. Shih and H. I. Lévy: 'Galactosemia screening of new born in Massachusetts'. *N.E.M.J.* **284**, 733—737 (1971).
18. R. Mycrowitz and R. L. Proja: 'cDNA clone for the α chain of human β hexoaminidase, deficiency of α chain mRNA in Aschkenazi Tay-Sachs fibroblasts'. *P.N.A.S.* **81**, 5394—5398 (1984).
19. P. Meyer: *Hypertension artérielle.* Flammarion, Paris (1978).
20. S. Gershon and B. Shopson: *Lithium: its Role in Psychiatric Research and Treatment.* Plenum (1973).
21. R. H. S. Thompson: *Thiamine and the Metabolism of Pyruvate.* T.I.B.S., 460 (1983).
22. A report of the surgeon general: smoking and health, 1979. US department of health. Education and welfare public health service, office of the assistant secretary for health.
23. C. S. Lieber: 'The metabolism of alcohol'. *Sci. Am.* **234**, 25—33 (1976).
24. P. Leder: 'The genetics of antibody diversity'. *Sci. Am.* **246**, 72—83 (1981).
25. R. A. Lerner: 'Synthetic vaccines'. *Sci. Am.* **248**, 48—56 (1983).

Metallo-Organic Complexes and Carcinogenesis

PHAM V. HUONG
*Laboratoire de Spectroscopie Moléculaire et Cristalline (UA 124 CNRS),
Université de Bordeaux I, 351, Cours de la Libération, 33405 Talence, France.*

1. Introduction

In cancer research, two kinds of methods are currently practised. In the first series, chemicals or radiations are administered to animals or humans and after a while, a body count helps to establish the statistics of the disease or the cure rate and the effectiveness of a therapy (Figure 1). We call these the 'black box methods'. In the other manner, people try to get to know what could happen inside the black box by examining the disease at the molecular scale.

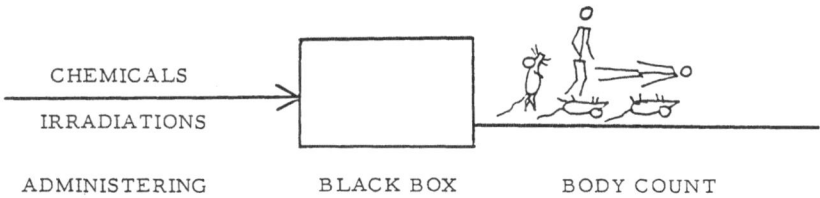

CANCER RESEARCH

I. BLACK BOX METHODS

     practical and effective

CHEMICALS

IRRADIATIONS

ADMINISTERING   BLACK BOX   BODY COUNT

II. MOLECULAR STRUCTURES AND INTERACTIONS STUDIES

Fig. 1. 'Black box methods' in cancer research.

Among the latter methods, Laser Raman spectroscopy which can identify the molecular vibrations of the chemical entities involved in the disease, can help to identify their nature, their interaction modes and eventually the localization of the interactions between biological sites and carcinogens or drugs [1].

2. Possibilities of Raman Spectroscopy

Among the molecular and non-destructive investigating methods, Laser Raman spectroscopy is revealed to be very suitable for the study of *live objects* [2, 3]. For

Jean Maruani (ed.), Molecules in Physics, Chemistry, and Biology, Vol. IV, 87—109.
© 1989 *by Kluwer Academic Publishers.*

instance, the vibrational bands characteristic of various nucleic bases in transfer-Ribo-Nucleic Acid [4] can be easily identified in its Raman spectrum (Figure 2).

When a molecule is in interaction, big changes can occur in its Raman vibrational spectrum especially for chemical groups involved in the interaction site. It is easy to recognize for instance, the S—H stretching vibration at 2526 cm^{-1} in new anti-cancer reduced glutathione (Figure 3). This function disappears completely in the oxidized form and gives place to the appearance of an S—S bridge [5] with a vibrational frequency located at 516 cm^{-1}.

By choosing appropriate laser exciting frequency, the *Resonance Raman effect* of a chromophore can be produced with an intensity enhancement which can reach, in many cases, one million times or more. In consequence, a sample concentration of one million times lower than usual can be sufficient to allow the vibrational spectrum to be recorded. Figure 4 illustrates such a possibility with the Raman spectrum of an experimental anti-cancer drug: an ellipticin salt. A similar chemical, first synthesized by Dat Xuong [6] is already commercialized by Sanofi — Institut Pasteur as an anti-breast-cancer preparation, under the name of Celliptium.

In sending the incident laser light through an optical microscope, one can illuminate a sample area as small as one micrometer square, and record the Raman spectrum of this limited surface. This spatial selectivity allows the recording of the various components in a heterogeneous sample. This Raman micro-spectroscopy can be applied with efficiency to the investigation of live objects. An example can be seen in Figure 5 where the Raman spectrum of a liver cell is given.

By combining Raman micro-spectroscopy and resonance Raman effect, one can tremendously increase the ability of this non-destructive and molecular technique and apply it to the investigation of biological media, in particular for the study of blue particles recently found in tumors and blood of leukemic and cancerous patients [7, 8].

3. Blue Particles in Cancer Bloods and Tumors

Blue and blue-green particles of micrometer size were recently observed in tumors and blood of cancerous and leukemic persons. All the sixty cancerous patients of various natures examined:

> Breast cancer,
> Lung cancer,
> Rectum cancer,
> Uterus cancer,
> Myeloid leukemias,
> etc.. . .

present blue and blue-green particles. These colored materials have no specific morphology under the optical (Figure 6) or electron (Figure 7) microscopes.

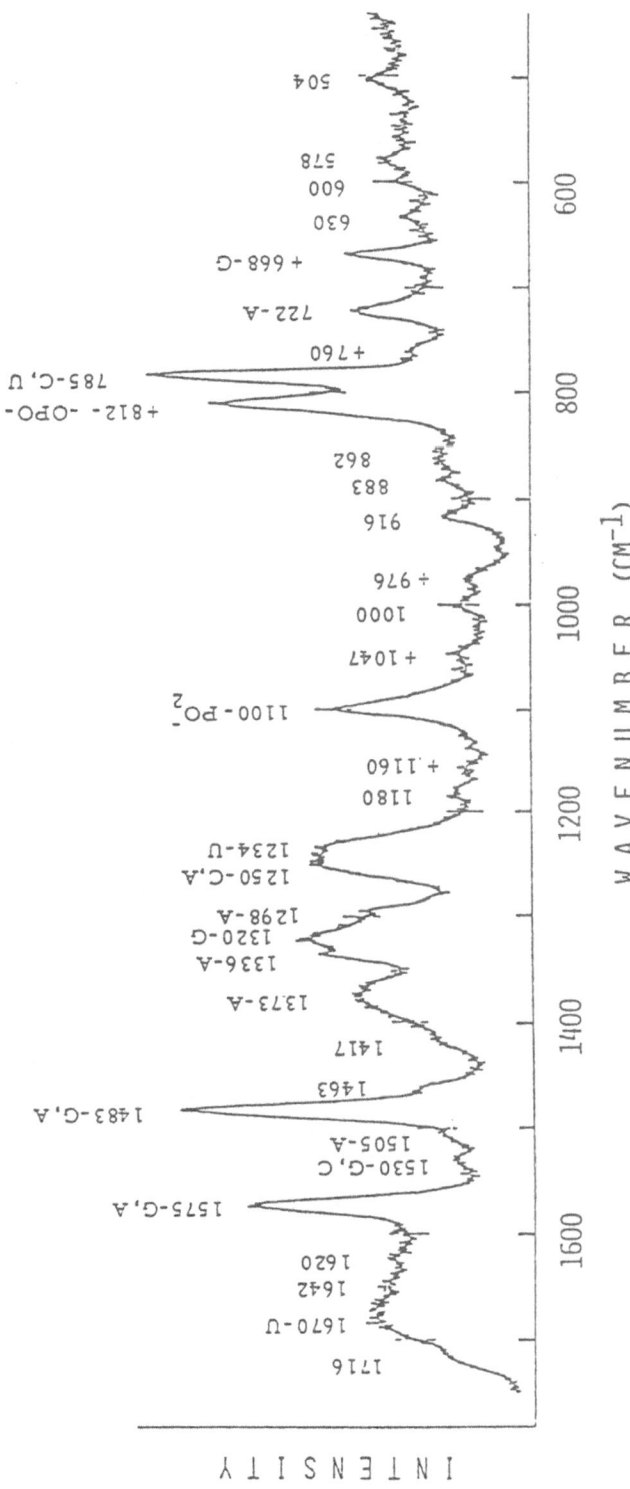

Fig. 2. Raman spectrum of an aqueous solution of yeast t-RNA.

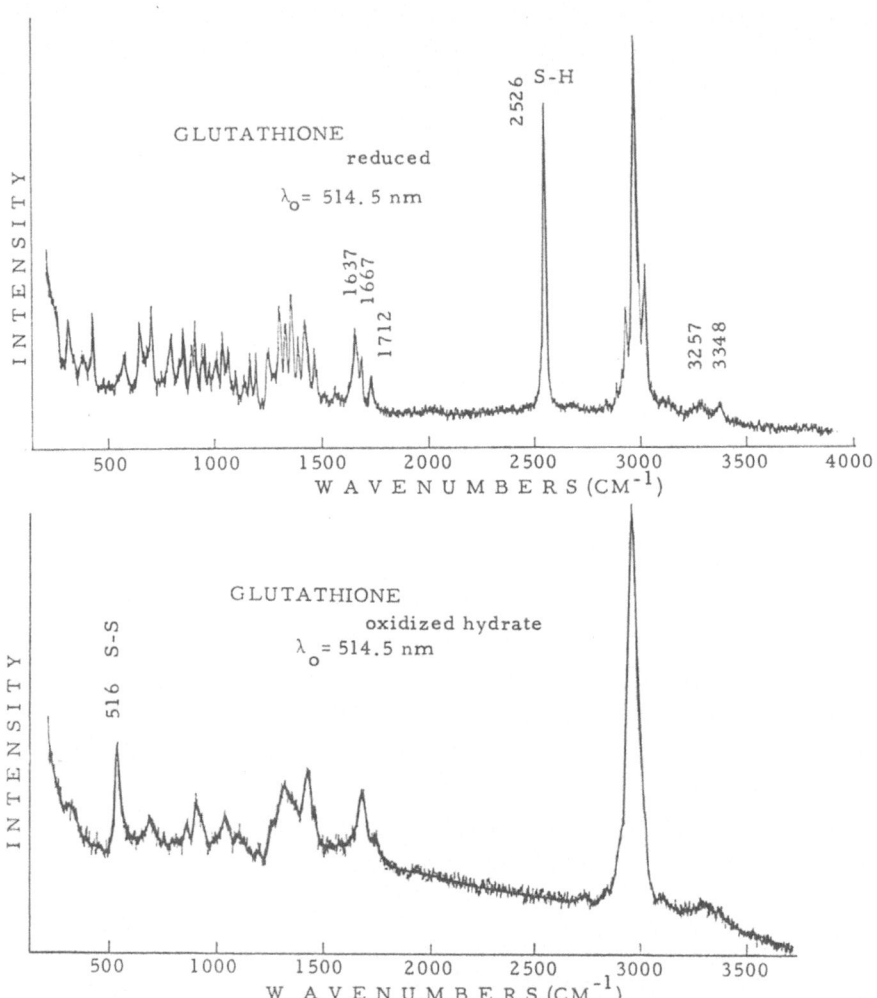

Fig. 3. Raman spectra of reduced and oxidized anticancer drug glutathione.

The sampling was conducted by smearing blood of leukemic or of generalized cancerous persons or by apposition of the tumoral piece cut by a surgeon. Other samples came from tapping (lung, sternum) or puncture (marrow).

When examining three samples from each patient, blue or blue-green particles of one micrometer to 20 micrometer size can be observed. Upon cultivation of the bacterial suspensions on a medium containing pre-warmed blood, the blue particles appear again, but in more abundant frequency.

Among fifty 'healthy' persons taken as reference, seven are found to give such colored particles. One of them was revealed later as cancerous (ovary); three others as having psoriasis, herpes and chronic staphylococcal disease. The last finding, clearly, shows that blue particles are not specific to cancer. Nevertheless,

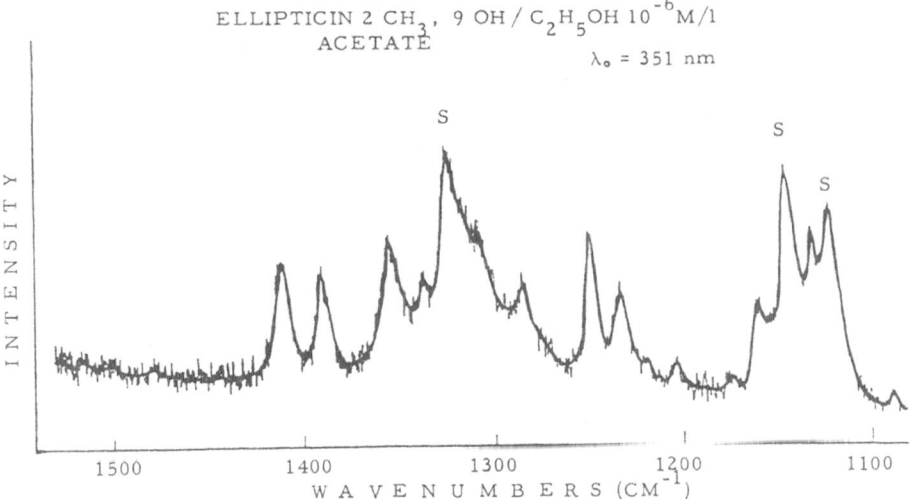

Fig. 4. Raman spectrum of an experimental anticancer drug, an ellipticin salt.

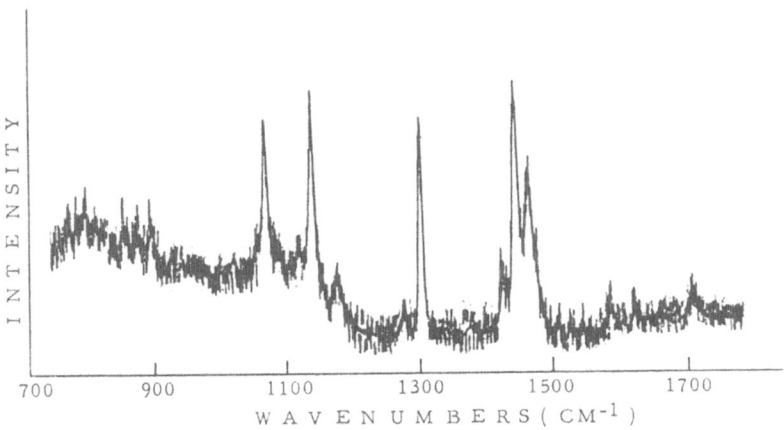

Fig. 5. Raman spectrum of a liver cell with a hepatisis due to use of 'Methotrexate'.

the fact that all sixty cancerous patients carry blue particles suggested that further investigations could be conducted.

So what we have done with new cancer samples and metallo-organic complexes modelling the interactions identified in cancerous media.

4. Micro-Raman Spectra of Cancer Blue Particles

Under the objectives of a micro-Raman spectrometer, all the blue particles do not

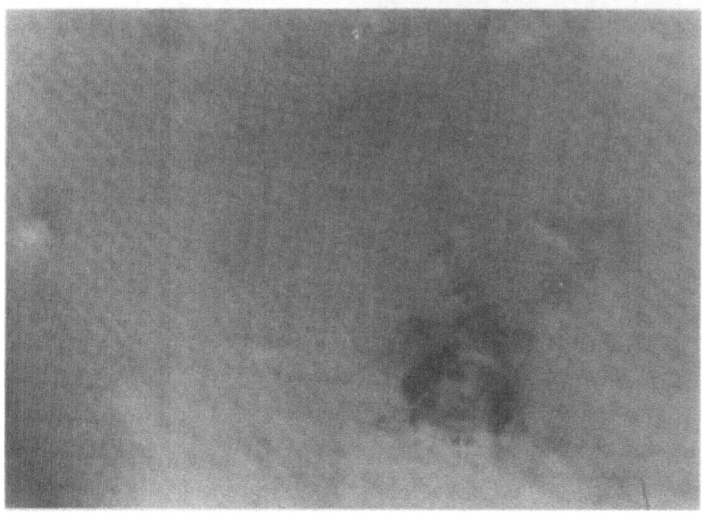

Fig. 6. Optical micro-photography of a blue particle from a mammary gland carcinoma opposition (breast cancer).

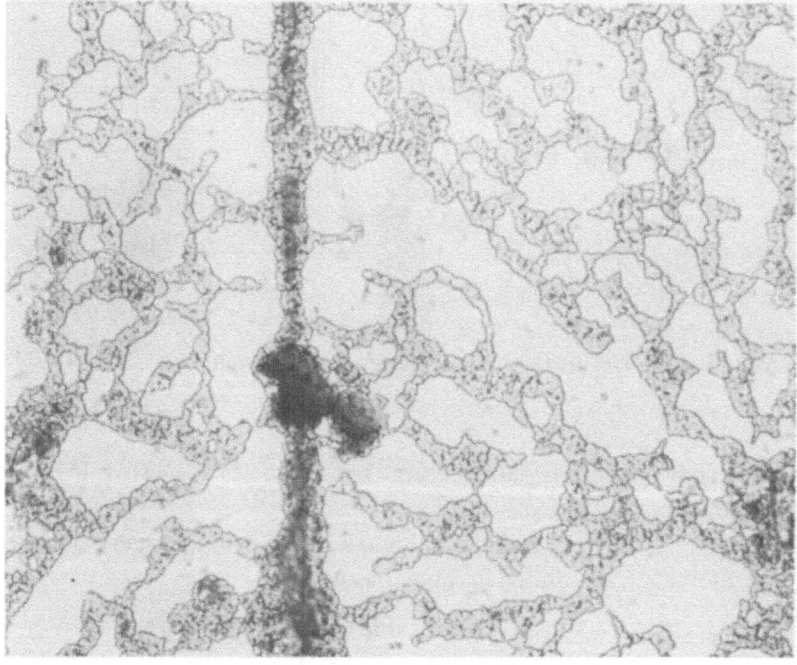

Fig. 7. Scanning electron micro-photography of a blue-green particle in a bacterial cultivation of blood from a lung cancer.

give a suitable spectral record. The ones located in blood of leukemia, of generalized cancer or in pleural fluid of lung cancer are generally easy to investigate. These colored materials appeared as crystal-like solids and give Raman spectra of good quality, void of any fluorescence.

That is also the case of the majority of blue particles spotted in samples cultivated upon gelose containing pre-warmed blood.

But some other samples, from breast cancer, rectum or uterus tissues, give bad Raman spectra, often overlapped with large fluorescence features. For the latter, a better spectral recording can sometimes be realized after washing the slide with some drops of water.

Figure 8 represents the Raman spectrum of such a blue particle, recorded by the micro-Raman technique described. One can recognize on the spectrum, between 500 and 1750 cm^{-1}, the spectral features currently found by vibrations of organic components.

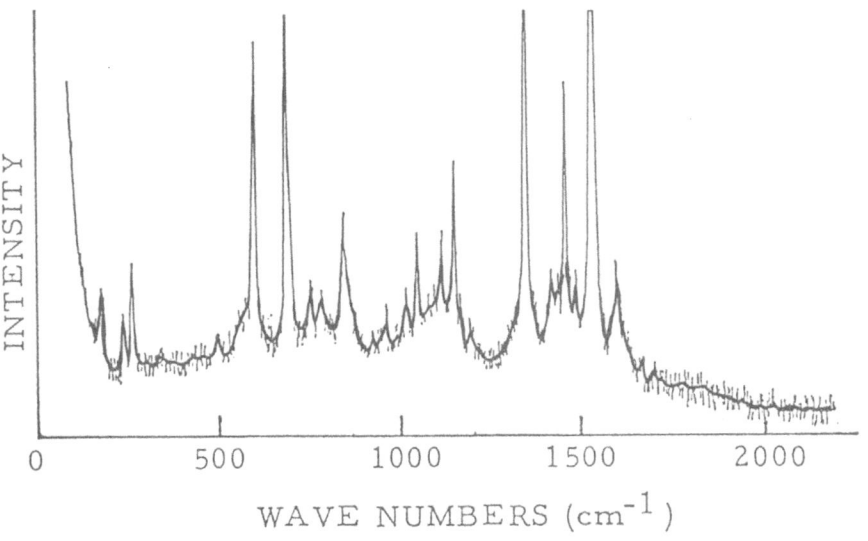

Fig. 8.   Raman spectrum of a blue particle from an ovary cancer (Exciting line 514.5 nm).

Below 500 cm^{-1}, many bands are also observed. This spectral region is characteristic of intermolecular vibrations, in particular the metal-ligand vibrations such as metal-oxygen, metal-nitrogen and metal-sulfur. When excited with a green laser line emitting at 514.5 nm, the low frequency Raman bands occur at 210, 265 and 440 cm^{-1}.

By choosing another laser line (457.9 nm) to approach the electronic level corresponding to metallic transition in the near ultra-violet, the intensity of some Raman bands in the range 200—450 cm^{-1} increases noticeably (Figure 9). This

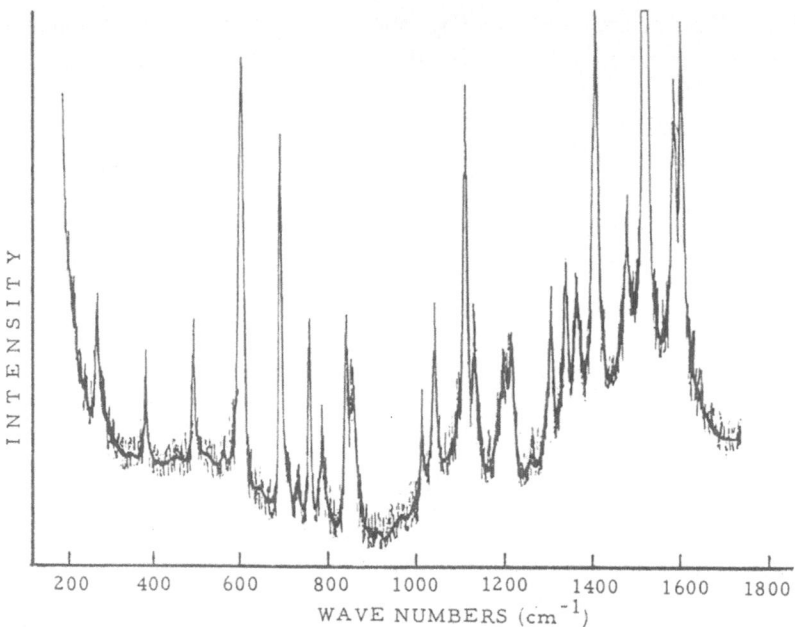

Fig. 9. Raman spectrum of a blue particle from gastric cancer (Exciting line 457.9 nm).

specific resonance effect confirms that the observed Raman bands are due to metal-ligand interactions.

This Raman spectrum is identical for nearly all blue particles examined, independently of the origin and the nature of the cancer. For the green particles observed, more rarely, in the same samples, the resonance micro-Raman spectrum is different from the blue one. Figure 10 reproduces the Raman spectrum of a green particle from the blood of a lung cancer. From any cancer, the green particles give the same Raman spectrum. If strong bands are observed in the spectral range where are usually located the nucleic bases vibrations [4, 9, 10] as well as other organic component vibrations; one strong line is also recorded at 670 cm^{-1}.

On the lower frequency side, two weak bands are also located at 260 and 350 cm^{-1} which can be assigned to metal-nitrogen and metal-sulfur vibrations by comparison to known complexes.

The nature of the metal has recently been identified by electron micro-spectroscopy [11]: copper is observed in the bulk of both blue and blue-green particles. In addition, electron micro-spectroscopy also reveals in the neighborhood of the colored particles the presence of spherules of copper and some other metals. Therefore, the observed metal ligands frequencies can be attributed to Cu—O, Cu—N and Cu—S vibrations.

Nevertheless, the assignment of the Raman lines of the organic part cannot be done without the study of model metallo-organic complexes.

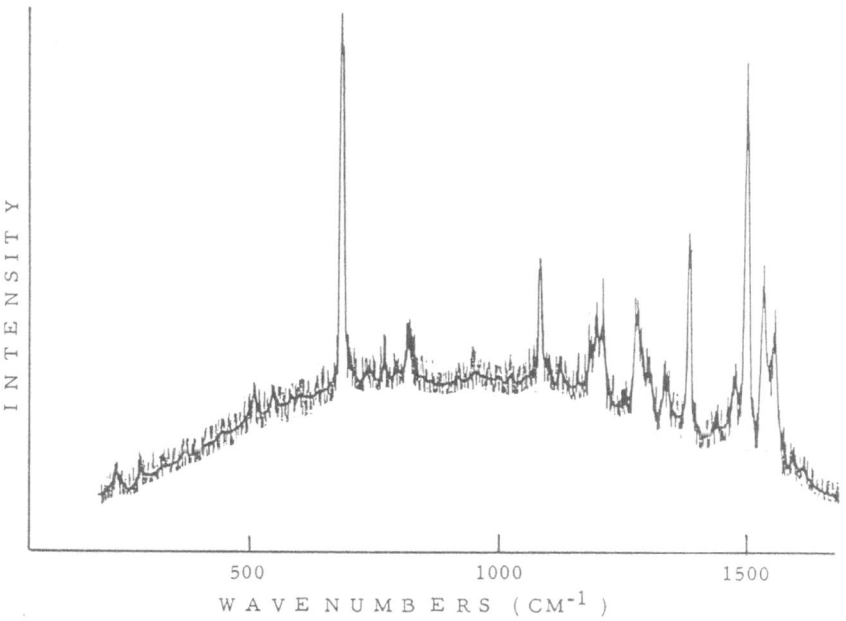

Fig. 10. Raman spectrum of a green particle of blood from a lung cancer.

5. Model Organo-Metallic Complexes

If the nature of the metal in the blue complexes found in tumors and cancer bloods is well established as copper, that of the organic part is not yet known.

The preliminary examination of the Raman bands of these colored particles indicates that the organic part could result from many bases. As in biological medium several bases sites exist, we tried to synthesize some main copper-bases models and to study their Raman spectra:

— copper-nucleotide complexes;
— copper-tyrosine complexes;
— copper-peptide complexes.

5.1. COPPER-NUCLEOTIDE COMPLEXES

The Raman spectra of nucleotides monophosphates complexes $Cu^{II}AMP$, $Cu^{II}CMP$, $Cu^{II}GMP$ and $Cu^{II}UMP$ (A: Adenines; C: Cytosines; G: Guanines, U: Uracines) (Figure 11), as well as polynucleotides $Cu^{II}Poly$-A, $Cu^{II}Poly$-C, $Cu^{II}Poly$-G and $Cu^{II}Poly$-U were recorded and analyzed in comparison to those of corresponding free bases.

For the copper complexes, the Raman vibrational frequencies of various chemical groups, when compared to the free bases, are not affected the same

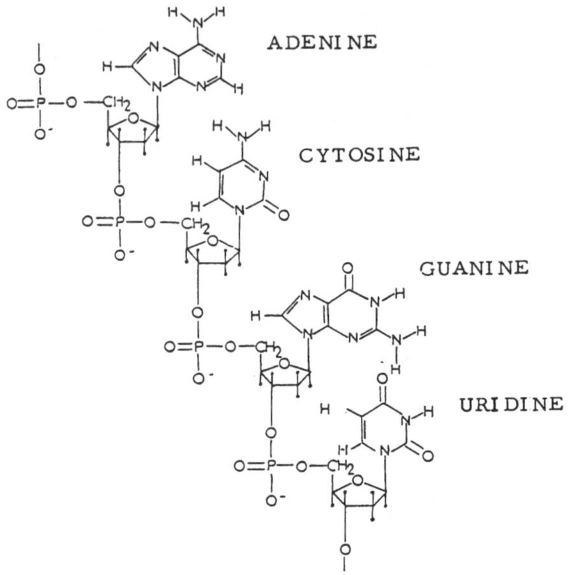

Fig. 11. A representative segment of DNA chain.

manner for all of them. With Adenosine and Guanosine Phosphate series (Figure 12), the band shifts are the most important for Raman bands characteristic of the immidazol cycle [10], while the frequencies corresponding to other chemical groups remain nearly unchanged. We can then deduce that in these series the interactions are localized on the N_7 nitrogen atom of the immidazole ring (Figure 13).

In addition, in the low frequency range, new bands appear which can be assigned to metal-ligands vibrations. With $Cu^{II}GMP$ complex, the Cu—N vibration is located at 374 cm^{-1}.

With the other series, $Cu^{II}UMP$ and $Cu^{II}CMP$ complexes, most changes in frequency were recorded for carbonyl stretchings. For UMP, $v(C{=}O)$ is located at 1671 cm^{-1}. This vibration is shifted toward 1635 cm^{-1} in the copper complex $Cu^{II}UMP$. We then deduced, for this series, that copper is bonded to the carbonyl group (Figure 14).

5.2. COPPER-TYROSINE COMPLEXES

Phenolate bridged binuclear Cu^{II} complexes (Figure 15) were studied by resonance Raman spectroscopy [12, 13]. The striking features are located in the region from 1250 to 1650 cm^{-1} which corresponds mostly to phenolate vibrations, especially strong bands at 1170, 1270, 1500 and 1600 cm^{-1} which are characteristic phenolate ring vibrations (Figure 16), and bands at 1288 cm^{-1} which is the stretching vibration $v(C{—}O)$ of phenolate.

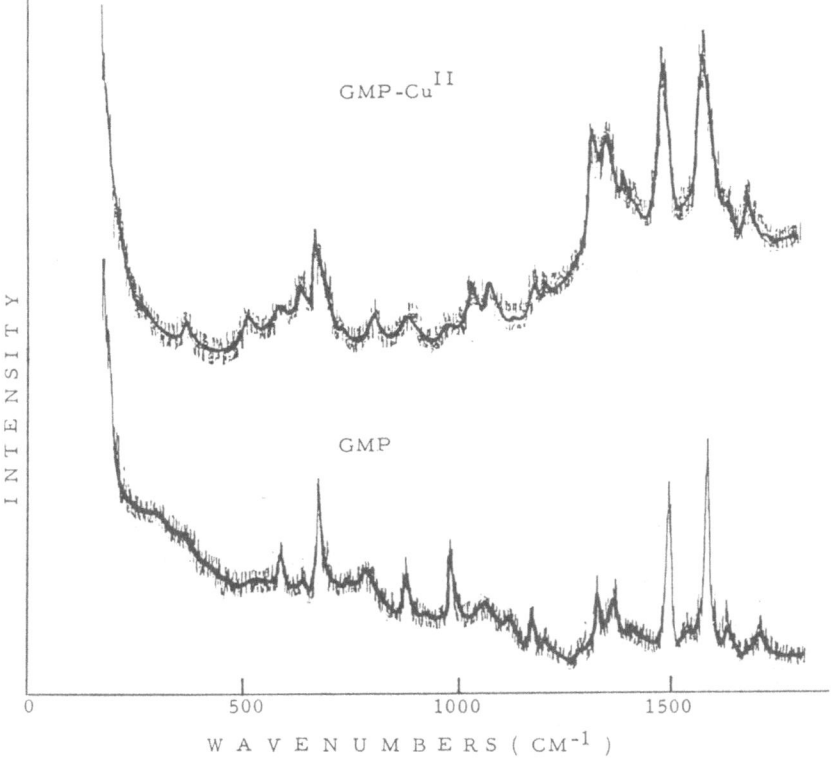

Fig. 12. Raman spectra of guanosinemonophosphate (GMP) and Cu^{II}GMP.

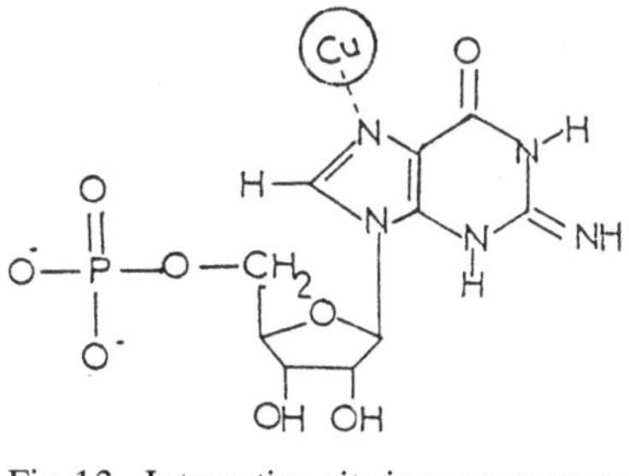

Fig. 13. Interaction site in copper-guanosinemonophosphate complex.

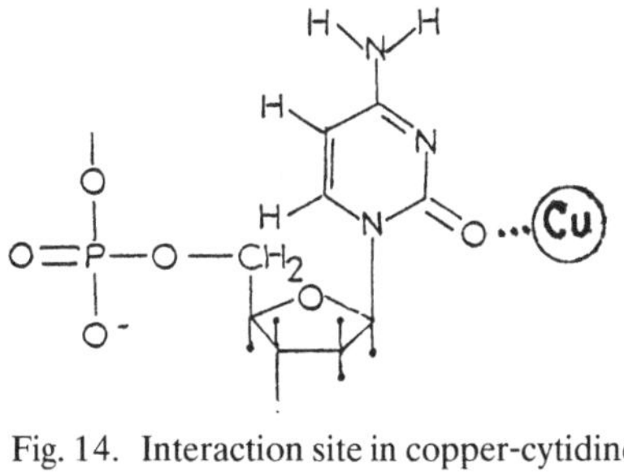

Fig. 14. Interaction site in copper-cytidinemonophosphate complex.

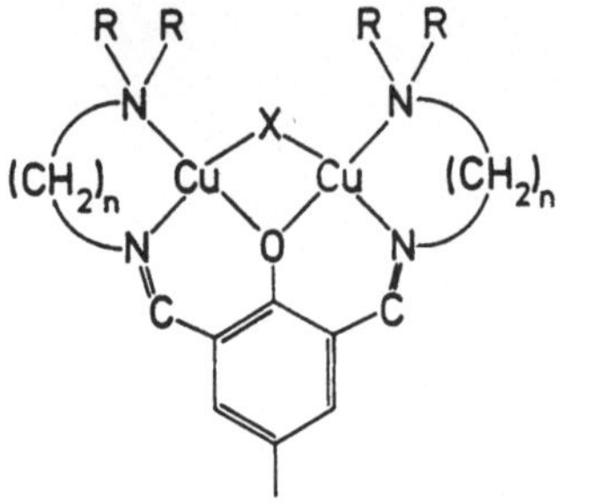

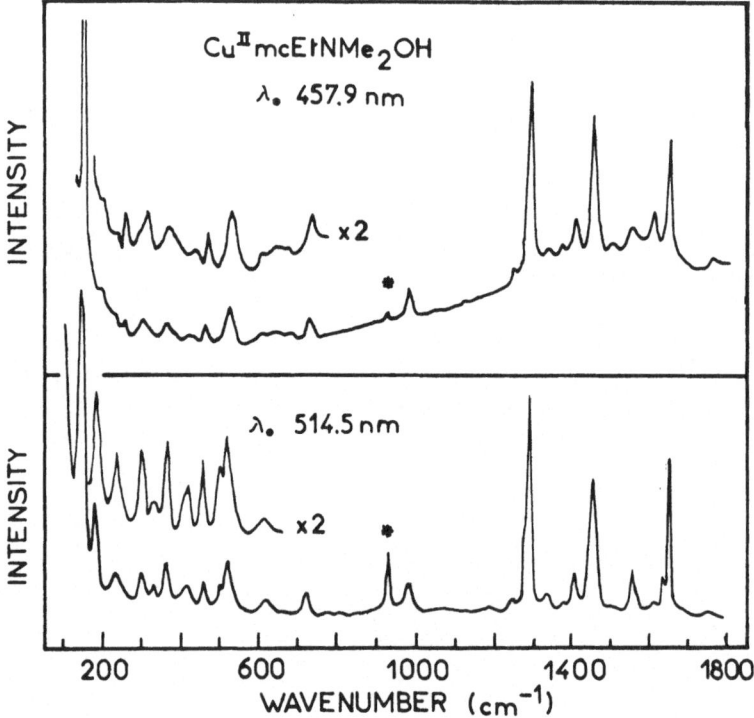

Fig. 15. Phenolate complexes.

Fig. 16. Resonance Raman spectra of phenolate complexes.

The weak bands lying from 200 to 430 cm^{-1} are assigned to copper-ligand vibrations.

We have to point out that in these complexes the copper atom is bonded to four ligands of completely different nature.

5.3. COPPER AMINO-ACID COMPLEXES

Among the representative components of proteins which could coordinate, in one of their conversion stages, to the metal, basic sites of amino-acids could be foreseen. We have chosen γ-amino-butyric acid (GABA).

The 'complex' $Cu^{II}GABA$ was prepared by mixing a dilute aqueous solution of GABA to an aqueous solution of copper dichloride $CuCl_2$. The precipitate was centrifugated then washed many times before final centrifugation and drying under vacuum.

It could be reminded that in the condensed state, GABA exists under the zwitterion form [14]

$$H_3N^+\text{—}(CH)_3\text{—}C\underset{O}{\overset{O}{\lessgtr}}\text{—}$$

and this form remains unchanged in aqueous medium [5].

The Raman spectrum of the copper salt of GABA is reported on Figure 17. By comparison to that of free GABA in aqueous solution [5], the appearance of new Raman bands in the 200—450 cm^{-1} range undoubtedly indicates the establishment of metal-ligand bonds. As the $\nu(COO^-)$ bands are noticeably affected, as well as other band representing C—N group, we naturally think that the interaction is located at the level of both the oxygen and nitrogen atoms. Therefore, the Raman band observed in this region is due to Cu—O and Cu—N bonds.

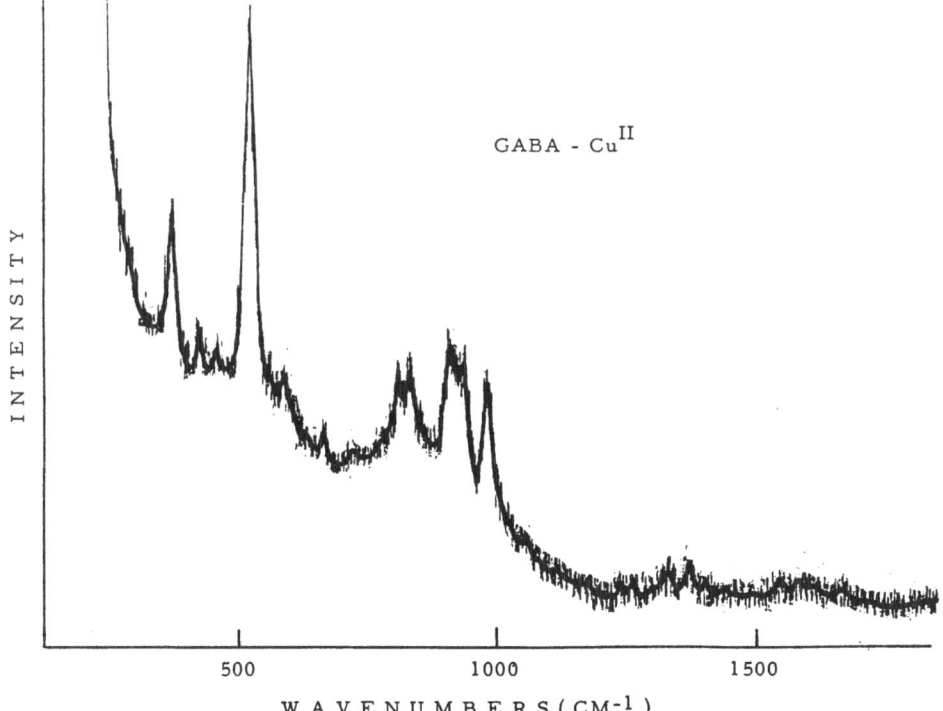

Fig. 17. Raman spectrum of copper-GABA 'complex'.

5.4. COPPER-PEPTIDE COMPLEXES

The presence of a strong Raman band at 670 cm^{-1} in the blue complexes spotted in cancer bloods and tissues suggests that a sulfur base could be involved. No compounds containing sulfur peptides are known to not give such vibrations [15—17].

Model sulfur-peptides were therefore synthesized and their resonance Raman spectra analyzed.

5.4.1. *Copper-Cysteine Complex*

Cysteine is a sulfur-containing peptide currently found in enzymatic reactions. Its Raman spectrum clearly shows the presence of S—H stretching band at 2560 cm^{-1} (Figure 18), the ν(C—S) at 690 cm^{-1} and ν(C=O) at 1740 cm^{-1}. In the spectrum of copper-cysteine complex, the ν(SH) band completely disappears, indicating that the sulfurized peptide is oxidized. In addition, a strong band

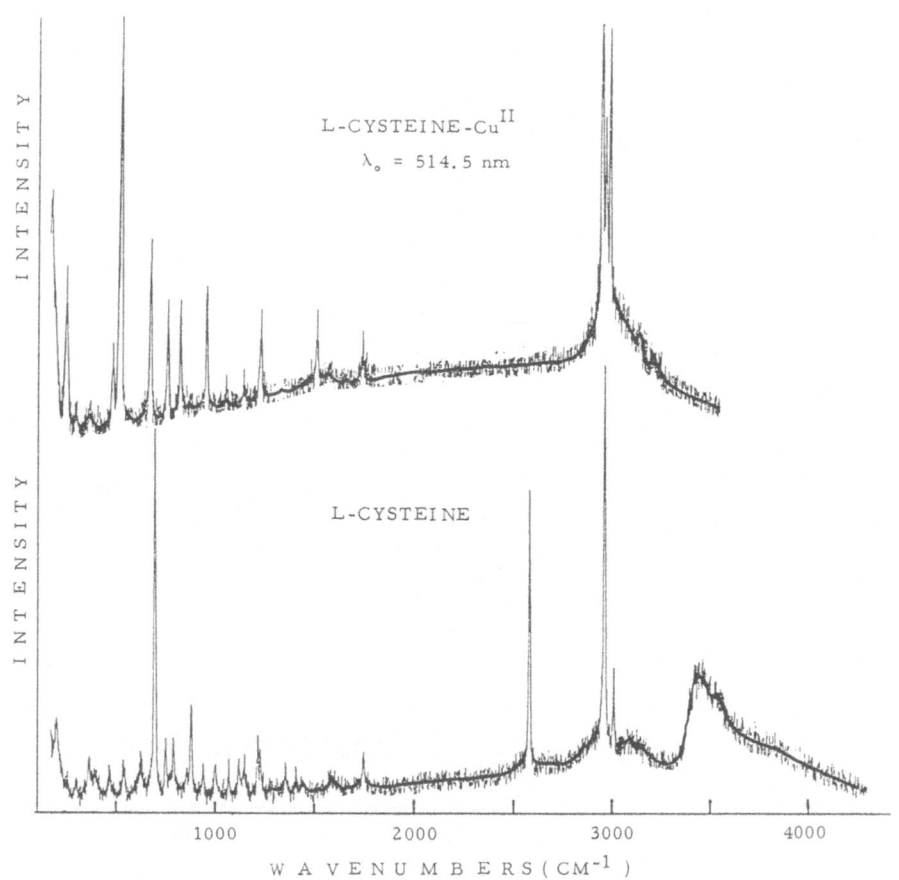

Fig. 18. Raman spectra of L-cysteine peptide and its copper[II] complex.

appears at 520 cm^{-1} which can be assigned to an S—S stretching vibration, as this frequency is very well localized with known S—S containing compounds [18]. We also notice that in the 'cysteine dimer', the C—S stretching vibration is lowered to 660 cm^{-1}, while the carbonyle ν(C=O) is shifted to 1720 cm^{-1}. These frequency shifts indicate that the C=O group and probably sulfur atoms are both involved in the complex to the copper metal. In addition, the new bands at 260 and 350 cm^{-1} strongly support the existence of Cu—O and Cu—S vibrations.

5.4.2. Copper-Glutathione Complex

Glutathione is a tripeptide, well known as an enzymatic reducing agent. Under the reduced form, its Raman spectrum clearly shows the presence of SH stretching vibration at 2526 cm^{-1}, ν(NH) and ν(NH$_2$) at 3200—3300 cm^{-1} region and the carbonyl vibrations at 1600—1750 cm^{-1} range (Figure 19). In the oxidized form, the Raman spectrum attests that the S—H group completely disappeared, while the new S—S stretching vibration appears at 516 cm^{-1}.

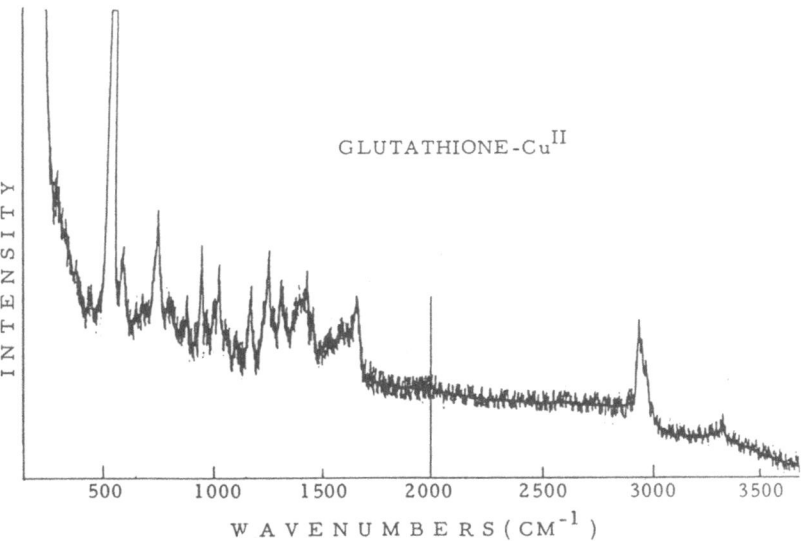

Fig. 19. Raman spectrum of copper^{II}-glutathione complex.

When reduced glutathione is complexed with copper II, no trace of ν(SH) is observed, meaning that the active hydrogen is removed. The reaction with copper dichloride has certainly triggered the 'dimerization' of glutathione. Nevertheless, the Raman spectrum of glutathione-Cu^{II} complex is different from that of oxidized glutathione.

The very strong band located at 750 cm^{-1} is still due to the C—S stretching vibration of the complex. The intensity enhancement of this band indicates that important electronic rearrangement might occur in the S atom level. This fact implies that S atom is directly involved in the ligand to copper metal. Its frequency is however much higher than that observed in copper-cysteine complex. The carbonyl $\nu(C{=}O)$ is also lowered, indicating that $C{=}O$ group is also site of the interaction. Finally the Raman band at 350 cm^{-1} confirms the establishment of metal-ligand bonds.

6. Comparative Analysis of the Raman Spectra

The analysis of the Raman spectra of the cancer blue and green particles in comparison to various model copper-organic complexes shows that some characteristic bands of nucleotide copper complexes and of cysteine-copper complex find some similarities with the Raman bands of the blue particles. At this stage of the study, the complete assignment of the Raman spectrum of the latter is not yet achieved. Nevertheless, we can already say that:

— more than one ligand intervenes in the blue complexes;
— both nucleic bases and proteins are among these ligands; especially sulfur containing bases are involved.

Anyhow, the fact that such organo-metallic complexes are evidenced can lead to advance some ideas concerning the disorder which can initiate this complex formation.

7. A Mechanism of Carcinogenesis

The discovery of blue particles in blood and tumors of cancerous and leukemic persons and animals and its partial identification help to understand how the disorder can occur in biosynthesis.

Of course, the intercalating disorder model (Figure 20) is among the most attractive existing theories on carcinogenesis. Nevertheless, from this mechanism, it is not easy to explain the selective activity of an anticancer drug.

The evidence of blue complexes suggests itself a simple mechanism of carcinogenesis.

7.1. METAL CONCENTRATION BY ABSORPTION AND BY METAL CARRIERS

In the healthy medium the amount of metal is low [18]. But the metal concentration can be increased by inhalation, ingestion [19], or transportation by metal carriers such as carcinogens. Many carcinogens contain in their molecule many strong basic sites. They can drain metals by coordination in their path and bring them toward the negative sites of nucleic and proteins bases (ex. aflatoxin B$_1$).

Aflatoxin B$_1$

Polynuclear or methylated carcinogens interact in fact when they assume the metabolized epoxy forms. The new negative epoxy sites play then the same role as metal carriers.

In many cases, the metals could also have been concentrated by bacteria, in particular in lung, rectum, etc.

All these metal concentration processes are in agreement with the observation of high concentrations of metal spherules in the neighborhood of blue particles found in cancer tumors. These processes constitute the *external causes* of the possible appearance of cancer.

7.2. DISORDERING COMPLEXES

When the metal concentration is locally high enough in the site of nucleic and proteinic bases, *blue complexes* can occur. Metals like copper have the possibility to coordinate with up to six different ligands [21].

As we have shown earlier, the complexes can be formed between copper and

INTERCALATING DISORDER

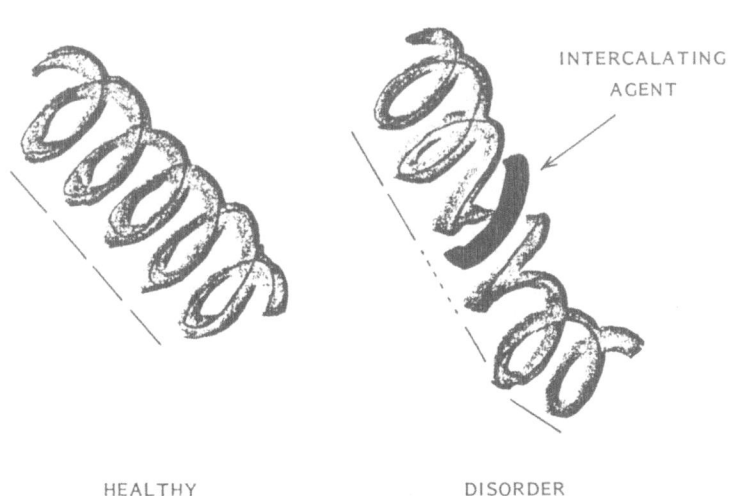

HEALTHY   DISORDER

Fig. 20. Intercalation model promoting disorder (double helix symbol).

nucleic and sulfur-peptides of the protein chain. This number is not exclusive. One or more ligands could be the carcinogens (metal carriers) themselves.

The fact that copper metal can fix up to six ligands confers to this element the possibility to release one or two ligands easily, then to share new ones easily. This property is similar to that of the central metal atom in big cycles, such as metallo-porphyrins [22, 23], metallo-corrines [2] or in other macrocycles.

In biological medium, a binding similar to that we showed earlier with model complexes, for instance to N-7 of a guanine base, could take out this nucleic base from the regularity of DNA chains and provoke the disorder base pairing and induce the miscoding. The regulation will then be no longer respected in the cell growth.

Copper is well known to induce *in vitro* aberrations in base pairing or con-formation of nucleic acids [24, 25]. Copper can also induce infidelity of DNA replication *in vitro* [26].

In organism, the disordering complexes (blue particles) creates a local disorder which is the triggering factor initiating the missing of codage and messages and can explain the origin of cancer (Figure 21) as the cell division is then no longer regulated.

A MECHANISM OF CARCINOGENESIS

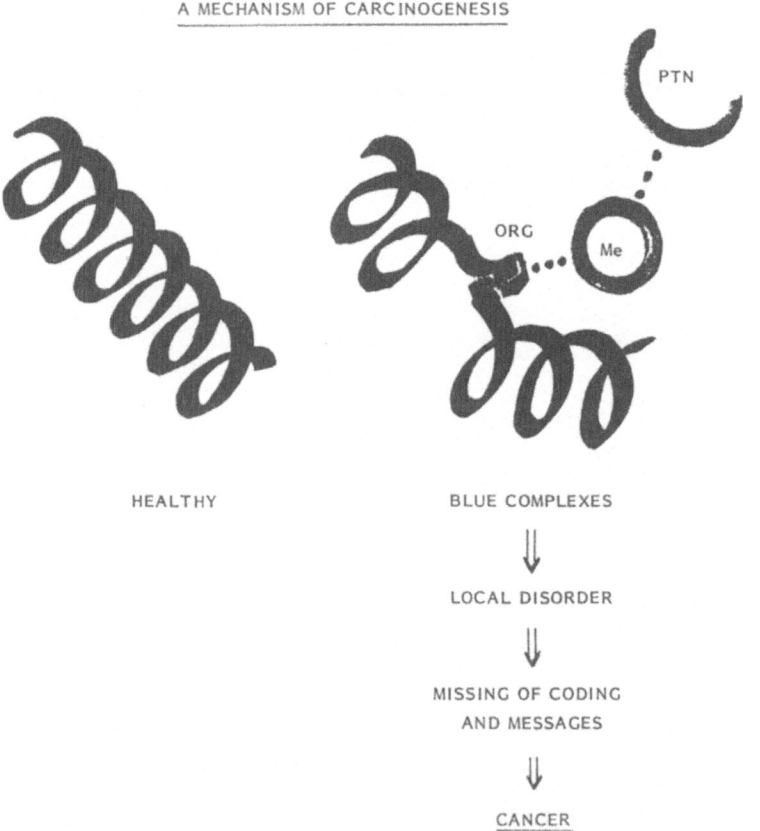

HEALTHY   BLUE COMPLEXES

⇓

LOCAL DISORDER

⇓

MISSING OF CODING
AND MESSAGES

⇓

CANCER

Fig. 21. Proposed mechanism of carcinogenesis (double helix symbol).

8. Selective Reactivity of Cancer Drugs

From existing theories, no satisfying explanation could be given concerning how a drug kills a cancerous cell and not a healthy cell.

The idea of a drug intercalating in the double helix may not be sustained, as this mechanism induces disorder and therefore cannot play the repairing role.

In our model of disordering complexes, the selective reactivity of a drug is obvious. In healthy medium, because of the low level of metal, no blue complex exists; while in cancerous media, metallo-organic complexes are present and thus can be destroyed by an eventual competing equilibrium induced by an anti-cancer drug.

9. A Way Toward Therapeutic Research

From the thermodynamic point of view, when a metallo-organic complex exists, one can introduce a competitive equilibrium to take out the metal (scheme).

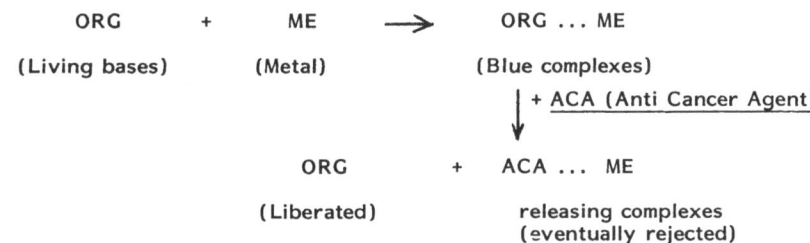

Scheme. A way toward therapeutical research.

If an agent, ACA, is basic enough and if its conformation is suitable, it can displace the equilibria in its favour to form what we call *releasing complexes*, and the lived organic part of the blue complexes could then be released, and the tumor growth therefore be inhibited.

10. A Way Toward Pharmaceutical Research

In respect to thermodynamic equilibria conditions, the anti-cancer-agent (ACA) might have a similar conformation than the ligand bases we found in the blue complexes. Although the total identification of the organic part in these complexes has not yet been achieved, we already have good confidence of the presence of nucleotides and sulfur-protein bases.

Of course, these agents might have slightly stronger basic (electron donor) properties than lived bases. These properties could be acquired with suitable substituents.

The ACA agents might not complex all the amount of metals in the organism, even locally; a minimum threshold of metals might be kept present.

10.1. MODIFIED NUCLEIC BASES

Although their selectivity is not yet perfectly obtained, some substituted nucleic bases, such as, fluoro-uracyl (Figure 22), could correspond to the ACA profile suggested by this theory.

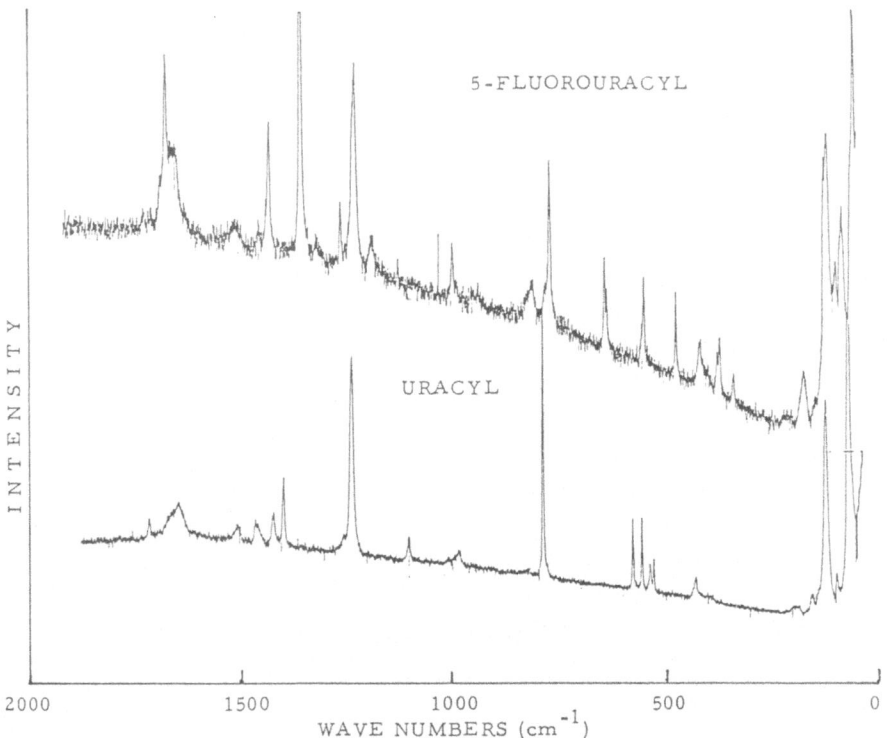

Fig. 22. Raman spectra of uracyl and 5-fluorouracyl, an anticancer drug.

10.2. SULFUR CONTAINING PEPTIDES

In the blue complexes, the Raman spectroscopic study reveals that one of the ligands surrounding the copper atom could be a sulfur-peptide like cysteine.

Among similar peptides, reduced glutathione merits a special attention. As we have seen earlier, glutathione is an enzymatic reducing agent. Its presence in organism is relatively high in liver but much less abundant in the other parts.

In cancerous patients, it is found that the amount of glutathione is lowered in the neighborhood of the tumors but its concentration is increased in the tumors and can reach the level of glutathione in liver [27]. There is in fact a migration of glutathione from healthy parts of organism toward the disease centers.

In the structural point of view, glutathione is both similar to cysteine and different. The differences are due to the presence of two 'substituants' which are glycocolle and glutamic acid. It could also correspond to the ACA profile suggested by the present theory on carcinogenesis.

We do think that glutathione can play the role of releasing-complexing-agent [28]. It can take out the metal from blue complexes.

From this analysis, a new insight could be formulated for the role of reduced glutathione in the organism. It could play there a defense game as a natural anti-cancer substance.

Therefore, glutathione, modified glutathione, and modified cysteine must be actively investigated as candidates in anti-cancer drug research.

A positive result obtained recently by Novi [29], seems to be very promising.

Reduced Glutathione

11. Conclusion

Metallo-organic blue particles have been found in blood and tumors of cancerous and leukemic people.

By comparison with the interaction modes in model complexes such as metallo-nucleotides and metallo-proteins, the nature of the blue complexes has been partially identified. This leads to a possible mechanism of carcinogenesis involving three steps.

1. Hazardous stage: favored by metal transport and metal concentration, due to:
       — absorption
       — metal carrying carcinogens
       — bacteria, etc. . . .

2. Disordering stage: with appearance of *decoding complexes*, initiating the missing of coding, message and regulation.

3. Hoping stage: with creation of *releasing complexes*, by competitive equilibrium able to dissociate the decoding complexes.

The nature of the nucleotides and sulfur containing peptides identified as ligand-bases in the blue complexes suggested that modified molecules with similar conformation and with stronger basicity could play the role of competitive complexing agents.

These results contribute, from the bio-molecular point of view, to help the orientation of cancer therapy and anti-cancer drug research.

Acknowledgements

The author wishes to express his warm thanks to S. P. Plouvier, D. Lespiaux, P. Lambert, R. Giege, J. Lorosch, P. Carmona, J. F. Goussot, C. Casahoursat, A. M. Cornut, and R. M. Escaich for their valuable assistance.

References

1. J. Lascombe and P. V. Huong, Eds.: *Raman Spectroscopy, Linear and Non-Linear*, Wiley, New-York (1982).
2. P. V. Huong: 'Chemical Applications of Resonance Raman Spectroscopy', in *Vibrational Spectra and Structure*, J. R. During, Ed., Elsevier, Amsterdam (1980).
3. P. V. Huong: 'New Analytical Possibilities of Raman and Resonance Raman Spectroscopy', in *Reviews on Analytical Chemistry*, L. Niinisto, Akademia Kiado, Budapest (1982).
4. P. V. Huong, E. Audry, R. Giege, D. Moras, J. C. Thierry, and M. B. Comarmond: 'Conformation change in yeast-transfer-Ribonucleic-Acid Aspartate'. *Biopolymers* **23**, 71 (1984).
5. P. V. Huong: 'Drug Analysis by Raman and Micro-Raman Spectroscopy'. *J. of Pharmaceutical and Biomedical Analysis* **4**, 811 (1986).
6. N. Vanbac, C. Moisand, A. Gouyette, G. Muzard, N. Datxuong, J. B. le Pecq, and C. Paoletti: 'Metabolism and disposition studies of ellipticines in animals'. *Cancer Treatment Reports* **64**, 879 (1980).
7. P. V. Huong and S. R. Plouvier: 'Blue particles in cancerous organs studied by Laser Micro-Raman Spectroscopy'. *J. Mol. Structure* **115**, 489 (1984).
8. S. R. Plouvier and P. V. Huong: 'Microbial chromophore materials in circulating blood identified by Laser Micro-Raman Spectroscopy'. *Biorheology*, Suppl. I, 345 (1984).
9. R. Giege, P. V. Huong, and D. Moras: 'Molecular dynamics in *t*-RNA. Crystallographic and Raman spectroscopic evidences'. *Spectrochim. Acta* **42A**, 378 (1986).
10. P. Carmona, P. V. Huong, and E. Gredilla: 'Raman spectroscopic study of the binding of copper to nucleotides', in *Raman Spectroscopy*, W. Peticolas and B. Hudson, Eds. (1986), p. 2/15.
11. P. V. Huong, S. R. Plouvier, and P. Lambert: 'Complexes organométalliques dans les tissus cancéreux', in *Spectroscopy of Biomolecules*, Alix, Bernard, Manfait, Eds., New York (1985), p. 946.
12. J. Lorosch, W. Haase, and P. V. Huong: 'Resonance Raman spectroscopic investigations on phenolate-bridged binuclear Cu complexes: a basis for the identification of the endogenous bridging ligand in hemocyanin'. *J. Inorg. Biochemistry* **27**, 53 (1986).
13. P. V. Huong, P. Carmona, and J. Lorosch: 'Metalloproteins and metallo-nucleotides', in *Raman Spectroscopy*, W. Peticolas and B. Hudson, Eds. (1986), p. 4/3.
14. P. V. Huong and J. C. Cornut: 'The interconversion of the zwitterion and uncharged form of γ-amino-butyric acid (GABA)'. *J. Chem. Phys.* **65**, 4748 (1976).
15. M. Avignon, P. V. Huong, J. Lascombe, M. Marraud, and J. Neel: 'Etude par spectroscopie infrarouge de la conformation de quelques composés peptidiques modèles'. *Biopolymers* **8**, 69 (1969).
16. M. Avignon and P. V. Huong: 'Une méthode de dosage des isomères de rotation des dipeptides en solution par spectroscopie infrarouge'. *Biopolymers* **9**, 427 (1970).
17. M. Marraud, J. Neel, M. Avignon, and P. V. Huong: 'Contribution à l'étude conformationnelle des composés peptidiques en solution'. *J. Chim. Phys.* **67**, 959 (1970).
18. E. Frieden: 'New perspectives on the essential trace elements'. *J. Chem. Education* **62**, 917 (1985).
19. F. W. Sunderman: 'Mechanisms of Metal Carcinogenesis', in *The Clinical Biochemistry of Cancer*, M. Fleisher Ed., AACC, Washington D.C. (1979).
20. W. A. Creasey: *Cancer*, Oxford University Press (1981).
21. K. D. Karlin and J. Zubieta, Eds.: *Biological and Inorganic Copper Chemistry*. Adenine Press, Guilderland N.Y. (1986).

22. P. V. Huong and J. C. Pommier: 'Une relation entre fréquences Raman de résonance et rayon du coeur de cycle des métalloporphyrines'. *Compt. Rend. Acad. Sci.* **285C**, 519 (1977).
23. U. Mioc, P. V. Huong, and H. Ledon: 'Effect of dimerization on the structure of molybdenium tetraphenylporphyrin'. *J. Mol. Structure* **142**, 481 (1986).
24. M. J. Murray and C. P. Flessel: 'Metal-polynucleotide interactions. A comparison of carcinogenic and non-carcinogenic metals *in vitro*'. *Biochim. Biophys. Acta* **425**, 256 (1976).
25. Y. P. Blagoi, U. A. Sorokin, V. A. Valew, and G. O. Gladchenko: 'Characteristics of the helix-coil transition in DNA in the region of GC and AT base-pairs. Relative stability inversion due to Cu^{2+} and Mn^{2+}'. *Dokl. Akad. Nauk. SSSR* **240**, 459 (1978).
26. L. A. Loeb, M. A. Sirover, and S. S. Agarwal: 'Infidelity of DNA synthesis as related to mutagenesis and carcinogenesis', in *Inorganic and Nutritional Aspects of Cancer*, G. N. Schrauzer, Ed., Plenum Press, New York (1978), p. 103.
27. J. W. Thompson and C. Voegtlin: 'Glutathione content of normal animals'. *J. Biol. Chem.* **70**, 793 (1926).
28. P. V. Huong and S. P. Plouvier: in *Molecular Structure and Spectra*, J. R. Durig, Ed., Elsevier, Amsterdam (1988).
29. A. M. Novi: 'Regression of Alflatoxin B_1 Induced Hepatocellular Carcinomas by Reduced Glutathione'. *Science* **212**, 541 (1981).

Topics in Biomolecular Physics

Vibrational Spectroscopy, Molecular Flexibility and Molecular Graphics

GÉRARD VERGOTEN
Laboratoire de Génie Biologique et Médical, INSERM U 279, Faculté de Pharmacie, Lille, France.

1. Vibrational Spectroscopy and the Theoretical Determination of Molecular Vibrations

There are three main experimental techniques from which vibrational frequencies may be obtained. They are neutron inelastic scattering, infrared absorption spectroscopy, and light inelastic (Raman) scattering. In the biological field, Raman spectroscopy seems a very attractive technique since water is the best solvent for studying solutions. For example in the solid state very low frequencies can be observed in Raman without much difficulties, using triple monochromators with very low stray light [1].

Raman studies of protein crystals are of great interest, since it has been shown that in general the crystal and solution structures are essentially the same [2].

Using multichannel detectors it is now possible to record very high quality spectra in short times. One of the main disadvantages of Raman spectroscopy, however, is due to the fluorescence of biological samples. Fluorescence is usually due to impurities which prevent the observation of Raman spectra. This difficulty can be reduced if we use Resonance Raman spectroscopy, which means that we use ultraviolet laser lines as exciting lines for the Raman effect. Once the spectra are obtained, we have to assign all observed bands or most of them. In the case of large molecules, the concept of group frequencies cannot be applied [3, 4]. Moreover we have to make a distinction between localized (high frequency) small amplitude vibrations and overall (low frequency) large amplitude vibrations. The spectral frontier between these two series of vibrations may be fixed at 200 cm^{-1}, which is the value of thermal energy kT at room temperature. The theoretical determination of (low frequency) molecular vibrations is reached using classical mechanics. This is clearly seen from the Heisenberg uncertainty principle, which relates the energy ΔE of one physical event and its time period ΔT:

$$\Delta E \cdot \Delta T \sim h.$$

Since we are concerned with energies of the same order than thermal energy, the former relation gives $T = 2 \times 10^{-14}$ s, which means a limiting dimension of 0.2 Å. So a classical determination of molecular vibrations is appropriate for macromolecular systems, even if energy levels are determined by quantum

Jean Maruani (ed.), Molecules in Physics, Chemistry, and Biology, Vol. IV, 113—131.
© *1989 by Kluwer Academic Publishers.*

mechanics. The theory of this determination is well established. It is mainly based on the $F \cdot G$ matrix eigenvalues and eigenvectors where F is the potential energy matrix and G the inverse kinetic energy matrix [5, 6].

If the molecular geometry is known, the G matrix is easily constructed. In order to evaluate the F matrix, a model of the force field has to be chosen. We are currently developing the local symmetry force field for a number of molecular classes.

2. The Biological Significance of Low-Frequency Molecular Vibrations

It is now well accepted that the static description of biological macromolecules as given by X-ray experiments is not satisfactory. In order to explain a number of experimental results, it is necessary to include both molecular flexibility and conformational adaptability. The phenomenon of protein flexibility is very complex and is characterized by a large variety of motions occurring over wide ranges of time $(10^{-4}-10^{-12}$ seconds). Flexibility takes place for example in aiding the access of ligands to active sites. A typical case is the binding of glucose to hexokinase where the protein molecule has to open and close in order for the enzyme to bind the whole molecule. Another example is given by immunoglobins where the mutual flexibility is an important factor occurring when the antigen-antibody complex is built. This is a molecular process which leads to a maximum efficiency of the antibody while modulating the spatial structure of the antigenic groups involved in the immunological response mechanism.

One of the best typical cases is given by citrate synthase [7]. This enzyme catalyses the reaction between acetylcoenzyme A and oxaloacetate to form citrate. This macromolecule is a dimer with identical subunits formed by 437 residues. This is a large globular protein consisting essentially of alpha helices. Two forms have been revealed using X-ray experiments: one open form and a closed form with respect to a hinge region. In the closed form the catalytic reaction takes place: condensation of oxaloacetate and acetyl coA to form citryl coA and hydrolysis of citryl coA to form citrate and coA. The open form corresponds to the free product and free substrate. The closing and opening of the hinge take place thousands of times per second during each catalytic cycle.

The last example given here is binding of O_2 to myoglobin and hemoglobin which requires some motions of the protein to allow access to the heme.

Low frequency vibrations of a macromolecule are of great importance since they involve larger scale movements.

Among the $(3N - 6)$ molecular vibrations for a N atomic molecules only those of low frequency (less than 200 cm^{-1} which represents the value of the thermal energy kT at room temperature) are of biological interest since their linear combination leads to the effective deformation of the molecule. An other characteristic point among molecular vibration is given by the calculation of populations on vibrational levels. At room temperature the fundamental level of a high frequency vibration (such as the well known amide I vibration) is 100% populated. On the contrary, for a very low frequency vibration (10 cm^{-1}) 20 levels are equally populated (5%) at room temperature. The spontaneous return of molecules from vibrational excited states to a fundamental level represents an appreciable source of energy which may be used by the macromolecule to undergo conformational transitions.

3. The Local Symmetry Force Field

The analytical expression of the potential energy V of an n-atom molecule is not known. The only certainty about it is that it must be a function of the $3n$ atomic displacement coordinates. The potential energy function is usually expanded in a Taylor series in terms of a set of displacement coordinates q_i, for small displacements:

$$V = V_0 + \sum_i f_i q_i + 1/2! \sum_{ij} f_{ij} q_i q_j + 1/3! \sum_{ijk} f_{ijk} q_i q_j q_k + \ldots .$$

The constants

$$f_i = \left(\frac{\partial V}{\partial q_i} \right)_0, \qquad f_{ij} = \left(\frac{\partial^2 V}{\partial q_i \partial q_j} \right)_0, \qquad f_{ijk} = \left(\frac{\partial^3 V}{\partial q_i \partial q_j \partial q_k} \right)_0$$

are the linear, quadratic and cubic force constants, respectively.

Since we are always concerned with variations of the potential energy with the displacements coordinates, its origin may be chosen as $V_0 = 0$. Furthermore the equilibrium configuration of the molecule must be at a minimum of potential energy. Provided that the q_i are all independent, we have

$$f_i = \left(\frac{\partial V}{\partial q_i} \right)_0 = 0.$$

For small displacements of the atoms around their equilibrium position, we can neglect the cubic and higher terms in the Taylor expansion of V so that

$$V = 1/2 \sum_{ij} f_{ij} q_i q_j.$$

The use of the harmonic potential function in the equations of motions leads to solutions giving harmonic molecular vibrations the frequencies of which may be compared with experimental data obtained in the Infrared and Raman spectra. From the molecular potential energy function a wide range physical properties may also be investigated. They are obtained from molecular mechanics calculations [8, 9]. A number of molecular mechanics computer programs are available.

They are for example MM2/MMP2 [10], QCFF/PI + MCA [11, 12], AMBER [13], and CHARMM [14]. All these programs use the internal coordinates (bond stretchings, bond angle bendings, torsions) as basis coordinates which are usually not independent. This means that not all linear terms in the potential energy function expansion vanish so that the equilibrium configuration of the molecule is not at a minimum of the energy. If the number m of internal coordinates is greater than $(3n - 6)$ (internal degrees of freedom) intramolecular tension force constants are added to the classical force constants to handle the remaining linear terms (see [15] and [16]). Intramolecular tension force constants are not easy to determine when rings and hydrogen bonds are present in the molecule. In order to avoid all the difficulties in treating redundancies, we use local symmetry coordinates as basis independent coordinates and then the local symmetry force field. In this case force constants are characteristic of one atomic group instead of one bond or one angle. Let us take two examples.

The first one will be the methyl CH_3X and methylene CH_2XY groups where X and Y are any kinds of atoms (e.g. carbon atoms).

The CH_3XY group has the local symmetry C_{3v}, the hydrogen located at the trans position relative to Y is numbered as 1. The H atoms numbered 2 and 3 are defined in the clockwise direction viewing the atoms along the bond X—C, with the atom X nearer to the observer than C. The H_1CXY defines the reference plane.

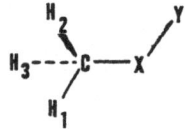

For this five atomic group we have to define 9 independent local symmetry coordinates among the 6 angle bending and 4 bond stretching internal coordinates.

Let us call a_{12} the variation of angle bending H_1CH_2, b_{1X} the variation of angle bending H_1CX, r_1 the variation of the bond stretching CH_1 and R the variation of the bond stretching CX internal coordinates.

Since the 6 angles cannot increase at the same time, the redundancy condition is obtained:

$$S(I) = (6)^{-1/2}(a_{12} + a_{13} + a_{23} + b_{1X} + b_{2X} + b_{3X}).$$

The nine independent local symmetry coordinates are given below.

CH_3 total symmetry stretching:

$$TS = (3)^{-1/2}(r_1 + r_2 + r_3) \qquad \text{(in plane)}.$$

CH_3 degenerate stretching:

$$DS = (6)^{-1/2}(2 \cdot r_1 - r_2 - r_3) \qquad \text{(in plane)}.$$

CH_3 degenerate stretching:

$$DS' = (2)^{-1/2}(r_2 - r_3) \qquad \text{(out of plane).}$$

CH_3 symmetric deformation:

$$SD = (6)^{-1/2}(a_{12} + a_{13} + a_{23} - b_{1X} - b_{2X} - b_{3X}).$$

CH_3 degenerate deformation:

$$DD = (6)^{-1/2}(2 \cdot a_{23} - a_{12} - a_{13}) \qquad \text{(in plane).}$$

CH_3 degenerate deformation:

$$DD' = (2)^{-1/2}(a_{12} - a_{13}) \qquad \text{(out of plane).}$$

CH_3 degenerate rocking:

$$DR = (6)^{-1/2}(2 \cdot b_{1X} - b_{2X} - b_{3X}) \qquad \text{(in plane).}$$

CH_3 degenerate rocking:

$$DR' = (2)^{-1/2}(b_{2X} - b_{3X}) \qquad \text{(out of plane).}$$

CX stretching:

$$CX = R.$$

For the CH_2XY group with C_{2v} symmetry, we also have to define 9 independent local symmetry coordinates among the 10 internal coordinates.

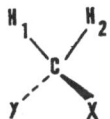

The sign of the vibrations corresponding to rocking, wagging or twisting will be positive if, when viewing the XCH_2Y group from the direction perpendicular to the HCH plane, the atom X being nearer to the observer, the HCH rotates in the clockwise direction (for rocking), the H atoms move away from the X atom in order to increase the XCH angles (for wagging) or the H atom on the right comes nearer and the H atom on the left moves away (for twisting).

Let us take a the variation of bending H_1CH_2 internal coordinate, b_{1X}, b_{2Y} the variations of bending H_1CX and H_2CY internal coordinate respectively.

The redundancy conditions among the internal coordinates is

$$S(I) = (6)^{-1/2}(a + b_{1X} + b_{2Y} + b_{2X} + b_{1Y} + XCY).$$

We are now able to define the 9 independent local symmetry coordinates classified as 'in plane' or 'out of plane' for some of them with respect to the XCY plane.

CH_2 symmetric stretching:

$$SS = (2)^{-1/2}(r_1 + r_2).$$

CH_2 antisymmetric stretching:

$$AS = (2)^{-1/2}(r_1 - r_2).$$

CH_2 scissor:

$$SC = (20)^{-1/2}(4 \cdot a - b_{1X} - b_{1Y} - b_{2X} - b_{2Y}) \qquad \text{(in plane)}.$$

CH_2 wagging:

$$WA = (2)^{-1/2}(b_{1X} + b_{2X} - b_{1Y} - b_{2Y}) \qquad \text{(in plane)}.$$

CH_2 twisting:

$$TW = (2)^{-1/2}(b_{1X} - b_{2X} - b_{1Y} + b_{2Y}) \qquad \text{(out of plane)}.$$

CH_2 rocking:

$$RO = (2)^{-1/2}(b_{1X} - b_{2X} + b_{1Y} - b_{2Y}) \qquad \text{(out of plane)}.$$

CY stretching:

$$RX.$$

CY stretching:

$$RY.$$

XCY bending:

$$XCY.$$

In order to determine the local symmetry force field, we usually start with a classical mode of the energy function, for example the Urey–Bradley–Shimanouchi (UBS) force field [17]. We use the following UBS force constants:

stretching (mdyne/Å)

$$K(CC) = 2.563$$
$$K(CH)(CH_3) = 4.301$$
$$K(CH)(CH_2) = 3.936$$

bending, repulsion (mdyne/Å)

$$H(CCC) = 0.287$$
$$F(CCC) = 0.369$$
$$H(HCH)(CH_3) = 0.378$$
$$F(HCH)(CH_3) = 0.195$$

$H(HCH)(CH_2) = 0.332$

$F(HCH)(CH_2) = 0.279$

$H(CCH)(CH_3) = 0.208$

$F(CCH)(CH_3) = 0.385$

$H(CCH)(CH_2) = 0.191$

$F(CCH)(CH_2) = 0.537$

torsion (mdyne.Å)

torsion$(CH_2—CH_3) = 0.086$

torsion$(CH_2—CH_2) = 0.107$

interaction force constants (mdyne.Å) (see [17])

$trans = 0.170$

$trans' = 0.033$

$gauche = 0.014$

$gauche' = 0.086$

Coriolis CH, CH: $p = -0.116$

CH_2 deformation: $\ell = 0.011$.

In order to eliminate in the Taylor expansion of the potential energy function the linear term due to redundancy, we have to add to second order terms internal tension force constants (mdyne.Å) [17]:

$X(CH_3) = 0.025$

$X(CH_2) = 0.054$

this force constant applies to the following expression [17]:

$$3 \cdot (2)^{1/2}/8((a)^2 + (b_{1X})^2 + (b_{2X})^2 + (b_{1Y})^2 + (b_{2Y})^2 + (XCY)^2) + (2)^{1/2}/2(a \cdot b_{1X} + a \cdot b_{2X} + a \cdot b_{1Y} + a \cdot b_{2Y} + a \cdot XCY + b_{1X} \cdot b_{2X} + b_{1X} \cdot b_{1Y} + b_{1X} \cdot b_{2Y} + b_{1X} \cdot XCY + b_{2X} \cdot b_{1Y} + b_{2X} \cdot b_{2Y} + b_{1X} \cdot XCY + b_{1Y} \cdot b_{2Y} + b_{1Y} \cdot XCY + b_{2Y} \cdot XCY)$$

for a perfect tetrahedral $—CH_2$-group.

Transforming the preceding force field with the local symmetry coordinates, we obtain the local symmetry force constants which are specific of the corresponding atomic groups and may be transferred with them when building molecules from atomic fragments. It is clearly seen that these force constants take redundancies into account. We are giving some of them for the CH_3- and $—CH_2$-groups:

CH_3-:

$$F(SD) = 0.550\, H(HCH) + 0.238\, F(HCH) + 0.839\, H(CCH) + 0.322\, F(CCH) + 0.530\, X(CH_3) = 0.584 \text{ mdyne.Å}$$

$F(DD)$ = 1.188 $H(HCH)$ + 0.475 $F(HCH)$ − 0.177 $X(CH_3)$ = 0.534 mdyne.Å

$F(DR)$ = 1.679 $H(CCH)$ + 0.645 $F(CCH)$ − 0.177 $X(CH_3)$ = 0.593 mdyne.Å.

—CH_2-:

$F(SC)$ = 0.950 $H(HCH)$ + 0.380 $F(HCH)$ + 0.336 $H(CCH)$ + 0.129 $F(CCH)$ − 0.318 $X(CH_2)$ = 0.538 mdyne.Å

$F(WA)$ = 1.679 $H(CCH)$ + 0.645 $F(CCH)$ + 0.530 $X(CH_2)$ − 2ℓ = 0.674 mdyne.Å

$F(RO)$ = 1.679 $H(CCH)$ + 0.645 $F(CCH)$ + 0.530 $X(CH_2)$ + 2ℓ = 0.717 mdyne.Å

$F(TW)$ = 1.679 $H(CCH)$ + 0.645 $F(CCH)$ + 0.884 $X(CH_2)$ = 0.619 mdyne.Å.

We also give the local symmetry force constants relative to the two stretching local symmetry coordinates:

$F(SS)$ = 4.735 mdyne/Å
$F(AS)$ = 4.576 mdyne/Å.

The second example is concerned with the planar vibrations of the benzene ring (without hydrogens). This six membered ring belongs to the symmetry point group D_{6h}. 18 internal coordinates are defined. They consist of 6 bond stretching coordinates, 6 angle bending coordinates and 6 torsion coordinates. The internal degrees of vibrational freedom are $3 \times 6 - 6 = 12$ in which we find 9 in plane vibrations and 3 out of plane vibrations.

Among the twelve planar internal coordinates, only 9 are independent so we have to determine 3 redundancy relations, according to the irreducible representations of the D_{6h} group. They are:

$A1_g$ SI = RCC/$(6)^{1/2}$(CCC612 + CCC123 + CCC234 + CCC345 + CCC456 + CCC561)

$E1_u(a)$ SII = −1/2(CC12 + 2·CC23 + CC34 − CC45 − 2·CC56 − CC61) + RCC·$(3)^{1/2}$/2(−CCC123 − CCC234 + CCC456 + CCC561)

$E1_u(b)$ SIII = $(3)^{1/2}$/2(CC12 − CC34 − CC45 + CC61) + RCC/2(−CCC123 + CCC234 + CCC456 − CCC561)

where CC12 is the variation in the bond length between carbon 1 and carbon 2, CCC123 is the variation in the bond angle between carbon 1, carbon 2 and carbon 3 and RCC the equilibrium value of the carbon-carbon bond length.

Taking these three redundancy relations into account, we may define the nine independent 'local' symmetry planar coordinates of the benzene ring. There are given below according to the D symmetry species:

$E2_g(a)$ $S1a = 1/(12)^{1/2}(CC12 - 2 \cdot CC23 + CC34 + CC45 - 2 \cdot CC56 + CC61)$

$E2_g(b)$ $S1b = 1/(2)^{1/2}(CC12 - CC34 + CC45 - CC61)$

$E1_u(a)$ $S2a = 1/(10)^{1/2}(CC12 - CC34 - CC45 + CC61) - RCC/(20)^{1/2}(2 \cdot CCC612 + CCC123 - CCC234 - 2 \cdot CCC345 - CCC456 + CCC561)$

$E1_u(b)$ $S2b = 1/(30)^{1/2}(CC12 + 2 \cdot CC23 + CC34 - CC45 - 2 \cdot CC56 - CC61) - RCC(3)^{1/2}/(20)^{1/2}(CCC123 + CCC243 - CCC456 - CCC561)$

$B2_u$ $S3 = 1/(6)^{1/2}(CC12 - CC23 + CC34 - CC45 + CC56 - CC61)$

$B1_u$ $S4 = RCC/(6)^{1/2}(CCC612 - CCC123 + CCC234 - CCC345 + CCC456 - CCC561)$

$A1_g$ $S5 = 1/(6)^{1/2}(CC12 + CC23 + CC34 + CC45 + CC56 + CC61)$

$E2_g(a)$ $S6a = RCC/(12)^{1/2}(2 \cdot CCC612 - CCC123 - CCC234 + 2 \cdot CCC345 - CCC456 - CCC561)$

$E2_g(b)$ $S6b = RCC/(2)^{1/2}(CCC123 - CCC234 + CCC456 - CCC561)$.

To these 6 local symmetry coordinates (including doubly degenerate coordinates) correspond 6 diagonal local symmetry force constants F_1 to F_6. One additional off diagonal force constant F_7 has to be added to fit the experimental spectral data. This last constant is in interaction constant between local symmetry coordinates $S1a$ to $S6a$ (also between $S1b$ and $S6b$).

Under these conditions the in plane potential energy harmonic function takes the expression

$$2V = F_1 \cdot ((S1a)^2 + (S1b)^2) + F_2 \cdot ((S2a)^2 + (S2b)^2) + F_3 \cdot (S3)^2 + F_4 \cdot (S4)^2 + F_5 \cdot (S5)^2 + F_6 \cdot ((S6a)^2 + (S6b)^2) + 2 \cdot F_7 \cdot (S1A \cdot S6A + S1b \cdot S6b)$$

where $F_1 = 6.6882$, $F_2 = 7.3971$, $F_3 = 4.0490$, $F_4 = 1.1719$, $F_5 = 6.9660$, $F_6 = 1.1443$ and $F_7 = 0.1316$.

The units are: mdyne/Å for stretching diagonal force constants, mdyne.Å for bending diagonal force constants and mdyne for stretching-bending interaction force constants.

The following table gives the observed and calculated frequencies, the assignments in terms of local symmetry coordinates and the potential energy distribution among the local symmetry force constants.

obs. freq. (cm⁻¹)	1596	1596	1486	1486	1310
calc. freq. (cm⁻¹)	1595.43	1595.43	1486	1486	1309.94
assignments	$S1b(S6b)$	$S1a(S6a)$	$S2b$	$S2a$	$S3$
F_1	91.5	91.5	0	0	0
F_2	0	0	100	100	0
F_3	0	0	0	0	100
F_4	0	0	0	0	0
F_5	0	0	0	0	0
F_6	0	0	0	0	0
F_7	0	0	0	0	0

obs. freq. (cm⁻¹)	1010	992	606	606
calc. freq. (cm⁻¹)	1008.9	992	606.6	606.6
assignments	$S4$	$S5$	$S6b$	$S6a$
F_1	0	0	8.7	8.7
F_2	0	0	0	0
F_3	0	0	0	0
F_4	100	0	0	0
F_5	0	100	0	0
F_6	0	0	88.6	88.6
F_7	0	0	2.7	2.7

From this table it appears that the description of vibrational frequencies in terms of local symmetry coordinates is very simple that means that local symmetry coordinates are very close to normal modes. From the present calculation, using the Lx matrix description, it is possible to have a direct visualization of the 'in plane' vibrations of the benzene ring. They are shown in the following nine figures. For frequencies higher than 1000 cm⁻¹, the magnification coefficient was 20; for frequencies less than 1000 cm⁻¹, this factor was 10. We only have to treat redundancies once and then transfer local symmetry force constants of small atomic groups in order to build more complex molecules.

We are currently developing the local symmetry force fields of the 20 amino-acids and the 5 nucleic acid bases. They will be used to determine internal vibrations of peptides, proteins and nucleic acids using both geometry and force field data bases. For these classes of molecules, additional interaction force constants may be added if necessary (Van der Waals interactions, hydrogen bonds,...).

4. Molecular Graphics and a Direct Visualization of Molecular Vibrations

Computer graphics is a new applied science. Computer graphics may be defined as creation of non-numerical and non-alphabetical images on computer screens. It gives new possibilities to all scientists concerned with molecules. Computer graphics also makes possible the construction of macromolecules dedicated to control and regulation of biological mechanisms.

Molecular structures can be manipulated in a more sophisticated way than can be done using hand-held molecular models. Molecular graphics is now widely used for pharmacophore identification and computer assisted drug design [18, 19]. The use of molecular graphics is of great help for understanding low frequency molecular vibrations. Color vector graphics technology makes possible the display in pseudo real time of vibrating macromolecules. We are currently using a computer graphics system which consists of an Evans and Sutherland PS300 graphics display [20], linked to a Digital Equipment VAX 11/780 minicomputer [21]. A number of graphics software packages are available. We are using the MANOSK program [22].

It is not easy to imagine molecular vibrations of complex molecules using the potential energy distribution among internal coordinates (bond stretching, bond angle bendings and torsions). To overcome this difficulty, the use of molecular graphics is very attractive.

To visualize molecular vibrations, the following matrix relations are needed:

$$R = BX = LQ$$

where X is the Cartesian coordinate column vector, R the internal coordinate column vector and Q the normal coordinate column vector.

We must define the inverse transformation $X = AR$ with $BA = E$ where E is the identity matrix.

Under these conditions, we obtain the inverse kinetic energy matrix G:

$$BM^{-1}\tilde{B} = G$$

where M is a diagonal matrix of order $3n$ (n the number of atoms) whose elements are the inverse of the atomic masses, $\tilde{B}$ is the transposed matrix of B.

Multiplying the former relation by G^{-1} we obtain:

$$BM^{-1}\tilde{B}G^{-1} = GG^{-1} = E = BA.$$

This gives: $A = M^{-1}\tilde{B}G^{-1}$.

Since $X = L_x Q = ALQ = (M^{-1}\tilde{B}G^{-1}L)Q$, we have the expression of the L_x matrix:

$$L_x = M^{-1}\tilde{B}G^{-1}L.$$

After solving the secular equation $GFL = L\Lambda$, a new definition of the L_x matrix is obtained:

$$L_x = M^{-1}\tilde{B}FL\Lambda^{-1}.$$

During the normal vibration a, the motion of atom i is given in terms of time dependent Cartesian coordinates x_i^a, y_i^a, z_i^a as follows:

$$x_i^a = x_i^0 + AD(L_x)_{x_i}^a \cdot \sin \varphi$$

$$y_i^a = y_i^0 + AD(L_x)_{y_i}^a \cdot \sin \varphi$$

$$z_i^a = z_i^0 + AD(L_x)_{z_i}^a \cdot \sin \varphi$$

where x_i^0, y_i^0, z_i^0 are the Cartesian coordinates of atom i at the equilibrium position, $(L_x)_{x_i}^a$, $(L_x)_{y_i}^a$, $(L_x)_{z_i}^a$ are the elements of the L_x matrix, $\varphi = 2\pi \subseteq \bar{v}_a t$ (where $\bar{v}_a$ is the wavenumber of the vibration a and t the time), $D = 6.8465(T)^{1/2}/\bar{v}_a$ is used to make all physical quantities dimensionally homogeneous, T is the absolute temperature, and A a magnification factor.

These three equations are obtained if the total energy of the system is assumed to be kT, the unit of length being the angström, amu the unit of mass.

The motion is obtained by giving the values $-90°$, $-80°$, . . . , $+90°$ to φ (Figs. 1—9).

Nineteen intermediate positions of the vibrating molecule are thus obtained. The refresh capabilities of the graphic display are then used to make motion pictures.

5. Low Frequency Molecular Vibrations of the Z-Form of DNA

A typical example of a low frequency vibration determination of biological substances currently carried out in our group is given by the hexanucleotide d(CpGpCpGpCpG) which has been shown to adopt the Z conformation with 12 base pairs per turn of the left handed double helix (see [4] and references therein). The complete normal coordinate treatment of the crystal has been performed and 1185 normal modes of vibration thus determined. The equilibrium position of the atoms in the crystal is shown in Figures 10 and 11 (see [4] for the atom correspondence). The lowest vibrational frequencies are calculated for the totally symmetric $A1$ species at 2.1 and 3.8 cm^{-1}.

The corresponding extreme atomic displacements ($\varphi = \pm 90°$) are given in Figures 12 and 13 for the 2.1 cm^{-1} vibration and Figures 14 and 15 for the 3.8 cm^{-1} vibration. The first one (2.1 cm^{-1}) corresponds to a transverse motion (perpendicular to the helix axis or lateral breathing of the helix). During this vibration the guanine is rejected out of the overall shape while the cytosine base stays inside. This particular motion is in good agreement with former experimental observations [23]. The second lowest frequency vibration (3.8 cm^{-1}) is a longitudinal motion (in the direction of the helix axis). Lateral shape does not change. The G and C bases rotate around axes perpendicular to the helix axis. Such a motion may make possible intercalation of planar aromatic substances.

The magnification factor used to display these two vibrations is 10. The low frequency vibrations strongly contribute to the molecular flexibility.

The overall deformation is obtained if we linearly combine the molecular vibrations with frequency less than 50 cm^{-1}.

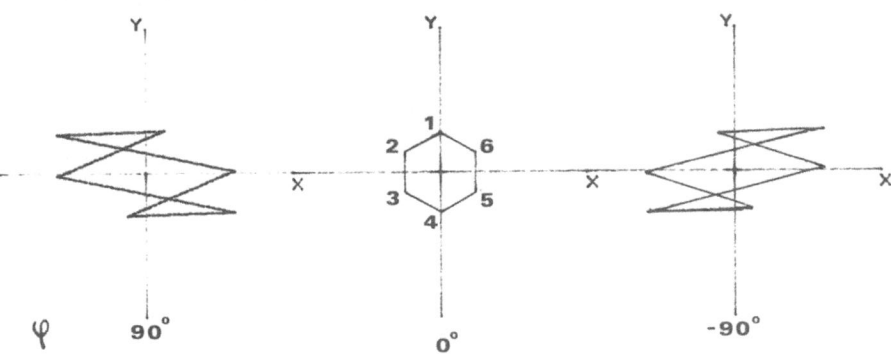

Fig. 1. Extreme atomic displacements in the 1595.43 cm⁻¹ vibration.

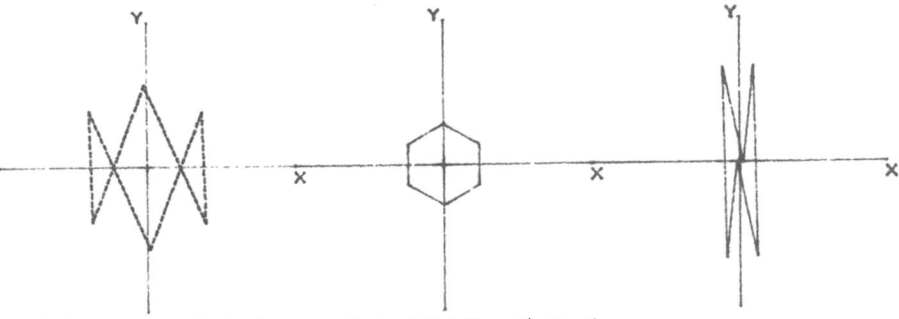

Fig. 2. Extreme atomic displacements in the 1595.43 cm⁻¹ vibration.

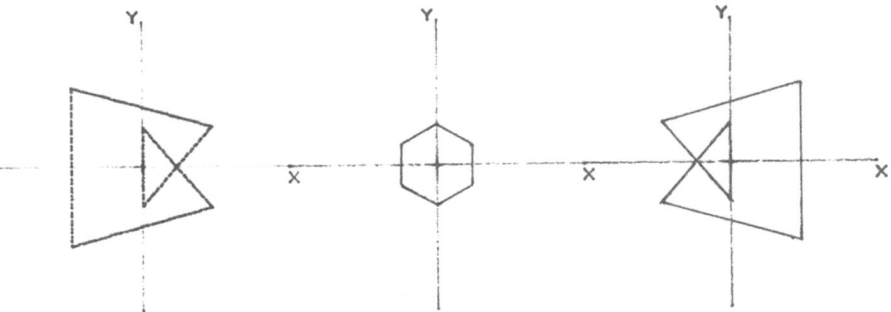

Fig. 3. Extreme atomic displacements in the 1485.99 cm⁻¹ vibration.

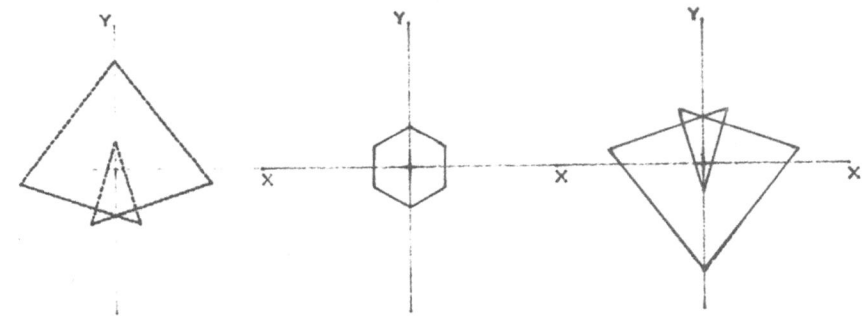

Fig. 4. Extreme atomic displacements in the 1485.99 cm⁻¹ vibration.

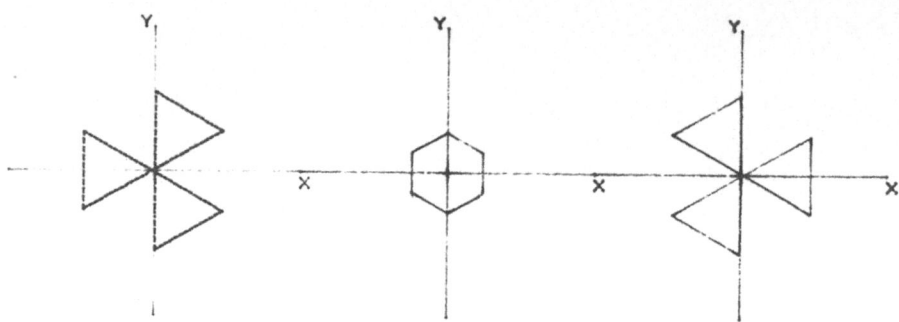

Fig. 5. Extreme atomic displacements in the 1309.94 cm^{-1} vibration.

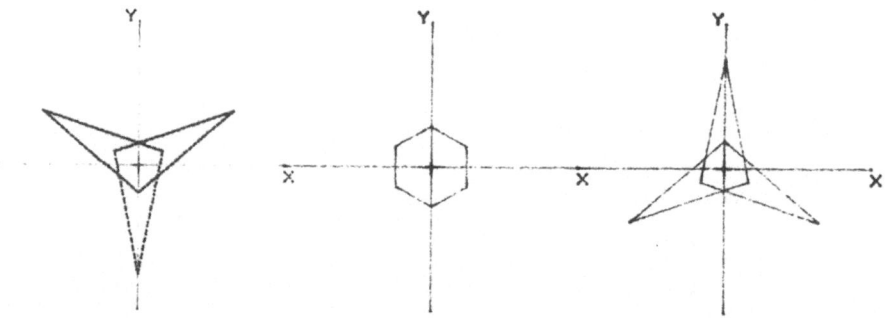

Fig. 6. Extreme atomic displacements in the 1008.90 cm^{-1} vibration.

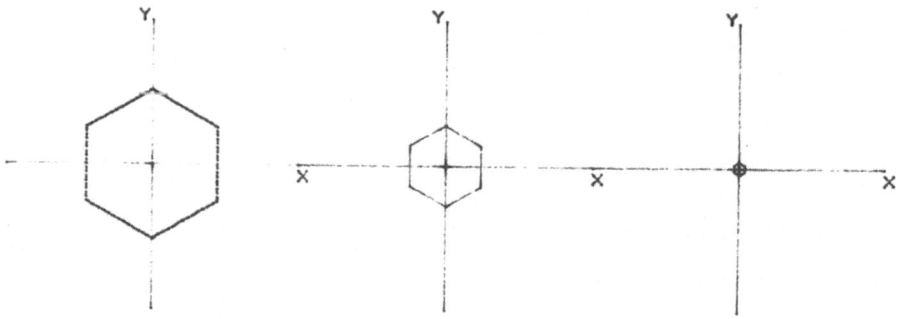

Fig. 7. Extreme atomic displacements in the 991.99 cm^{-1} vibration.

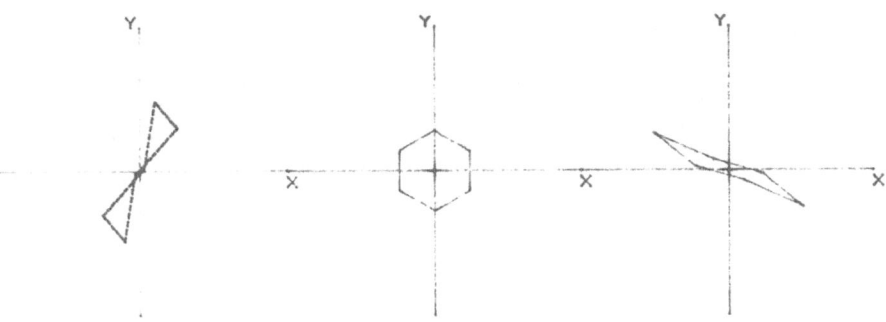

Fig. 8. Extreme atomic displacements in the 606.63 cm⁻¹ vibration.

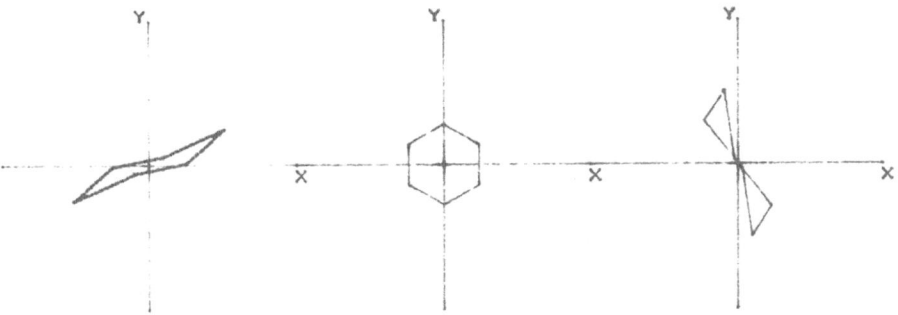

Fig. 9. Extreme atomic displacements in the 606.61 cm⁻¹ vibration.

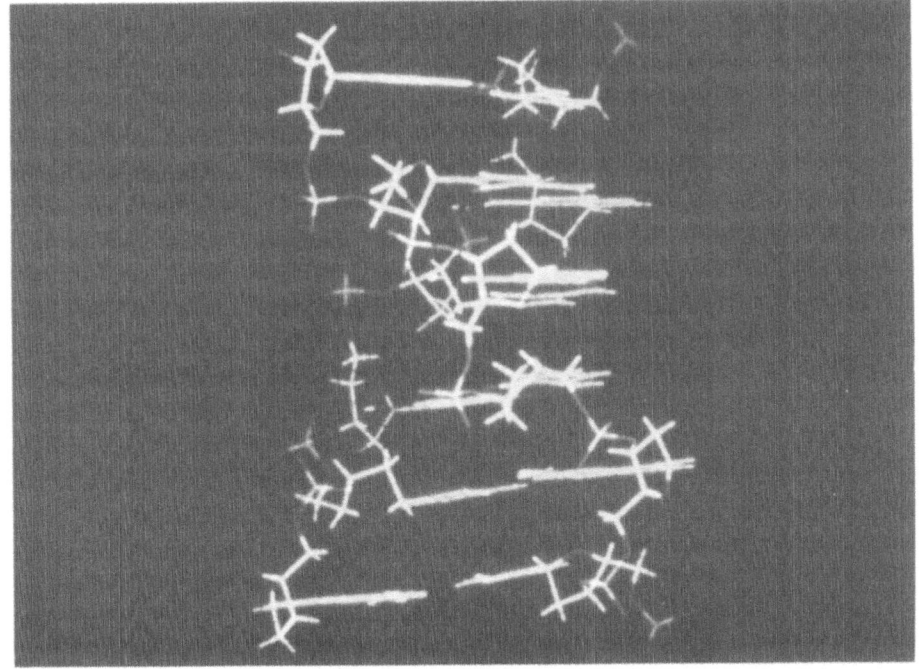

Fig. 10.

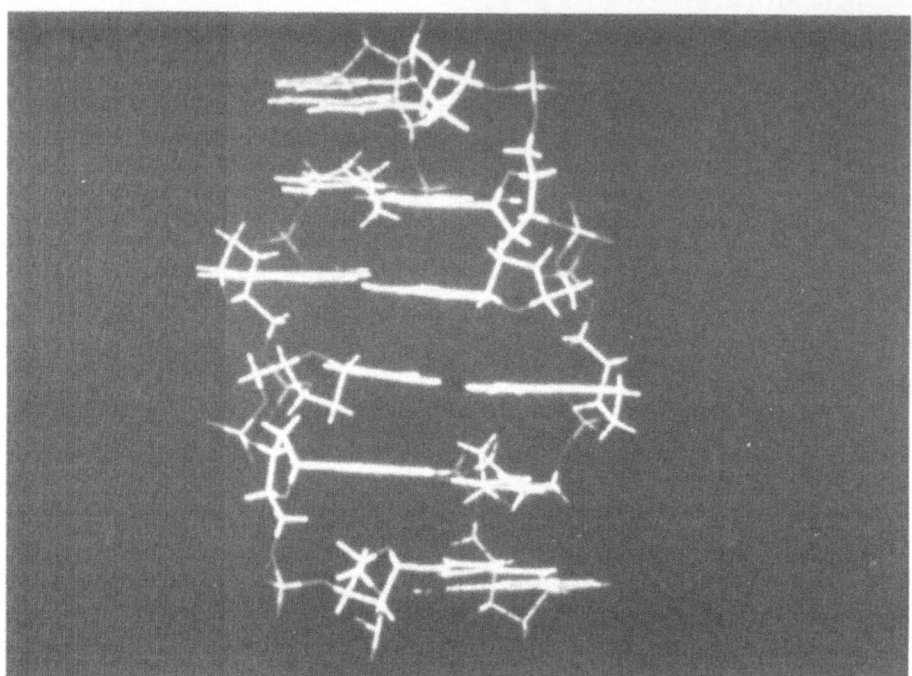

Figs. 10—11. Crystal structure of the CpG hexamer.

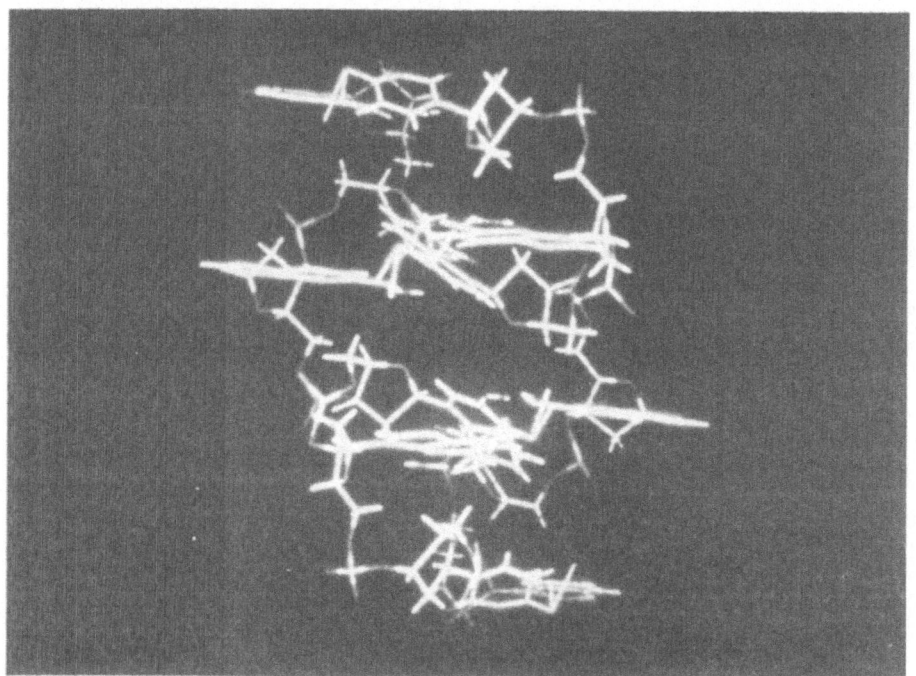

Fig. 12.

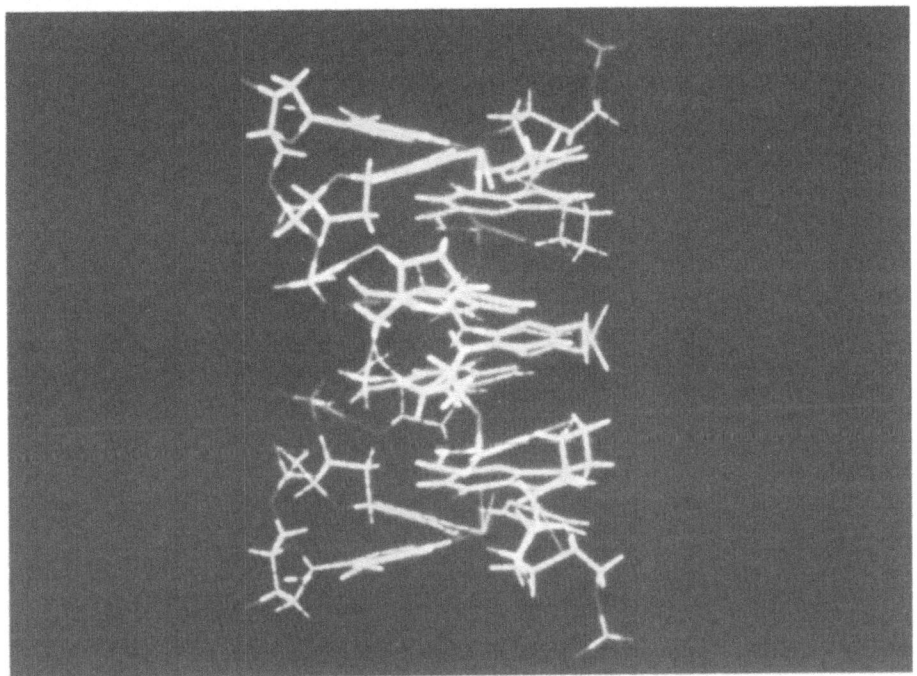

Figs. 12—13. Extreme atomic displacements in the 2.1 cm^{-1} vibration.

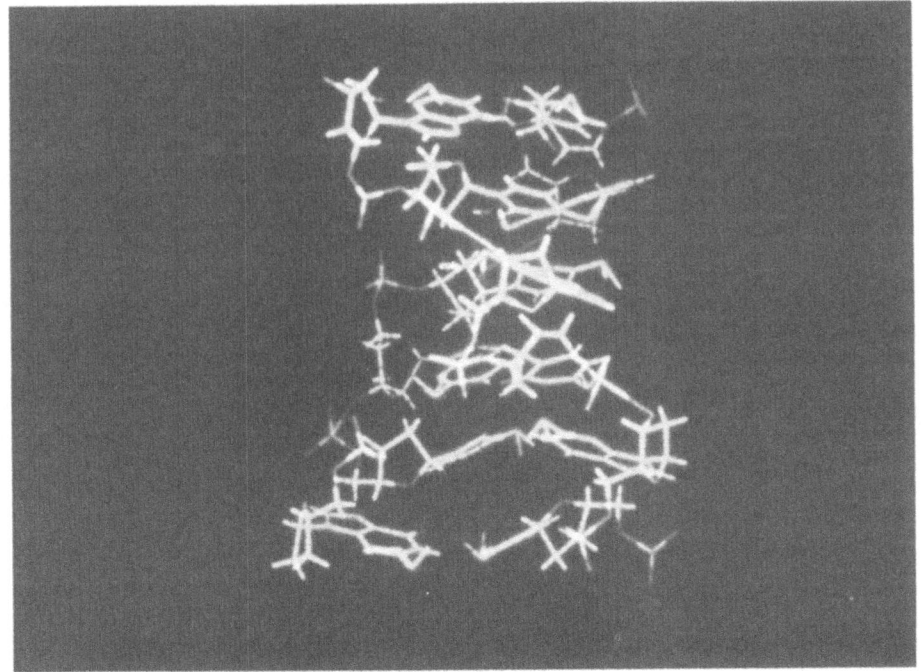

Fig. 14.

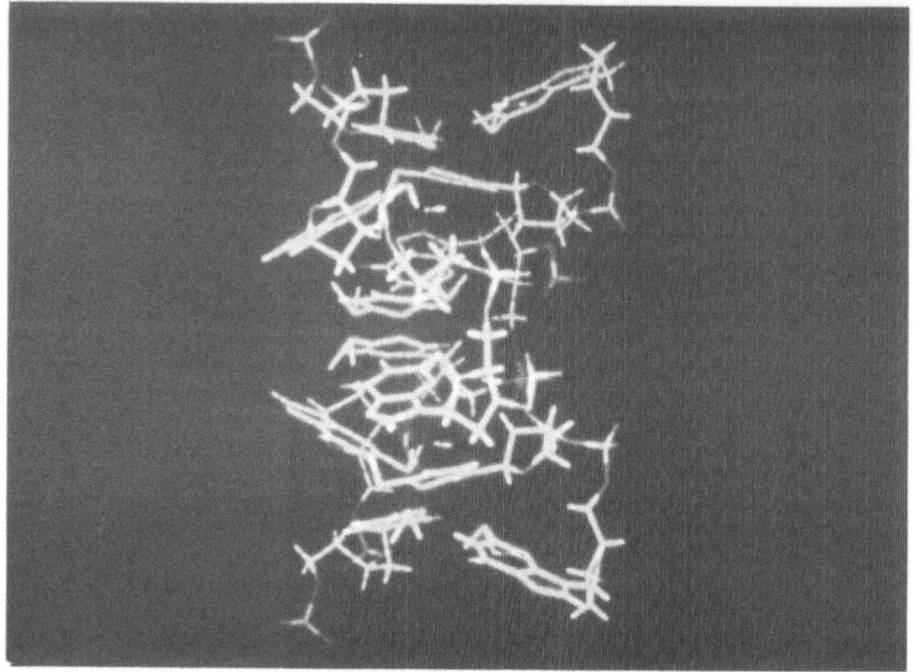

Figs. 14—15. Extreme atomic displacements in the 3.8 cm^{-1} vibration.

6. Conclusions and Perspectives

Recent progress in computer technology makes possible the theoretical determination of internal vibrations of large molecular systems. The availability of monochromators with very low stray light has led to experimental observation of very low frequency Raman bands. The development of computer graphics is of great help in understanding the motions, since it is now possible to visualize directly the calculated molecular vibrations. One can then use flexible molecular species in order to better understand biological functions.

References

1. G. Vergoten, G. Fleury, and Y. Moschetto: in *Advances in Infrared and Raman Spectroscopy*, volume 4, chapter 5, eds R. J. H. Clark and R. E. Hester, Heyden (1977).
2. A. Fersht: in *Enzyme Structure and Mechanism*, p. 32, W. H. Freeman and Company (1985).
3. T. Shimanouchi: *Pure Appl. Chem.* **36**, 93—107 (1973).
4. G. Vergoten, P. Lagant, W. L. Petitcolas, Y. Moschetto, I. Morize, M. C. Vaney, and J. P. Mornon: *Journal of Molecular Graphics* **4**, 187 (1986).
5. E. B. Wilson, Jr., J. C. Decius, and P. C. Cross: in *Molecular Vibrations*, McGraw Hill, New-York (1955).
6. T. Shimanouchi: 'Molecular force field', in *Physical Chemistry*, volume 4, chapter 6, eds Eyring, Henderson, Jost, Academic Press, New York (1970).

7. R. Huber and W. S. Bennett, Jr.: *Biopolymers* **22**, 261—279 (1983).
8. H. Burkert and N. L. Allinger: *Molecular Mechanics*, American Chemical Society, Washington D.C. (1983).
9. N. L. Allinger: International School of Crystallography, 'Static and dynamic implications of precise structural informations', Erice, Italy (1985).
10. N. L. Allinger and Y. A. Yuh: MM2 *Quantum Chemistry Program Exchange* **13**, 395 (1981).
11. A. Warshel: *Computers and Chemistry*, vol. 1, pp. 195—202, Pergamon Press (1977).
12. A. Warshel and M. Levitt: *Quantum Chemistry Program Exchange* **247** (1974).
13. P. K. Weiner and P. A. Kollman: *Journal of Comput. Chem.* **2**, 287 (1981).
14. B. R. Brooks, R. E. Bruccoleri, B. D. Olafson, D. J. States, S. Swaminathan, and M. Karplus: *Journal of Comput. Chem.* **4**, 187 (1983).
15. S. Califano: in *Vibrational States*, chapter 8, John Wiley and Sons (1976).
16. J. P. Huvenne: Thèse Doctorat ès Sciences Physiques, Lille, France (1979).
17. T. Shimanouchi: *Pure and Applied Chemistry* **7**, 131 (1963).
18. J. P. Tollenaere, H. Moereels, and L. A. Raymaeken: *X-ray Crystallography and Drug Action*, International School of Crystallography, Erice, Italy, March 1983, eds. A. S. Horn and C. J. De Ranter, Clarendon Press, Oxford (1984)
19. A. J. Hopfinger: *Pharmacy International* **5**, 224—228 (1984).
20. Evans and Sutherland Computer Corporation, P.O.B. 8700, Salt Lake City, Utah, 84108 USA.
21. Digital Equipment Corporation, 146 Main Street, Maynard, Ma., 01754—2571, USA.
22. M. C. Vaney *et al.*: *Journal of Molecular Graphics* **3**, 123 (1986).
23. J. Ramstein, N. Vogt, and M. Leng: *Biochemistry* **24**, 3603—3609 (1985).

Collective Excitations in Biological Macromolecules: Photoelectron and Exciton Spectra of Polyene, Polypeptides and Polynucleotides

S. SUHAI
Molecular Biophysics Project, German Cancer Research Center, Im Neuenheimer Feld 280, D-6900 Heidelberg, Federal Republic of Germany.

1. Introduction

There is mounting theoretical and experimental evidence that the electrons of biological macromolecules behave like a highly cooperative system. Electronic delocalization effects play an important role in the optical and transport properties, in cooperative melting processes, in non-linear phenomena, etc. Computation of excited state wave functions shows, for instance, that the excited electron and the remaining hole (forming a bound exciton) have a large probability amplitude to move together at a distance of quite a few peptide units in polypeptides. Even in polynucleotides, where the intermolecular interactions are smaller, the binding energy of excitons originates to a non-negligible extent from delocalized charge-transfer components. A lot of measurements and calculations on organic molecular and polymer crystals, which are very similar in their chemical composition and geometrical structure to biopolymers, also prove that the proper description of electronic processes in these systems must take into account the delocalization of their wave function over a number of subunits.

Adequate quantum mechanical methods for the investigation of excited states in biological macromolecules must, therefore, combine theoretical and computational procedures of both molecular and solid state physics to be able to describe all fine details of the interactions at a given lattice site on the one hand (molecular aspect), but also to properly take into account the collective response of the system originating from the more or less regular spatial structure (solid state aspect) on the other hand. The major conceptual and computational difficulties in these investigations arise from the latter aspect of the problem, since biopolymers are by their very nature quasi-periodic systems. This means, from the point of view of optical activities, for instance, that, though the major part of the interaction between a polypeptide and an external radiation field takes place in the backbone (built as a regular chain of peptide units), the side groups of the amino acids still exert a perturbing influence, which has to be included either by perturbation theory or by other means if we want to obtain a proper correlation between molecular structure and optical properties. The same is true for polyene, where the

Jean Maruani (ed.), Molecules in Physics, Chemistry, and Biology, Vol. IV, 133–194.
© 1989 *by Kluwer Academic Publishers.*

regular *trans*-model is only the first step leading through structural modifications to the building unit of visual pigments.

Fortunately, a number of new theoretical tools have also been developed in the past years for aperiodic solids which can be generalized to the case of aperiodic biopolymers. The first step of such calculations is usually the computation of approximate periodic models which supply the input data to the (more involved) investigation of quasi-regular structures.

The present paper will give a review of first principles theoretical methods for the calculation of optical properties of polyene (β-carotene), polynucleotides and polypeptides. After discussion of various aspects of Hartree—Fock calculations on biopolymers as zeroth order approximation (Section 2), we will turn to the more sophisticated problem of the proper inclusion of electron correlation effects, using Møller—Plesset perturbation theory, and to the development of a theoretical framework for the interpretation of photoelectron spectra in polymers (electron polaron model, Section 3). Applications of these theoretical tools to the investigation of various ground state properties of polyene, polypeptides and polynucleotides demonstrate the possibilities and limitations of these methods, in Section 4. The Green's function theory of charge-transfer excitons and of exciton-photon interaction (with applications to the above systems) will be, finally, presented in Sections 5 and 6, respectively.

2. Hartree—Fock Calculations on Biological Macromolecules

Emphasis will be laid in this review on *a priori* or *ab initio* realizations of various computational schemes proposed for biological macromolecules. We have to recall at this point that the great advantage of the *ab initio* techniques in molecular as well as in polymer theory is that (within the framework of a given atomic basis set) all one- and two-electronic integrals are calculated exactly without recourse to any empirical information apart from the structural data. The above definition will have to be made more precise for polymers. Namely, as it will be shown later, to reach a meaningful precision for all physical properties in question, (i) integrals whose absolute value is smaller than a given threshold (usually 10^{-6}—10^{-7} a.u.) can be neglected and (ii) certain core-electron and electron-electron integrals describing interactions between very far monomeric units can also be neglected pairwise (even if the members of the pairs are above the prescribed threshold value).

The evaluation of the infinite lattice sums remains the most difficult step in *ab initio* polymer calculations. It can be performed in configuration space as well as in momentum space (or the two procedures can also be combined). Not enough experience has been accumulated in the literature to decide which approach is more favourable from the computational point of view. Theoretically, the first one would be preferable for semiconducting polymers while the second one seems to be more appropriate for metallic ones. In the calculations reviewed in this article, the configuration space approach has been used with appropriately truncated lattice sums.

The use of a one-particle picture (Hartree–Fock theory) is a reasonable starting point in polymer conformational studies though the Hartree–Fock absolute energies are, of course, in considerable error (correlation error). For conformational problems one is concerned, however, with energy differences, and it is well established that the correlation error in the Hartree–Fock energy remains relatively constant in the neighbourhood of the extreme points of the potential energy surface, thus allowing acceptable predictions for the geometry. Other quantities (like band gaps, optical transitions, etc.) may be much more sensitive to correlation effects. Therefore, only predictions concerning the relative values of these quantities for different structures seem to be significant at the Hartree–Fock level, and further calculations with explicit allowance for electron correlation are needed to obtain them more accurately.

Taking into account the capabilities of modern high speed computers equiped with special vector or array processors as well as advanced computational procedures, we can state that careful handling of lattice sums and efficient use of inter- and intracell localization of the wave function makes it possible to perform *a priori* studies even for polymers with large units like a whole nucleotide. These computations are able to economically supplement experimental work on these systems.

2.1. THE ONE-PARTICLE MODEL OF POLYMER MANY-ELECTRON THEORY

The introduction of the concept of one-electron polymer orbitals (PO's) considerably reduces difficulties associated with the many-electron nature of the polymer electronic structure problem. The Hartree–Fock (HF) solution represents the best possible description of a many-electron system with a one-determinantal wave function built from symmetry-adapted one-electron PO's (also called Bloch functions). The HF approach is, of course, only a first approximation to the many-particle problem, but it has many advantages both from practical and theoretical points of view.

(i)   The HF total energy per monomer unit in the polymer is variationally determined and admits quite realistic geometry optimizations, potential energy surface calculations (phonon spectra), etc.

(ii) The electronic charge distributions obtained by the HF method are very accurate and suitable not only for qualitative but also for quantitative discussions and interpretations of properties related to them.

(iii) The one-determinantal wave function of the method is precisely defined; therefore, it can serve as a reasonable starting point for the investigation of effects lying beyond the capabilities of the one-particle approach (correlation corrections, optical properties, etc.).

The HF PO method is especially efficient if the Bloch orbitals are calculated in the form of a linear combination of atomic orbitals (LCAO) [1, 2] since in this case the large amount of experience collected in the field of molecular quantum mechanics can be used in polymer HF studies. The atomic basis orbitals applied for the above mentioned expansion are usually optimized in atoms and molecules. They can be Slater-type exponential functions if the integrals are evaluated in momentum space [3] or Gaussian orbitals if one prefers to work in configuration space. The specific computational problems arising from the infinite periodic potential, which do not appear in molecular *ab initio* calculations, will be discussed later.

2.2. VARIATIONAL CALCULATION OF BLOCH-TYPE POLYMER ORBITALS

The method for the calculation of HF Bloch functions in polymers at the *ab initio* level has been discussed earlier in detail [4]; therefore, we repeat here only the basic expressions to allow a self-contained discussion of the procedures applied. We will work entirely in configuration space, writing down the N-electron polymer wave function Φ_{HF} as a Slater-determinant built from doubly filled Bloch functions, $\phi(\mathbf{r})$, in the form

$$\Phi_{HF} = (N!)^{-1/2} \det[\ldots \phi_n^{\mathbf{k}}(\mathbf{r}_i)\alpha(\sigma_i)\phi_n^{\mathbf{k}}(\mathbf{r}_{i+1})\beta(\sigma_{i+1})\ldots]. \tag{1}$$

$\phi_n^{\mathbf{k}}(\mathbf{r})$ represents here a one-electron state with quasi-momentum $\mathbf{k}$ in band n. The Bloch functions themselves will be expressed as linear combinations of Bloch basis orbitals in the form of

$$\phi_n^{\mathbf{k}}(\mathbf{r}) = \sum_{a=1}^{\nu} C_{an}^{\mathbf{k}} \psi_a^{\mathbf{k}}(\mathbf{r}), \tag{2}$$

where ν is the number of basis orbitals per site, and the Bloch orbitals $\psi_a^{\mathbf{k}}(\mathbf{r})$ are symmetry-adapted combinations of the AO's:

$$\psi_a^{\mathbf{k}} = N_c^{-1/2} \sum_{j=1}^{N_c} \exp\{i\mathbf{k}\mathbf{R}_j\}\chi_a^j(\mathbf{r}). \tag{3}$$

N_c stands here for the number of elementary cells, and the AO $\chi_a^j(\mathbf{r}) = \chi_a(\mathbf{r} - \mathbf{R}_a - \mathbf{R}_j)$ is centred in the cell j at $\mathbf{R}_a + \mathbf{R}_j$. Applying the non-relativistic electronic Hamiltonian (in atomic units)

$$\hat{H} = -\frac{1}{2}\Delta_r - \sum_{h=1}^{N_c}\sum_{A=1}^{n_A}\frac{Z_A}{|\mathbf{r} - \mathbf{R}_h - \mathbf{R}_A|} + \sum_{\ell < m}\frac{1}{|\mathbf{r}_\ell - \mathbf{r}_m|}, \tag{4}$$

(Z_A being the core charge of the atom A at position $\mathbf{R}_A$), a Ritz variation of the expectation value $\langle\Phi_{HF}|\hat{H}|\Phi_{HF}\rangle$ leads to the polymer HF equation

$$\hat{F}\phi_j^{\mathbf{k}} = \varepsilon_j^{\mathbf{k}}\phi_j^{\mathbf{k}}, \tag{5}$$

with the Fock-operator

$$\hat{F}(\mathbf{r}) = -\frac{1}{2}\Delta_{\mathbf{r}} - \sum_{h=1}^{N_c}\sum_{A=1}^{n_A}\frac{Z_A}{|\mathbf{r} - \mathbf{R}_h - \mathbf{R}_A|} +$$

$$+ \int d^3\mathbf{r}'\,\frac{\rho(\mathbf{r}',\mathbf{r}')}{|\mathbf{r} - \mathbf{r}'|} - \int d^3\mathbf{r}'\,\frac{\rho(\mathbf{r},\mathbf{r}')}{|\mathbf{r} - \mathbf{r}'|}\,\hat{P}(\mathbf{r},\mathbf{r}'), \tag{6}$$

where the permutation operator $\hat{P}$ interchanges the variables $\mathbf{r}$ and $\mathbf{r}'$, respectively, before integration for the function standing behind it. The first-order density matrix $\rho(\mathbf{r},\mathbf{r}')$ is constructed from the occupied Bloch orbitals as

$$\rho(\mathbf{r},\mathbf{r}') = \sum_{n}^{occ.}\sum_{\mathbf{k}}^{\{BZ\}}\phi_n^{\mathbf{k}}(\mathbf{r})\,[\phi_n^{\mathbf{k}}(\mathbf{r}')]^*. \tag{7}$$

Substitution of the LCAO expansion of the Bloch orbitals transforms Eq. (5) to the complex pseudo-eigenvalue problem

$$\mathbf{F}^{\mathbf{k}}\mathbf{C}_n^{\mathbf{k}} = \varepsilon_n^{\mathbf{k}}\mathbf{S}^{\mathbf{k}}\mathbf{C}_n^{\mathbf{k}}, \tag{8}$$

with

$$F_{ab}^{\mathbf{k}} = \sum_j \exp\{i\mathbf{k}\mathbf{R}_j\}f_{ab}^{0j}, \tag{9a}$$

$$S_{ab}^{\mathbf{k}} = \sum_j \exp\{i\mathbf{k}\mathbf{R}_j\}s_{ab}^{0j}. \tag{9b}$$

Here the atomic overlap integrals between the reference cell and cell j are defined by $s_{ab}^{0j} = \langle\chi_a^0(\mathbf{r})|\chi_b^j(\mathbf{r})\rangle$, and the corresponding Fock matrix elements have the form

$$f_{ab}^{0j} = h_{ab}^{0j} + g_{ab}^{0j}, \tag{10a}$$

$$h_{ab}^{0j} = -\frac{1}{2}\langle\chi_a^0(\mathbf{r})|\Delta_{\mathbf{r}}|\chi_b^j(\mathbf{r})\rangle -$$

$$- \sum_{h=1}^{N_c}\sum_{A=1}^{n_A}\left\langle\chi_a^0(\mathbf{r})\left|\frac{Z_A}{|\mathbf{r} - \mathbf{R}_h - \mathbf{R}_A|}\right|\chi_b^j(\mathbf{r})\right\rangle, \tag{10b}$$

$$g_{ab}^{0j} = \sum_h\sum_\ell\sum_c\sum_d p_{cd}^{h\ell}\left[2\times\left(\begin{array}{cc}0j & h\ell\\ab & cd\end{array}\right) - \left(\begin{array}{cc}0h & j\ell\\ac & bd\end{array}\right)\right]. \tag{10c}$$

The electronic density matrix elements $p_{cd}^{h\ell}$ connecting the orbitals c and d in cells h and ℓ, respectively, are calculated by summation over all occupied states $(\mathbf{k}, n)$

in the first Brillouin zone (BZ):

$$p_{cd}^{h\ell} = \sum_{k}^{\{BZ\}} \sum_{n}^{occ.} \exp\{-i\mathbf{k}(\mathbf{R}_h - \mathbf{R}_\ell)\}[C_{cn}^{\mathbf{k}}]^* C_{dn}^{\mathbf{k}}, \tag{11}$$

while the two electron integrals in Eq. (10c) are defined by

$$\begin{pmatrix} 0j & h\ell \\ ab & cd \end{pmatrix} \equiv \int d^3\mathbf{r}_1 \int d^3\mathbf{r}_2 \chi_a^0(\mathbf{r}_1)^* \chi_b^j(\mathbf{r}_1) \frac{1}{|\mathbf{r}_1 - \mathbf{r}_2|} \chi_c^h(\mathbf{r}_2)^* \chi_d^\ell(\mathbf{r}_2). \tag{12}$$

It should be noted here that the condition of charge neutrality in the polymer, whose satisfaction plays an important role in actual calculations, requires

$$2 \sum_j \sum_a \sum_b p_{ab}^{0j} s_{ab}^{0j} = \sum_A Z_A. \tag{13}$$

Finally, the total energy per monomer unit is obtained from

$$E = \frac{1}{2} \sum_j \sum_a \sum_b (h_{ab}^{0j} + f_{ab}^{0j}) p_{ab}^{0j} + \frac{1}{2} \sum_j \sum_A \sum_B{}' \frac{Z_A Z_B}{|\mathbf{R}_A - \mathbf{R}_j - \mathbf{R}_B|}, \tag{14}$$

where the $B = A$ term is excluded from the summation over B if j refers to the reference cell. It should be noted here that the Bloch form of the one-electron orbitals automatically implies translational symmetry. In the case of polymers, this symmetry operation can be combined with a simultaneous rotation around the polymer axis (helix operation). It can be shown that, if the AO's are properly transformed in the translated and rotated elementary cells [5, 6], the above described formalism can still be applied.

2.3. TRUNCATION OF INFINITE LATTICE SUMS IN POLYMERS

We can see from Eq. (10c) that the major difficulties in applying the HF-PO method arise at the calculation of the bielectronic integrals whose number is proportional to v^4 for each upper (cell) index triplet $\{j, h, \ell\}$. The method can be reasonably applied only if the three infinite sums in Eqs. (9a), (10b) and (10c) can be truncated with a relatively short radius around the reference cell. Fortunately, even in this case only a very small portion of the bielectronic integrals have a physically significant value. Therefore, instead of actually calculating v^4 integrals for a $\{j, h, \ell\}$-group, our program examines first, on the basis of a Mullikan-type integral approximation, whether the absolute value of a certain integral is larger than the prescribed threshold. Our integral program package is based on the Gaussian 76 program system which is specially efficient if the s- and p-type Gaussians share the same exponent. In this case, for an (s, p_x, p_y, p_z)-shell one can decide in one step whether any of the 256 possible integrals will make a significant contribution. A p-type function is rotated for this purpose in the case of each shell into a maximum overlap position, and the largest possible integral is tested by the Mullikan method for significance. For the majority of cell index triplets with different elements, one has to actually calculate less than one percent of the bielectronic integrals in this way.

Returning to the problem of cell index truncation, we can see from Eqs. (8)–(11) that there are six physical quantities which will have to be properly truncated in this procedure:

(i)   the overlap integrals s_{ab}^{0j},

(ii) the matrix elements of the kinetic energy:

$$t_{ab}^{0j} = -\tfrac{1}{2}\langle \chi_a^0(\mathbf{r})|\Delta_{\mathbf{r}}|\chi_b^j(\mathbf{r})\rangle, \tag{15}$$

(iii) the electron-core interaction:

$$z_{ab}^{0j} = - \sum_{h=1}^{N_c} \sum_{A=1}^{n_A} \left\langle \chi_a^0(\mathbf{r}) \left| \frac{Z_A}{|\mathbf{r} - \mathbf{R}_h - \mathbf{R}_A|} \right| \chi_b^j(\mathbf{r}) \right\rangle, \tag{16}$$

(iv) the electron-electron Coulomb repulsion:

$$c_{ab}^{0j} = \sum_h \sum_\ell \sum_c \sum_d 2p_{cd}^{h\ell} \left(\begin{matrix} 0j \\ ab \end{matrix} \middle| \begin{matrix} h\ell \\ cd \end{matrix} \right), \tag{17}$$

(v) the non-local exchange potential:

$$e_{ab}^{0j} = - \sum_h \sum_\ell \sum_c \sum_d p_{cd}^{h\ell} \left(\begin{matrix} 0h \\ ac \end{matrix} \middle| \begin{matrix} j\ell \\ bd \end{matrix} \right), \tag{18}$$

(vi) and the core-core repulsion:

$$E_{n,n} = \frac{1}{2} \sum_j \sum_A \sum_B{}' \frac{Z_A Z_B}{|\mathbf{R}_A - \mathbf{R}_j - \mathbf{R}_B|}. \tag{19}$$

The first five quantities contribute, besides to the total energy, also to the Fock matrices and, therefore, also play a crucial role in determining the energy band structure.

For AO's of either Slater or Gaussian type, the overlap and kinetic energy integrals decay exponentially, and they can be safely truncated after a few neighbors. It is more complicated to truncate the contributions from Eqs. (16) and (17). Both of them behave separately as $\ln(r)$ at large distances; their sum is, as an alternating series, only conditionally convergent. Defining the operators

$$\hat{c}(\mathbf{r}) = \sum_h \sum_\ell \sum_c \sum_d 2p_{cd}^{h\ell} \int d\mathbf{r}' \, \frac{\chi_c^h(\mathbf{r}')\chi_d^\ell(\mathbf{r}')}{|\mathbf{r} - \mathbf{r}'|}, \tag{20}$$

$$\hat{z}(\mathbf{r}) = -\sum_h \sum_A \frac{Z_A}{|\mathbf{r} - \mathbf{R}_h - \mathbf{R}_A|}, \tag{21}$$

we obtain the core attraction and electron repulsion contributions to each Fock matrix element in the form

$$\langle \chi_a^0(\mathbf{r}) | \hat{c}(\mathbf{r}) + \hat{z}(\mathbf{r}) | \chi_b^j(\mathbf{r}) \rangle. \tag{22}$$

It follows from these considerations that, for a physically correct truncation of the infinite lattice sums, the following criteria will have to be fulfilled.

(i)   The sum over h in operators $\hat{c}$ and $\hat{z}$ will have to be truncated with the same radius in such a way that the elements of $\hat{c} + \hat{z}$ have an accuracy of 10^{-6}– 10^{-7} a.u.

(ii) The sum over ℓ in $\hat{c}$ will have to reach beyond the sum over h in such a way that all contributions to the electronic charge of the cell h are properly taken into account. This is necessary, since the electron population of a given cell is built up according to Eq. (13) as a sum over a certain number of neighbors. If one would use the same radius (measured from the reference cell) for the lattice sums over ℓ and h, the tail of the electron distribution belonging to the cell at $\mathbf{R}_h$, but collected from the cells $\mathbf{R}_{h+1}$, $\mathbf{R}_{h+2}$ etc., would be neglected. This loss of electronic charge would, of course, violate the electrostatic balance of the crystal.

(iii) The matrix elements of the non-local HF exchange potential stand alone without any balancing counterpart of one-electron integrals. They can, therefore, be truncated only if their sum has converged with a prescribed accuracy. The summation radius to be used here is expected to be different for semiconducting and metallic systems, since the long range tail of the exchange decays exponentially in the former case but it follows only a power law for metals [7, 8]. According to our experience, however, there is no significant difference between the two types of polymers *in the short and medium range* and too early a cut-off of the exchange either leads to serious convergence difficulties in the self-consistent field procedure or to non-physical solutions of the HF equations. We will show in the next paragraph how relatively small errors in the computation of the Fock matrix elements can be amplified in the course of the self-consistent procedure and become the source of disastrous instabilities and of non-physical results. The example of alternating and equidistant *trans*-polyene representing a semiconducting and a metallic system, respectively, should only demonstrate here the above mentioned similarity in the short and medium range.

Figure 1 shows the dependence of the Fock matrix element $\langle \chi_\pi^0 | \hat{F} | \chi_{\pi'}^j \rangle$ in the case of the alternating structure (the first $2p_z$ orbital of π symmetry is centered in the reference cell, the other one of the same symmetry in the cell at a distance of j CH units). (In polyene, due to the symmetry of the elementary cell, the above

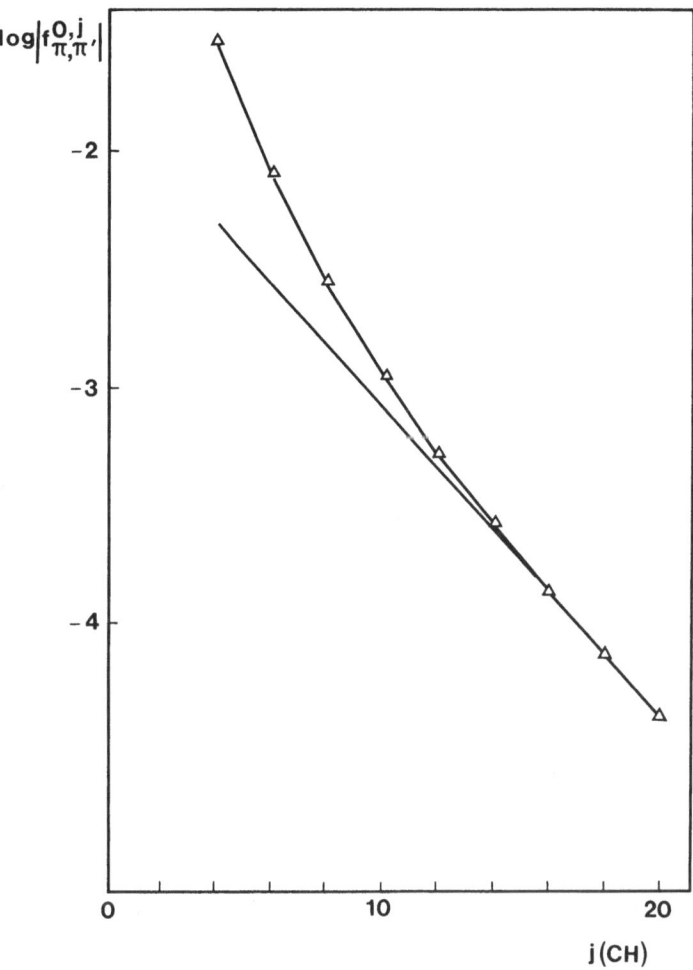

Fig. 1. The matrix element of the Fock operator, $f^{0,j}_{\pi,\pi'} = \langle \phi^0_\pi | \hat{F} | \phi^j_{\pi'} \rangle$, shown here for alternating *trans*-polyene as function of the distance of the two π-type atomic orbitals. j is measured in CH units, f is given in a.u.

mentioned electrostatic contributions compensate each other very fast, and the Fock matrix has longer range contributions practically only from the exchange potential.) We can see that the matrix element decays exponentially after the distance of 14—16 units but it has a much slower decay before. The contributions in the region of the 6th to 10th neighbours are of the order of 10^{-2}—10^{-3} a.u., and they originate nearly exclusively from the exchange.

From Figure 2, we can see that the situation is even somewhat worse in the case of the equidistant (metallic) case where the exchange decays only with a power law.

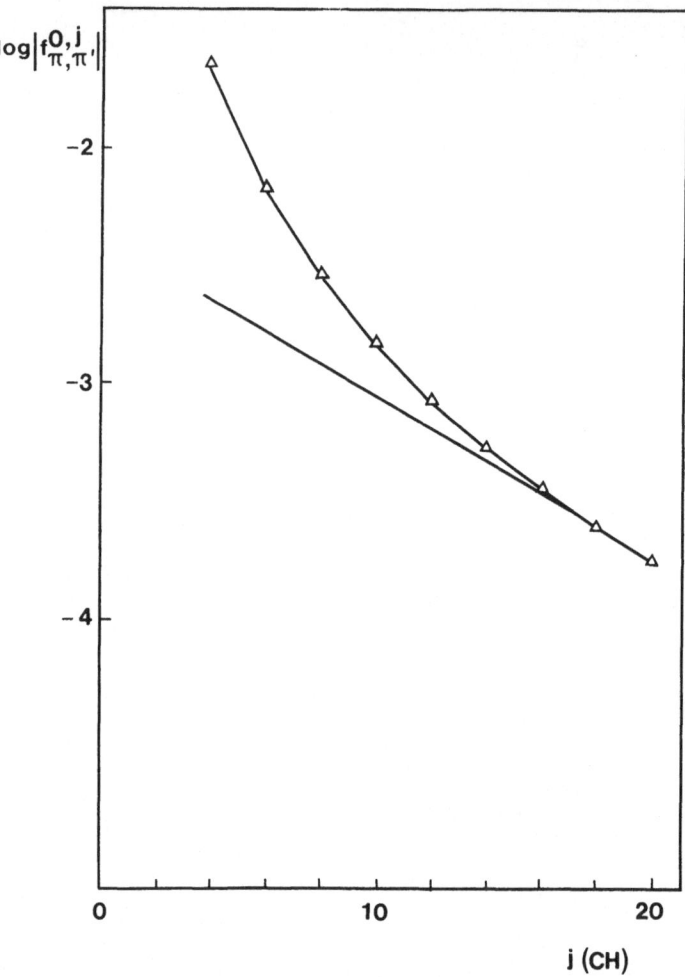

Fig. 2. The matrix element of the Fock operator, $f_{\pi,\pi'}^{0,j} = \langle \phi_\pi^0 | \hat{F} | \phi_{\pi'}^j \rangle$, shown here for equidistant *trans*-polyene as function of the distance of the two π-type atomic orbitals. j is measured in CH units, f is given in a.u.

(iv) In the case of elementary cells containing more atoms, the truncation can be performed only concerning the lower (AO) indices to preserve translational and point group symmetry. The number of terms included for a given atom as seen from the positive direction, must be equal to the number from the negative direction. This requirement can obviously not be satisfied if the whole cell is treated as an entity during the truncation procedure.

For a computationally efficient formulation of these requirements, let us introduce three different cut-off radii: d_s controlling the (exponential) decay of overlap-type quantities, d_c for the much slower decay of Coulomb-type interactions and d_e for the exchange terms. The choice of d_s is the easiest, since the order of magnitude of the overlap integrals reaches machine accuracy after a few neighbors. The overlap charge distributions $\chi_c(\mathbf{r})\chi_d(\mathbf{r})$ appear, however, besides kinetic energy integrals also in core attractions and electron-electron repulsions. Since the careful balance of the two latter quantities is important for keeping the crystal electrically neutral, let us consider here in more detail the consequence of their truncation. The AO $\chi_a^0 = \chi_a(\mathbf{r} - \mathbf{R}_A)$ is centered in the reference cell on atom A. Let us calculate the core attraction integral

$$- \left\langle \chi_a^0(\mathbf{r} - \mathbf{R}_A) \left| \frac{Z_c}{|\mathbf{r} - \mathbf{R}_h - \mathbf{R}_c|} \right| \chi_a^0(\mathbf{r} - \mathbf{R}_A) \right\rangle, \tag{23}$$

assuming that the distance between atoms A and C is just the Coulomb truncation radius: $|\mathbf{R}_A - \mathbf{R}_h - \mathbf{R}_c| = d_c$. This attraction is compensated in the polymer (besides the corresponding core-core repulsion) by the interelectronic repulsion between the charge distributions $\chi_a\chi_a$ and $p_{cd}^{h\ell}\chi_c^h\chi_d^\ell$, respectively, summed up over all c's ($c \in C$) to obtain the full charge of atom C according to Eq. (13). As a consequence, when checking the interorbital distances for the intercell integrals of the form $\binom{00}{aa}\binom{h\ell}{cd}$, we have to apply d_c for $|\mathbf{R}_A - \mathbf{R}_h - \mathbf{R}_c|$ but $d_c + d_s$ for $|\mathbf{R}_A - \mathbf{R}_\ell - \mathbf{R}_c|$. In other words, the nth neighbor approximation always means that for bielectronic integrals the neighbors $n + 1$, $n + 2$, etc. are also included. Neglecting these additional interactions introduces, due to the long range of the corresponding Coulomb integrals, a substantial error into the Fock matrices even for large values of n.

To obtain a quantitative estimate of the parameters d for a given system, one first has to check the behaviour of the intercellular overlap integrals for the given basis set and fix d_s. It is less straightforward to choose d_e properly. The simplest procedure is to make first the choice $d_e = d_c$ which overestimates the range of the exchange integrals. Since usually only one set of two-electron AO integrals is calculated (which is then used for the construction of both Coulomb- and exchange contributions to the Fock matrix elements), this choice means only a little superfluous effort during the SCF steps. It may be worthwhile, however, to optimize the value of d_e if, for example, a geometry optimization is performed for a given system.

Finally, the value of d_c is also system-dependent but the qualitative knowledge of the electronic structure of the elementary cell usually helps to estimate the difficulties one will encounter in the long-range part of the electrostatic potential. Namely, if the above discussed requirement for charge neutrality is fulfilled, the zeroth moment of the electron distribution is exactly compensated by the core charges of a given cell (in the sense of a multipole expansion) so that the long-range part of $\hat{c} + \hat{z}$ starts, in the worse case, with the dipole—dipole interaction, i.e. with a distance dependence of r^{-3}. Furthermore, if the polymer has an inversion center in the unit cell, the first long-range electrostatic term is the quadrupole—quadrupole interaction, proportional to r^{-5}. One will have to keep in mind, however, that these considerations are valid only for the really long-range region, where the charge distributions $\chi_a\chi_b$ and $\chi_c\chi_d$ do not overlap any more.

2.4. ERROR ANALYSIS FOR POLYMER HARTREE—FOCK CALCULATIONS

From Eqs. (10—12), we can immediately identify four possible sources of error entering the elements of the Fock matrix.

(i)   Numerical errors made in the evaluation of the individual mono- and bielectronic integrals: for current *ab initio* integral programs they lie in the region of 10^{-7} a.u. $\leqslant \delta_1 \leqslant 10^{-10}$ a.u.

(ii) Errors of the numerical integration in the *BZ* to obtain the matrices $p^{h\ell}$. There are different sophisticated methods to calculate highly accurate integrals for oscillatory functions and, according to our experience, the Gauss-Legendre procedure applied in our calculations provides the same accuracy as that obtained for the electronic integrals in (i).

(iii) Accumulating errors δ_2 due to the five-fold summations performed for each element of $\mathbf{F(k)}$. In principle, we have for them $\delta_2 = (N_c^3 \nu^2)^{1/2}\delta_1$ from Eqs. (9—10). It turns out in practice, however, that the dependence on the number of interacting cells passes over after a few neighbors into a simple linear dependence. On the other hand, for increasing values of N_c, the magnitude of the density matrix elements as well as that of the majority of the repulsion integrals decreases rapidly. For polymers with a unit cell of medium size, using a double-zeta type basis set, one finds that the situation is always better than $\delta^2 \approx 100 \times \delta_1$, i.e. $10^{-5} \leqslant \delta_2 \leqslant 10^{-8}$.

(iv) The most serious errors arise in HF calculations on polymers due to inappropriate lattice sum truncations. The fulfilment of three requirements seems to be important in this respect: (a) preservation of the translational and full point group symmetry for any value of N_c, (b) maintenance of the electrostatic balance between attractive and repulsive contributions to $\mathbf{F(k)}$ and, finally, (c) N_c has to be large enough so that the non-local HF exchange converges properly. According to our experience, the violation of any of these requirements leads to errors in the Fock matrix elements which are much graver than those mentioned previously. Depending on the system and on the basis set applied, they may reach the region of 10^{-2}—10^{-3} a.u. These errors can lead to serious numerical difficulties in the calculations, and they can also be identified as the source of spurious linear dependences in polymers in the case of otherwise harmless atomic basis sets.

The numerical difficulties in HF polymer studies arise from the transformation of the matrices $\mathbf{F(k)}$ to an orthogonal basis to solve the generalized eigenvalue problem in Eq. (8). Different methods have been proposed for this purpose. Löwdin showed [9] that the method of canonical orthogonalization also provides a convenient means to remove nearly linearly dependent combinations of basis functions from the basis set. This particular method has recently been applied to the polymer HF-error analysis [10], but it can easily be shown that the results are general and also hold for other kinds of transformations. Let us consider a transformation $T(k)$ for which

$$T^\dagger(k)S(k)T(k) = I, \tag{24}$$

where I is the unit matrix. We can transform Eq. (8) into an eigenvalue problem by writing

$$\tilde{F}(k)d_i(k) = \varepsilon_i(k)d_i(k), \tag{25}$$

or

$$\tilde{F}(k)D(k) = D(k)E(k),$$
$$[E(k)]_{ij} = \delta_{ij}\varepsilon_i(k), \tag{26}$$

with

$$\tilde{F}(k) = T^\dagger(k)F(k)T(k), \tag{27}$$

$$C(k) = T(k)D(k). \tag{28}$$

The matrices $C(k)$ and $D(k)$ are built here from the columns $c_i(k)$ and $d_i(k)$, respectively. The transformation $T(k)$ is, of course, not unique, but all suitable T's having the property (24) are related by a unitary matrix. As mentioned before, we chose for our error analysis T derived in the following way:

$$S(k)V(k) = V(k)\Lambda(k),$$
$$\Lambda_{ij}(k) = \delta_{ij}\lambda_i(k), \tag{29}$$
$$T(k) = V(k)\Lambda^{-1/2}(k).$$

Let us further define the column vectors $t_i(k)$ by

$$[t_i(k)]_j = V_{ji}(k)\lambda_i^{-1/2}(k), \qquad i = 1,\ldots,n, \tag{30}$$

and assume that the eigenvalues and eigenvectors of $S(k)$ are ordered: $0 < \lambda_1(k) \leqslant \lambda_2(k) \leqslant \cdots \leqslant \lambda_n(k)$. Since the accuracy of single overlap integrals is limited only by the machine precision and, on the other hand, expansion (9b) converges exponentially with N_c, we can assume that the matrices $S(k)$ are precise and in Eq. (8) only $F(k)$ contains errors due to reasons discussed earlier. The actually calculated Fock matrices $F^c(k)$ contain a hypothetically exact part $F^e(k)$ and an error:

$$F^c(k) = F^e(k) + \Delta(k). \tag{31}$$

Let us consider the $F_{11}(k)$ element obtained by transforming with the vector $t_1(k)$

belonging to the lowest eigenvalue of $S(k)$:

$$\tilde{F}^c_{11}(k) = t^\dagger_1(k)\,[F^e(k) + \Delta(k)]t_1(k)$$
$$= \tilde{F}^e_{11}(k) + \tilde{\Delta}_{11}(k), \tag{32}$$

or in more detail:

$$\tilde{\Delta}_{11}(k) = \sum_i \sum_j V^*_{1i}(k)\Delta_{ij}(k)V_{j1}(k)\lambda_1^{-1}(k)$$

$$= \tilde{\Delta}_{11}(k)\lambda_1^{-1}(k). \tag{33}$$

To obtain an order of magnitude estimate, we can substitute $\tilde{F}^e_{11}(k)$ by the corresponding eigenvalue $\varepsilon_1(k)$ in Eq. (32). We observe here that physically reasonable precision is obtained only if $\varepsilon_1(k) > \tilde{\Delta}_{11}(k)$ or:

$$\varepsilon_1(k) > \tilde{\Delta}_{11}(k)/\lambda_1(k). \tag{34}$$

The value of $\varepsilon_1(k)$ can always be estimated here from the approximately known band positions and $\lambda_1(k)$ can be exactly calculated. We can, therefore, estimate for each calculation the maximum error allowed in the Fock matrix elements from the requirement $\tilde{\Delta}_{11}(k) < \varepsilon_1(k)\lambda_1(k)$, which may be quite small if $\lambda_1(k)$ is small. Table I shows the lowest eigenvalues of the overlap matrix S at three different points of the BZ, both in equidistant and alternating *trans*-polyene, respectively.

TABLE I. The lowest eigenvalues of the overlap matrix $S(k)$ in equidistant and alternating *trans*-polyene, respectively, calculated with the 4-31 G basis set in different regions of the Brillouin zone as a function of the number of included neighbours, N.

N	Equidistant			Alternating		
	$k = 0$	$k = \pi/2$	$k = \pi$	$k = 0$	$k = \pi/2$	$k = \pi$
1	$-0.222367-01$	negative	negative	$-0.116024-01$	negative	negative
2	$0.719364-02$	$0.304497-02$	$0.504960-02$	$0.680816-02$	$0.325322-02$	$0.496967-02$
3	$0.714002-02$	$0.304257-02$	$0.506680-02$	$0.675710-02$	$0.325114-02$	$0.499119-02$
4	$0.714002-02$	$0.304256-02$	$0.506680-02$	$0.675710-02$	$0.325114-02$	$0.499119-02$

If we take 10^{-1} as the order of magnitude for ε, we can see that the early cut-off of the exchange, as discussed previously, would lead to serious errors according to Eq. (34). In fact, we could not obtain convergent HF solutions in the case of polyenes, for instance, if the cut-off radius was shorter than 10 CH units.

2.5. TRANSFORMATION TO AN OPTIMALLY LOCALIZED WANNIER FUNCTION BASIS

In most further applications of the wave functions obtained in HF-PO studies (calculation of correlation and excitonic effects, impurity levels, etc.), the use of

Wannier functions [11] instead of the original Bloch functions seems to be very promising. The connection between the two basis sets is given by the transformation

$$w_n^\ell \equiv w_n(\mathbf{r} - \mathbf{R}_\ell) = N_c^{-1/2} \sum_{\mathbf{k}} \exp(-i\mathbf{k}\mathbf{R}_\ell)\phi_n^{\mathbf{k}}(\mathbf{r}), \qquad (35)$$

where the Wannier function $w_n^\ell(\mathbf{r}) = w_n(\mathbf{r} - \mathbf{R}_\ell)$ is centred around the monomer at $\mathbf{R}_\ell$. From the point of view of the accuracy and economy of the above mentioned calculations, the extension of w_n is of great importance. As is well known, on the other hand (for a review see [12]), there still remains a residual degree of freedom in the Bloch functions, represented by the phase transformation

$$\tilde{\phi}_n^{\mathbf{k}}(\mathbf{r}) = \phi_n^{\mathbf{k}}(\mathbf{r}) \exp[i\lambda_n(\mathbf{k})], \qquad (36)$$

where $\lambda_n(\mathbf{k})$ can be any analytic function of $\mathbf{k}$ possessing the symmetry of the BZ. The phase factor $\exp\{i\lambda_n(\mathbf{k})\}$ can be used to predetermine certain properties of the Wannier functions obtained by Eq. (35). These properties are partly in conflict; therefore, one has to consider which of them is more important from the point of view of further calculations.

   If the transformed phase is written in the form $\tilde{\lambda}_n(\mathbf{k}) = \lambda_n^{orig.}(\mathbf{k}) + \lambda_n(\mathbf{k})$, where $\lambda_n^{orig.}(\mathbf{k})$ stands for the phase of the Bloch function obtained during the band-structure calculations, the following statements hold. (i) The antisymmetrical choice $\tilde{\lambda}_n(\mathbf{k}) = -\tilde{\lambda}_n(-\mathbf{k})$ makes the Wannier functions real. (ii) The symmetrical choice $\tilde{\lambda}_n(-\mathbf{k}) = \tilde{\lambda}_n(-\mathbf{k})$ preserves point group symmetries present in the Bloch functions in addition to the translational symmetry. (iii) One can apply variational procedures to determine that functional form of $\tilde{\lambda}_n(\mathbf{k})$ which will minimize the spatial extension of the Wannier functions.

   Different criteria can be used for the latter purpose. We found that maximization of the expectation values

$$I(\lambda_n^{\mathbf{k}}) = \int_{\{\Omega\}} d^3\mathbf{r}\, w_n^*(\mathbf{r}) z^2 w_n(\mathbf{r}), \qquad (37)$$

for a small region Ω around the reference cell ($\mathbf{R}_\ell = 0$, z is the co-ordinate parallel to the polymer axis) also provides efficient localization with reasonable computational efforts at *ab initio* level. It is evident that, making choices (i) and (ii), there does not remain any room for a further optimization of $\lambda_n^{\mathbf{k}}$. On the other hand, one of these properties can always be combined with (iii). Actually, we found that the combined use of (i) and (iii) provides the best starting point for the calculation of optical properties and localized impurity levels. After transformation (35), the Wannier functions are obtained in the same AO basis in which the Bloch functions are calculated:

$$w_n(\mathbf{r} - \mathbf{R}) = \sum_h \sum_p d_{pn}^h \chi_p(\mathbf{r} - \mathbf{R} - \mathbf{R}_h). \qquad (38)$$

The variational optimization of the phase factors reduces first the sum h over neighbouring cells. For a given cell, the Wannier function components $\Sigma_p d^h_{pm}(\mathbf{r} - \mathbf{R} - \mathbf{R}_h)$ are similar to molecular orbitals (MO's). It is not surprising, therefore, that methods worked out for MO localization also provide excellent tools in polymer theory. As an example, we show here the tails of the Wannier functions of π- and σ-symmetry, respectively, obtained for alternating polyene, using the previously discussed localization procedure (Figs. 3 and 4).

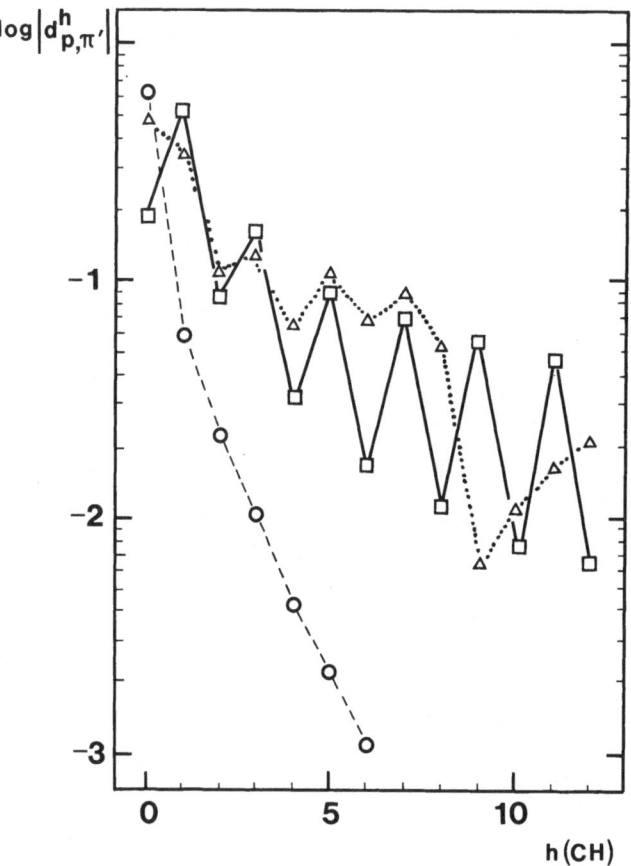

Fig. 3. Tail of the valence band Wannier function of π-symmetry in alternating *trans* polyene obtained using the original phase (solid line), a homogenized phase ($\lambda(\mathbf{k}) = 0$, dotted line) and a variationally optimized phase (dashed line), respectively, during the transformation of the Bloch functions.

3. Beyond Hartree–Fock: Perturbation Theory of Electron Correlation in Biopolymers

A large amount of experience collected in recent years in the field of molecular and solid state theory teaches us that Hartree–Fock-based procedures can never give satisfactory solutions for problems related to excited states. The aim of this

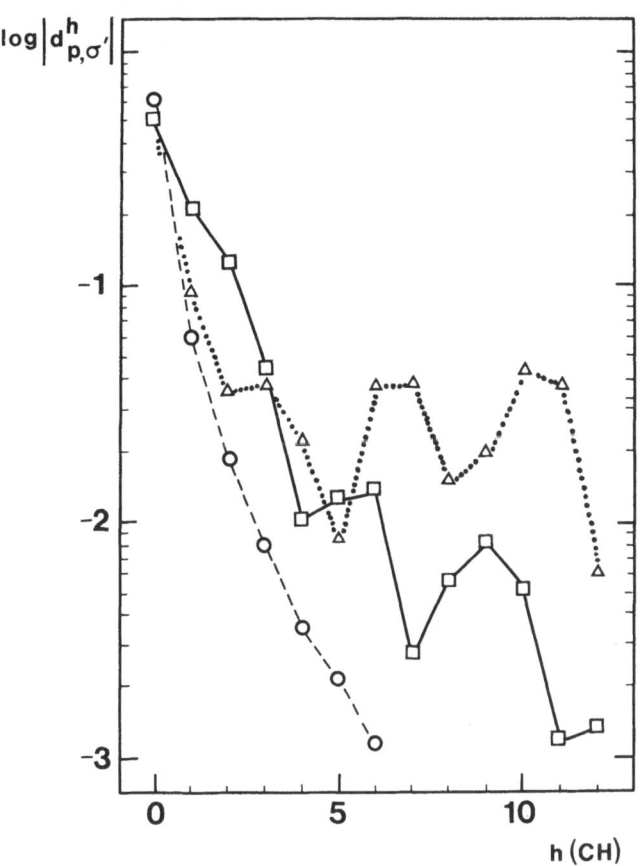

Fig. 4. Tail of the valence band Wannier function of σ-symmetry in alternating *trans*-polyene obtained using the original phase (solid line), a homogenized phase ($\lambda(\mathbf{k}) = 0$, dotted line) and a variationally optimized phase (dashed line), respectively, during the transformation of the Bloch functions.

paragraph is to demonstrate that the combination of the electron-polaron model with theoretical methods worked out to calculate correlation effects in atomic and molecular systems also provides a framework for the investigation of such effects in large biological macromolecules. Basically, different variants of two methods are mostly applied to smaller systems: configuration-interaction (CI) and perturbation theory (PT). CI can be applied only with truncation to larger systems (it will usually be restricted to single and double excitations). It is, however, in this form not size-consistent, i.e. its application even to an ensemble of isolated units leads to non-additive results. This feature makes it inapplicable to macromolecules.

On the other hand, many-body perturbation theoretical (MBPT) methods, based on Rayleigh-Schrödinger (RS) PT, define their total energy by the linked-cluster expansion, ensuring proper size dependence for each order of PT (see, for example, [13]). Their efficiency heavily depends on the choice of the zeroth-order Hamiltonian, $\hat{H}_0$. Various possibilities for the choice of $\hat{H}_0$ for correlation energy calculations in polymers have recently been compared [14—17]. For biopolymers, the application of the Møller-Plesset (MP) partitioning scheme, treating the full Hamiltonian as a perturbed spin-restricted HF (RHF) $\hat{H}_0$, proved to be the most successful one.

3.1. CALCULATION OF THE GROUND STATE CORRELATION ENERGY

Following Møller and Plesset [18], we choose the unperturbed Hamiltonian $\hat{H}_0$ as the sum of the Fock operators. Besides Φ_{HF}, we can construct further eigenfunctions of $\hat{H}_0$ by replacing some of the occupied orbitals in Φ_{HF} by virtual orbitals which we also obtain from Eq. (5). Introducing the fermion creation and annihilation operators $c_I^\dagger$ and c_I, respectively (with the compound index $I \equiv \{i, \mathbf{k}\}$) and labeling the states such that $I, J, \ldots$ stand for occupied levels in the ground-state configuration, while $A, B, \ldots$ refer to virtual levels, we can classify the eigenfunctions of $\hat{H}$ as singly, doubly, etc., excited configurations:

$$\Phi_0 = \Phi_{HF}, \tag{39}$$

$$\Phi_I^A = c_A^\dagger c_I \Phi_{HF}, \tag{40}$$

$$\Phi_{IJ}^{AB} = c_A^\dagger c_B^\dagger c_I c_J \Phi_{HF}, \tag{41}$$

$$\vdots$$

The corresponding eigenvalues of $\hat{H}_0$ are

$$E_0 = \sum_K^{(occ)} \varepsilon_K, \tag{42}$$

$$E_I^A = \sum_{K \neq I}^{(occ)} \varepsilon_K + \varepsilon_A, \tag{43}$$

$$E_{IJ}^{AB} = \sum_{\substack{K \neq I \\ K \neq J}}^{(occ)} \varepsilon_K + \varepsilon_A + \varepsilon_B, \tag{44}$$

$$\vdots$$

Expanding the eigenfunction Ψ^λ and the corresponding eigenvalue E^λ of $\hat{H}^\lambda = \hat{H}_0 + \lambda \hat{Q}$ (where $\hat{Q} = \hat{H} - \hat{H}_0$ is the perturbation operator) according to the RS PT in the form

$$\Psi^\lambda = \Phi_0 + \lambda \Psi_1 + \lambda^2 \Psi_2 + \ldots, \tag{45}$$

$$E^\lambda = E_0 + \lambda E_1 + \lambda^2 E_2 + \ldots, \tag{46}$$

and terminating these series at second order, we observe that the only matrix elements of $\hat{Q}$ that must be calculated are of the type $\langle \Phi_0 | \hat{Q} | \Phi_{IJ}^{AB} \rangle$ (due to Brillouin's theorem and of the fact that $\hat{Q}$ contains only two-electron operators). Setting $\lambda = 1$, we obtain in first order the term missing from E_0 to the HF energy, i.e.

$$E_{HF} = E_0 + E_1 = E_0 + \langle \Phi_0 | \hat{Q} | \Phi_0 \rangle, \tag{47}$$

while in higher orders we get correlation corrections to it. The full correlation energy is defined as $E_{corr} = E - E_{HF}$. Its value in second order is obtained as

$$E_2 = \sum_I \sum_J \sum_A \sum_B{}' \frac{\langle \Phi_0 | \hat{Q} | \Phi_{IJ}^{AB} \rangle \langle \Phi_{IJ}^{AB} | \hat{Q} | \Phi_0 \rangle}{E_0 - E_{IJ}^{AB}} \tag{48}$$

(the prime on the summations ensures that each double substitution is counted only once). Using the Slater-Condon rules to expand the matrix elements in Bloch functions, we get

$$E_2 = \sum_I \sum_J \sum_A \sum_B{}' \frac{|\langle \phi_I(1)\phi_J(2) | (1 - \hat{P}_{12}) r_{12}^{-1} | \phi_A(1)\phi_B(2) \rangle|^2}{\varepsilon_I + \varepsilon_J - \varepsilon_A - \varepsilon_B}, \tag{49}$$

where $\hat{P}_{12}$ is again the permutation operator for variables $\mathbf{r}_1$ and $\mathbf{r}_2$, respectively.

One of the major advantages of second-order PT in correlation studies is the fact that in this case only matrix elements between the ground state and doubly excited configurations will have to be computed. In the next order, however, the knowledge of the elements $\langle \Phi_{IJ}^{AB} | \hat{Q} | \Phi_{KL}^{CD} \rangle$ is also required, whose list may exceed that of the former elements by orders of magnitude, since even with the use of basis sets of moderate size, the virtual orbitals are much more numerous than the occupied ones. It seems, therefore, that even applying sophisticated techniques for the evaluation of the above mentioned matrix elements, the determination of the higher order correlation corrections will not be an easy task for biopolymers with medium-sized elementary cells. On the other hand, several applications of this technique for different polymers have shown that, from the point of view of the proper description of physical and chemical properties, the full correlation energy does not play an essential role. It is more important to reach a certain saturation for these properties themselves as functions of the correlation energy.

Another important feature, that makes second order PT especially attractive for polymer applications, is the fact that E_2 can be thought of as a sum of separate electron-pair contributions if it is written in the form

$$E_2 = \sum_I \sum_J{}' \varepsilon_{IJ}, \tag{50}$$

where the pair-correlation energies ε_{IJ} are defined by the comparison of Eqs. (49) and (50). In this way, the particle concept can also be preserved beyond the HF theory, and one can define quasiparticle states (and energy bands) that incorporate correlation effects at the given level of approximation. As is well known, further-more, the energy is invariant in all orders of MPPT with respect to unitary transformations among the occupied spin orbitls in Φ_{HF} (since $\hat{H}_0$ itself is invariant in the same sense). The individual terms ε_{IJ} in E_2, on the other hand, are not invariant and they may turn out to be quite different for, e.g., canonical (Bloch) orbitals or for localized ones. This degree of freedom permits us to transform the Bloch basis by the appropriate choice of the free phase (as discussed previously) into a set of Wannier functions for which the intraorbital correlation contributions are maximized. The same procedure may also help to compute ε_{IJ}'s that are transferable between related polymers and to facilitate the inclusion of correlation effects in aperiodic polymers.

3.2. ELECTRONIC POLARONS AND QUASIPARTICLE ENERGY BANDS: INTERPRETATION OF PHOTOELECTRON SPECTRA IN POLYMERS

The physical content of the HF energy bands is given by Koopmans's theorem which states that the orbital energies calculated by Eq. (5) for the conduction and valence bands, respectively, are equal to the corresponding electron affinities and ionization potentials:

$$\varepsilon_c^{\mathbf{k}_c}(HF) = E_{HF}^{(N+1)} - E_{HF}^{(N)}, \tag{51}$$

$$\varepsilon_{\ell}^{\mathbf{k}_{\ell}}(HF) = E_{HF}^{(N)} - E_{HF}^{(N-1)}. \tag{52}$$

If we assume that during the excitation of an electron the distribution of the other electrons is not basically changed, i.e., no relaxation takes place (which is quite a reasonable assumption for a large polymeric system), and, furthermore, if the excited electron and the remaining hole are infinitely separated (which is the case for Bloch electrons), then the excitation is a simple one-electron transition over the single-particle energy-band gap:

$$\Delta\varepsilon_g(HF) = \varepsilon_c^{\mathbf{k}_c}(HF) - \varepsilon_{\ell}^{\mathbf{k}_{\ell}}(HF). \tag{53}$$

Following Toyozawa's suggestion [19], we can also retain the above picture if we go beyond the HF model. By analogy to Eqs. (51) and (52), we define quasi-particle (QP) states using, instead of the HF total energies, the correlated energies from the preceding section:

$$\varepsilon_c^{\mathbf{k}_c}(QP) = E^{(N+1)} - E^{(N)}, \tag{54}$$

$$\varepsilon_{\ell}^{\mathbf{k}_{\ell}}(QP) = E^{(N)} - E^{(N-1)}. \tag{55}$$

By writing for the total energy $E = E_{HF} + E_2$, we obtain

$$\varepsilon_c^{\mathbf{k}_c}(QP) = \varepsilon_c^{\mathbf{k}_c}(HF) + E_2^{(N+1)} - E_2^{(N)}, \tag{56}$$

$$\varepsilon_{\ell}^{\mathbf{k}_{\ell}}(QP) = \varepsilon_{\ell}^{\mathbf{k}_{\ell}}(HF) + E_2^{(N)} - E_2^{(N-1)}. \tag{57}$$

In order to interpret these expressions, we use the decomposition of E_2 into the sum of independent pair correlations. We note that the application of the results obtained for the N-particle system involves a further approximation here since the $(N + 1)$- and $(N - 1)$-particle states are not closed-shell configurations. Thus, Brillouin's theorem does not exactly apply and, therefore, also single-particle excitations should be included in PT. It can be shown, however, that the contribution of these singly excited configurations is negligible as compared to the doubly excited ones which are included here.

A further problem should also be mentioned, though it is a general problem in band-structure theory: Namely, the ϕ's used to construct the $(N + 1)$- and $(N - 1)$-particle states are eigenfunctions of the N-particle Fock operator instead of belonging to the $(N + 1)$- or $(N - 1)$-particle Fock operators, respectively. In principle, one should again use an open-shell formalism for their determination, but it can be hoped for the infinite system (and the overall success of band theory supports this view) that the error made is not a serious one.

Using the convention that, e.g., $E^{(N-1)L}$ denotes the correlation energy of the $(N - 1)$-particle state which we receive by removing an electron from the valence-band state L, we obtain from Eq. (50):

$$E_2^{(N)} = \sum_I \sum_J{}' \varepsilon_{IJ}^{(N)} = \sum_{I \neq L} \sum_{J \neq L}{}' \varepsilon_{IJ}^{(N)} + \sum_{I \neq L} \varepsilon_{IL}^{(N)}, \tag{58}$$

$$E_2^{(N-1)L} = \sum_{I \neq L} \sum_{J \neq L}{}' \varepsilon_{IJ}^{(N-1)L}, \tag{59}$$

$$E_2^{(N+1)C} = \sum_{\substack{I \\ (C)}} \sum_{\substack{J \\ (C)}}{}' \varepsilon_{IJ}^{(N+1)C} = \sum_I \sum_J{}' \varepsilon_{IJ}^{(N+1)C} + \sum_I \varepsilon_{IC}^{(N+1)C}. \tag{60}$$

The first two summations in Eq. (60) also involve the extra occupied conduction-band state C. Substituting these expressions into Eqs. (56) and (57), we get for the quasiparticle energies

$$\varepsilon_C(QP) = \varepsilon_C(HF) + \sum_I \varepsilon_{IC}^{(N+1)C} + \sum_I \sum_J [\varepsilon_{IJ}^{(N+1)C} - \varepsilon_{IJ}^{(N)}], \tag{61}$$

$$\varepsilon_L(QP) = \varepsilon_L(HF) + \sum_{I \neq L} \varepsilon_{IL}^{(N)} + \sum_{I \neq L} \sum_{J \neq L} [\varepsilon_{IJ}^{(N)} - \varepsilon_{IJ}^{(N-1)L}]. \tag{62}$$

Following the proposal of Pantelides et al. [20], we can interpret the correction terms appearing in addition to the HF band energies in these equations as electron and hole self-energies, $\Sigma(e)$ and $\Sigma(h)$, respectively. For this purpose, we introduce the notation

$$\varepsilon_C(QP) = \varepsilon_C(HF) + \Sigma_C^{(N+1)}(e) + \Sigma_C^{(N+1)}(h), \tag{63}$$

$$\varepsilon_L(QP) = \varepsilon_L(HF) + \Sigma_L^{(N)}(e) + \Sigma_L^{(N)}(h). \tag{64}$$

Recalling Toyozawa's electronic-polaron model, we can identify the origin of these self-energy corrections as a cloud of virtual excitons dressing the *bare* HF particles (in complete analogy to the lattice polaron problem, where virtual optical phonons accompany the polarizing particle).

With this physical model in mind we can visualize the formation of an electronic polaron according to Eq. (61) in two steps. First, the extra particle put into the HF conduction-band state $\phi_c^{k_c}$ must establish its new correlation bonds with the other N particles present in the $(N + 1)$-particle system $\Phi^{(N+1)C}$, giving rise to the electronic self-energy correction $\Sigma_c^{(N+1)}(e)$. At the same time, owing to the occupation of the previously empty state $\phi_c^{k_c}$, the pair correlations $\varepsilon_{IJ}(I \neq C, J \neq C)$ are reduced in $\Phi^{(N+1)C}$ as compared to $\Phi^{(N)}$, since scatterings to $\phi_c^{k_c}$ are excluded. From the point of view of $\Phi^{(N+1)}$, the terms $\varepsilon_{IJ}^{(N+1)C} - \varepsilon_{IJ}^{(N)}$ in Eq. (61) destabilize the hole in $\phi_c^{k_c}$ (before its occupation with the extra electron); their sum, therefore, results in a hole self-energy $\Sigma_c^{(N+1)}(h)$. The inclusion of these correlation effects plays a very important role in the theoretical description of photoionization processes. We will demonstrate below that the experimental photoelectron spectra of polymers cannot be reasonably interpreted without such correction terms to the single particle energy levels.

4. Calculation of Ground State Energy Levels in Polyene, Polypeptides and Polynucleotides

Before turning to the discussion of ground state energy calculations in the case of biopolymers with chemically complicated building elements, we give a short overview of the experiences accumulated in our group in the past years in the field of polymer many-electron theory. As an example, we cite results obtained for *trans*-polyene, the structurally simplest biopolymer. This part of the paper should also give the reader a general impression about the difficulties and possibilities in this field and help him judge the different methods and approximations also subsequently applied to polypeptides and polynucleotides. Since, however, one is confronted with very heavy computational problems in applying the method to these systems, it is inevitable to choose a polymer built from smaller subunits for a flexible test of the theoretical tools.

The major, partly interrelated, questions to be answered are: how efficient is the MP PT when truncating after second order, how far will have to be extended the atomic basis to reasonably describe correlation, how important are virtual excitations into higher conduction bands, and, finally, how much of correlation energy will have to be included to obtain proper saturation in other properties, more relevant from a physical or chemical point of view.

The major computational effort in PT is the transformation of the bielectronic integrals, which are calculated for the HF problem in terms of the AO basis, into the Bloch basis. In second order, not considering **k**-dependence, this would be a relatively simple task since the number of operations needed would be proportional only to v^5. The difficulty arises, however, from the fact that the matrix elements in Eq. (49) must be calculated in a momentum-dependent manner, i.e. one would have to perform the above transformation for a larger number of representative points in the Brillouin zone which would be an impractical time-consuming procedure (especially for larger atomic basis sets).

We can see, on the other hand, that the problem is not the entire k-dependence in the Bloch functions (BF's), but only the part entering through the LCAO coefficients [c.f. Eqs. (2, 3)] into the fourfold summation over the AO indices when we form integrals of the type $\langle \phi_I \phi_J | r_{12}^{-1} | \phi_A \phi_B \rangle$. Thus, we can avoid these difficulties by expanding the Bloch orbitals in terms of Wannier functions (WF's), applying the inverse to the transformation defined by Eq. (35). Substituting the Bloch functions in this form into Eq. (49), we can calculate the products of the k-dependent exponential factors outside of the AO integral transformations. The computational problem has been further reduced in our calculations by an efficient selection procedure applied before the actual transformation of the integrals from the AO basis to the WF's. It is evidently important to keep the tail of the WF's as short as possible, as discussed before.

The next problem to be discussed is the size of the atomic basis set. Physical intuition tells us that it should play an even more important role in correlation studies than at the HF level since polarization functions are obviously needed to make it possible that the electrons avoid each other. It is nearly impossible to study this problem for polypeptides or polynucleotides; therefore, polyene again provides a very useful model system. The role of high-lying virtual bands in perturbation theory of correlation effects is an equally important problem. A lot of calculations performed in the past neglected virtual excitations into these bands, using the argument that the denominator in Eq. (49) is comparatively larger for them than for bands around the Fermi level. This problem is closely related to the question concerning the role of short-range correlations missing from the original Toyozawa model [19]. Short-range in configuration space means large momentum for the virtual excitons, i.e., thinking in an extended zone scheme, excitations to conduction bands with a larger positive energy. For alternating and equidistant polyene, the energy of virtual excitations has been cut at different values [16] to study the role of correlation energy left out. As the results presented below show, not only is a substantial part of E_2 lost in this way, but also the physically very important energy difference between the semiconducting and metallic phases is very poorly described.

4.1. FIRST PRINCIPLES ELECTRONIC STRUCTURE CALCULATIONS FOR THE GROUND STATE OF *TRANS*-POLYENE

Trans-polyenes with their extended conjugated π-electron system have been the subject of both theoretical and experimental investigations for many decades, particularly because of their intense absorption bands in the optical region [21,

22]. Upon excitation, these molecules can undergo a variety of chemical reactions which makes them very suitable as radiation energy gathering compounds in biological systems. Examples for such processes are the chlorophylls involved in the photosynthesis of bacteria and higher plants and the carotenoid pigments of plants and animals; the latter include 11-cis-retinal, which is the chromophore of the visual pigment in a wide variety of animals and is also found in the purple membrane of bacteria. For a satisfactory understanding of the spectra and the photochemical reactions of polyenes, a detailed knowledge of their electronic states is necessary. Recent interest in this material has been stimulated by the enormous increase in the electrical conductivity of the crystalline form of polyene upon doping [23].

Different aspects of the polyene problem have been investigated from the theoretical point of view in a large number of semiempirical and ab initio calculations from which we can cite only some representative ones [24—33]. In our two recent publications; the bond alternation [34] and the band-to-band transition energy (fundamental gap) [35] have been calculated from first principles also including electron correlation within the previously discussed scheme. An off-diagonal charge-density wave of the equidistant polyene structure has been identified at the HF level as a precursor of the bond alternate state which has been found by subsequent systematic optimization of the lattice parameters [34]. Electron correlation substantially reduces the Peierls-type lattice distortion and the corresponding energy barrier between the semiconducting (bond alternate) and metallic (equidistant) polymers, respectively, but extrapolation of the results to full correlation shows that the bond alternating ground state also persist in this case.

To demonstrate these effects is somewhat more detail, we present some results obtained using basis sets of increasing size. Table II describes the composition of these five sets.

TABLE II. Atomic basis function sets used in the LCAO expansion of Bloch orbitals.

Notation	Carbon atomic set	Hydrogen atomic set
STO-3G	$6s3p/2s1p$	$3s/1s$
4-31G	$8s4p/3s2p$	$4s/2s$
6-31G	$10s4p/3s2p$	$4s/2s$
6-31G*	$10s4p1d/3s2p1d$	$4s/2s$
6-31G**	$10s4p1d/3s2p1d$	$4s1p/2s1p$

Table III shows the HF and correlated $(E = E_{HF} + E_2)$ total energies per elementary cell of trans-polyene as obtained for the optimum of the bond alternation Δ_R (the four independent structural parameters shown in Figure 5 had been optimized using the same basis sets). We can see that the value of Δ_R is well converged at both levels of approximation with respect to the basis-set extension, and its correlated value (if projected on the polymer axis) is in good agreement with the experimental one [36].

TABLE III. Optimized single-double bond-length difference in *trans*-polyene as obtained with the HF approximation and using second-order Møller-Plesset PT ($E = E_{HF} + E_2$). The total energies per elementary C_2H_2-unit belong to the optimum bond alternation for a given basis set.

Basis set	ΔR_{HF} (Å)	E_{HF} (a.u.)	ΔR_E (Å)	E (a.u.)
STO-3G	0.156	−75.947121	0.122	−76.065389
4-31G	0.112	−76.776843	0.091	−76.955733
6-31G	0.109	−76.859247	0.085	−77.036179
6-31G*	0.103	−76.886912/	0.084	−77.137486
6-31G**	0.103	−76.8926757	0.084	−77.168100

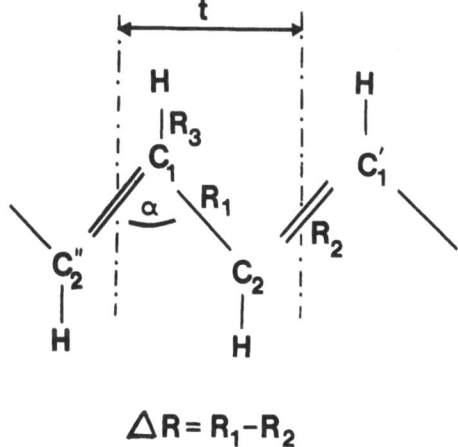

Fig. 5. The geometrical structure of the unit cell of *trans*-polyene showing the definition of the parameters optimized in the calculations.

We can observe at the same time that, though E_{HF} has reasonably converged, a considerable part of E_2 is still missing. This is physically understandable since our *spd* sets still do not have enough oscillations to make it possible for the electrons to avoid each other effectively. For this purpose, further basis functions with higher angular momenta would be required that could increase (through their nodes) the kinetic energy and thus result in a larger correlation contribution by virtue of the virial theorem. Comparison of the influence of the atomic basis for E_{HF} and E_2 is facilitated by Figure 6. We can see that the magnitude of the slope for the E_2-*vs*-E_{HF} curve gets larger and larger in the region of better basis sets.

As mentioned before, the role of short-range correlation effects is especially interesting from the point of view of electronic-polaron theory. To investigate the contribution of high-lying conduction bands associated with larger virtual exciton momentum, we introduced in a test calculation thresholds for the maximum excitation energy in calculating E_2. The results, presented in Table IV, show that high-energy excitations, despite the larger energy denominators associated with

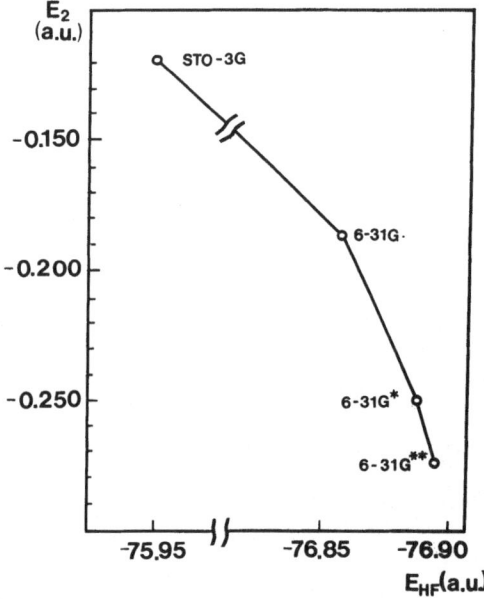

Fig. 6. Correlation energy per C_2H_2 formula unit vs HF energy per C_2H_2 formula unit in alternating *trans*-polyene using different basis sets (second-order Møller-Plesset PT, RHF scheme).

TABLE IV. Dependence of the correlation energy (E_2) in alternating and equidistant polyene on the number of conduction bands included in second order Møller-Plesset PT. $E_{exc,\,max}$ is the upper limit for conduction band states that are used to form virtual exciton pairs (results obtained with a 4-31G basis set).

$E_{exc,\,max}$ (a.u.)	Number of conduction bands included	$E_2^{alt.}$ (a.u.)	$E_2^{equid.}$ (a.u.)	$E_2^{alt.} - E_2^{equid.}$ (kcal/mol)
0.51	4	−0.035851	−0.040146	2.695
0.88	6	−0.069262	−0.072908	2.324
1.05	8	−0.104408	−0.107464	1.921
1.26	11	−0.151770	−0.154257	1.561
1.95	15	−0.178890	−0.181135	1.409

them, make substantial contributions to the correlation energy. It is, furthermore, interesting to observe, that energy differences are also strongly influenced by these effects. As we can see from the table, the prediction of the correlation contribution to the alternating-to-equidistant phase transition energy in *trans*-polyene would be in considerable error on the basis of low-lying excitations. The extension of the virtual space is, of course, closely related to the previously discussed basis-set problem and again underlines the need for *f, g,* etc. orbitals to obtain proper short-range correlations.

One would like to know, of course, the percentage of the full valence-shell correlation energy included in E_2 with the best *spd* basis used. We can get an approximate answer to this question if we recall that the valence-shell correlation energy of an acetylene unit was estimated to be about -10 eV [37]; therefore, our best energy (-7.5 eV) should cover 70—75% of the total value. On the other hand, since various π-electron Hamiltonians have been very extensively applied in the past to the polyene problem [38], it is also of interest to ask what part of the correlation originates from purely π-electron interactions. (Both the highest filled and lowest unfilled bands have π symmetry. This question is, therefore, decisive for doping and for conductive properties.) Since, due to symmetry, the σ- and π-type Bloch functions can be completely separated, we can evaluate the $\sigma - \sigma$, $\pi - \pi$ and $\sigma - \pi$ contributions to E_2 individually. Almost independently of the atomic basis set, we found that E_2 contributes only 15—20%, showing that for this kind of polymers the whole valence shell must be treated as an entity; a simple model separating bands with π-type symmetry would not work.

The central part of this work is the evaluation of the self-energy corrections to the HF band energies using Eqs. (63) and (64). These quantities are collected in Table V. The Σ_c's are calculated at the bottom of the lowest conduction band,

TABLE V. Different physical quantities contributing to the formation of the quasiparticle energy-band gap in alternating *trans*-polyene: one-particle energies, $\varepsilon(HF)$, electron and hole self-energies, $\Sigma(e)$ and $\Sigma(h)$, respectively, and quasiparticle energies, $\varepsilon(QP)$. All quantities are given in eV.

Quantity	STO-3G	6-31G	6-31G**
$\varepsilon_{c, min}(HF)$	3.719	-0.806	-1.322
$\Sigma_{c, min}^{(N+1)}(h)$	0.112	0.193	0.274
$\Sigma_{c, min}^{(N+1)}(e)$	-0.452	-0.623	-0.938
$\varepsilon_{c, min}(QP)$	3.379	-1.236	-1.996
$\varepsilon_{v, max}(HF)$	-4.563	-5.732	-5.749
$\Sigma_{v, max}^{(N)}(h)$	0.583	0.896	1.497
$\Sigma_{v, max}^{(N)}(e)$	-0.312	-0.436	-0.724
$\varepsilon_{v, max}(QP)$	-4.293	-5.361	-4.976

while the Σ_v's refer to the top of the highest valence band. These results substantiate for all basis sets the conclusions of the previous formal analysis, though the quantitative details are different. We can see that the shifts result for both bands from a positive and a negative term, but as a net effect the conduction band is shifted downward while the valence band is shifted upward.

We can follow the formation of the polyene energy-band gap in four consecutive steps as shown in Figure 7. The uppermost curve is obtained at the HF level using the fixed geometry obtained for the minimal basis. The extension of the atomic basis does not in this case significantly infuence the value of $\Delta\varepsilon_g(HF)$. On the other hand, it is reduced by ~ 4 eV if the bond alternation is optimized (at the HF level) with each basis set (second curve from the top). Nearly ~ 0.7 eV further reduction is obtained if the structural optimization is performed with the correlated wave function (third curve from the top). The lowest curve shows the energy-band gap values after polaron formation, $\Delta\varepsilon_g(QP)$. It can be seen here that the amount of correlation, included at this stage, plays a predominant role: The polaron effect is more than twice as large in the case of the extended *spd* basis than for the minimal one.

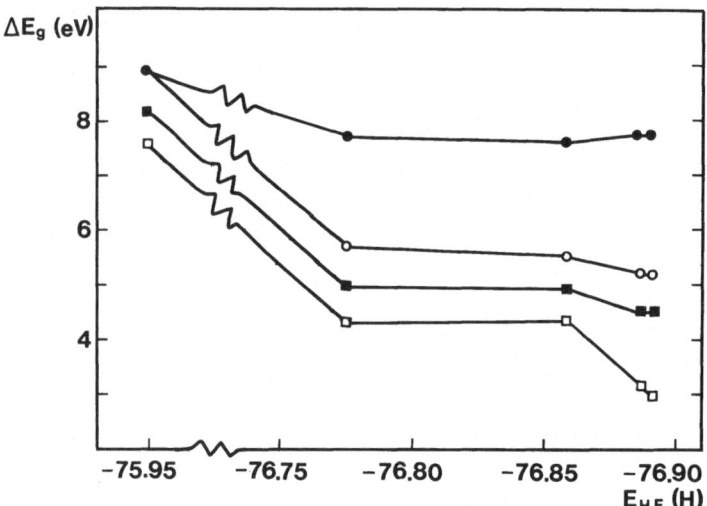

Fig. 7. Energy gap of *trans*-polyene *vs* HF energy per C_2H_2 formula unit obtained using the previously defined five different atomic basis sets. Methods of calculation (from top to bottom): HF with fixed bond alternation; HF with optimized bond alternation; HF + MPPT with optimized bond alternation but using $\Delta E_g = \varepsilon_{c, min}(HF) - \varepsilon_{v, max}(HF)$; electronic polaron model with $\Delta E_g = \varepsilon_{c, min}(QP) - \varepsilon_{v, max}(QP)$.

The best value obtained with this method is $\Delta\varepsilon_g = 2.98$ eV. For the ionization potential we get at the same time 4.976 eV which can reasonably be compared to the experimental value of 4.6 eV obtained from photoelectron spectroscopy [39, 40]. Since even our best wave function contains only about 75% of the total valence-shell correlation, it would be interesting to extrapolate the obtained value $\Delta\varepsilon_g(QP)$ for the case of full correlation.

Figure 8 shows that the estimated theoretical value would lie at about 2.5 eV, i.e. about half an electron volt higher than the position of the first peak in the absorption spectrum of pure *trans*-polyene [39]. The tail of the experimental spectrum reaches to 1.4—1.5 eV, probably due to structural disorder present in the crystalline samples. This tail is, of course, not represented in the above fully periodic calculation, for which the one-dimensional density of states curve has a sharp maximum at the band edge. Our theoretical result must, therefore, be compared with the position of the corresponding peak in experiment which lies at about 2 eV.

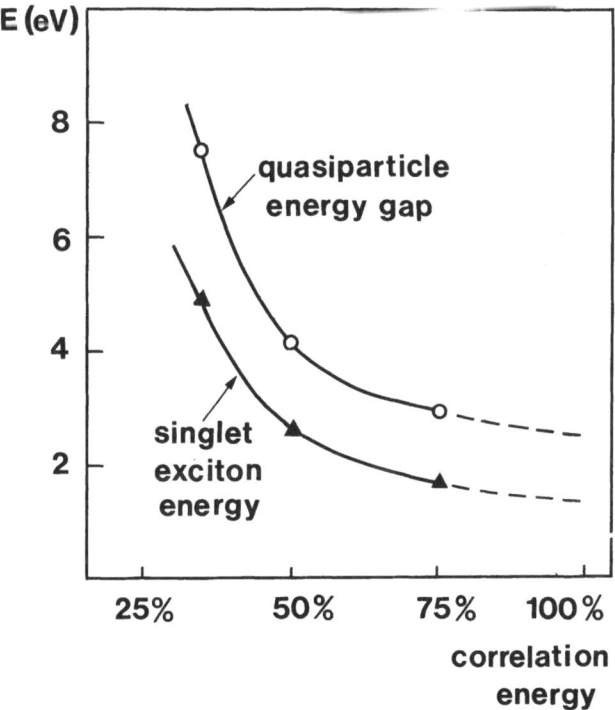

Fig. 8. Dependence of the quasiparticle energy gap of *trans*-polyene on the estimated amount of correlation accounted for in the particular calculation. The figure shows the corresponding excitonic energies as well. Their computation will be discussed in a subsequent paragraph.

4.2. CORRELATION CORRECTIONS TO THE HARTREE—FOCK PICTURE IN POLYPEPTIDES: QUASIPARTICLE ENERGY LEVELS

The geometrical structure of the α-helix used in the calculations reported here was taken from the results of the Scheraga group [41]. Since, as we have seen in the case of polyene, the electron correlation effects in polymers are very sensitive to the size of the atomic basis set, we also used in the HF part the largest possible set permitted by our computational facilities in view of the size of the unit cell of polyalanine $(6-31\,G^*)$. The proper truncation of the infinite lattice sums is, also in the case of a polypeptide chain, a very sensitive point of the HF calculation. Since the unit cell here has a dipole moment, the electrostatic interactions also contain contributions of long-range character besides the non-local exchange. Our test calculations, using first a minimal basis set, have shown that, in order to obtain convergence of *ab initio* quality, interactions with five neighbouring units will have to be taken into account in both directions along the helical backbone for each peptide unit. Using this truncation scheme with a careful balance between one- and two-electron contributions, we did not, in fact, observe any numerical instabilities which are usually encountered if improper approximations introduce errors into the Fock matrices of infinite systems.

We calculated the electronic structure of two α-helical polypeptide chains. The first system was a simple backbone without side groups (modelling polyglycine); the second one was polyalanine. In calculating the correlation contribution to the total energy per peptide unit, again only virtual excitations starting from valence band states were included, since in this work we were not interested in processes involving core electrons. On the other hand, the virtual space was used to its full extent. In this way, we obtained -0.641746 a.u. for the correlation energy in the second order of MP-PT (-29.2 mH per valence electron) for the backbone model and -0.806458 a.u. (-28.8 mH per valence electron) for polyalanine, respectively. Adding this contribution to the HF energy, we obtained -207.500687 a.u. as the total energy per glycine unit in the polyglycine model and -246.596061 a.u. per alanine residue in alpha-polyalanine, respectively. The change of these energies during conformational transitions is, in fact, physically more relevant than their absolute value. Our studies of proton motion in hydrogen bonds of polypeptides revealed, for instance, the importance of including correlation effects in the corresponding potential energy surfaces. On the other hand, the ingredients of the correlation energy, the individual pair-correlations, play a fundamental role in the theory of optical effects in these systems, as we will show in a subsequent paragraph.

The next step of the construction of the quasiparticle states, for which the self-energy correction terms are needed. Tables VI and VII show these quantities only at the corresponding band extrema for the two polymers under investigation. The HF eigenvalues were, in fact, calculated, at twenty inequivalent points of the Brillouin zone. Since the calculation of the self-energies is more time-consuming, they were computed explicitly at five points in each band and fitted by a trigonometric polynomial to get a fine mesh for the numerical integration needed for the computation of Green's function in the exciton calculation. We can see from these tables that the main consequence of correlation is again the shift of both the conduction and valence bands, down- and upwards, respectively. As a secondary effect, a band-narrowing also appears by analogy with the Franck-

TABLE VI. Different physical quantities contributing to the formation of the quasiparticle energy-band gap in polypeptides: one-particle energies $\varepsilon(HF)$, electron and hole self-energies, $\Sigma(e)$ and $\Sigma(h)$, respectively, and quasiparticle energies $\varepsilon(QP)$. Results obtained using a 6-31G* atomic basis set and second-order Møller-Plesset perturbation theory. All quantities are given in eV.

Quantity	α-helix backbone		α-polyalanine	
	val. band	cond. band	val. band	cond. band
$\varepsilon_{max}(HF)$	−10.428	3.363	−10.036	3.428
$\varepsilon_{min}(HF)$	−10.922	2.746	−10.508	2.746
$\Sigma_{bandmax.}(e)$	−0.646	−2.209	−0.744	−2.225
$\Sigma_{bandmin.}(e)$	−0.686	−2.086	−0.762	−2.052
$\Sigma_{bandmax.}(h)$	2.193	0.272	2.264	0.275
$\Sigma_{bandmin.}(h)$	2.312	0.267	2.352	0.240
$\varepsilon_{max}(QP)$	−8.881	1.426	−8.516	1.478
$\varepsilon_{min}(QP)$	−9.296	0.927	−8.918	1.008

TABLE VII. Valence and conduction band widths ($\delta\varepsilon_v$, $\delta\varepsilon_c$) and fundamental gap values (ΔE_g) for the simple α-helix backbone and for α-polyalanine, calculated at the Hartree–Fock (HF) and correlated quasiparticle (QP) levels, respectively. Results obtained using a 6-31G* atomic basis set and second-order Møller-Plesset perturbation theory. All quantities are given in eV.

Quantity	α-helix backbone		α-polyalanine	
	HF	QP	HF	QP
$\delta\varepsilon_v$	0.494	0.415	0.472	0.402
$\delta\varepsilon_c$	0.617	0.499	0.608	0.470
ΔE_g	13.174	9.812	12.856	9.524

Condon factor of phonon polaron theory. Furthermore, it is interesting to observe that the deviation between the glycine and alanine results only amounts to a few percent for all quantities: a useful hint for the future that side-chain effects could be reasonably calculated by perturbation theory.

The substantial reduction of the fundamental gap from ~ 13 eV (HF particles) to 9.5–9.8 eV (electron polarons) is a genuine correlation effect which is evidently important for a proper description of spectral properties.

4.3. GROUND STATE ELECTRONIC STRUCTURE OF POLYNUCLEOTIDES

The first *ab initio* Hartree–Fock energy band-structure calculations have been performed for polynucleotides at the minimal basis set level using spherical

Gaussian [42] and STO-NG-type atomic basis functions [43, 44, 46, 47], respectively. The extension of these basis sets not only permits the investigation of all nucleotide base stacks but also the use of a whole nucleotide (base + sugar + phosphate group) as elementary unit. It is also possible to study in this way important base-backbone interaction phenomena like the charge transfer between the sugar-phosphate (SP) group and the nucleotide bases themselves. Since, however, meaningful correlation corrections cannot be introduced at this level, the resulting energy bands can only be used for the comparison of different properties of polynucleotides and for the estimation of their physicochemical properties.

In the first calculations reported here, an STO-3G atomic basis set [44] and an *ab initio* polymer program have been used, which also takes into account the necessary rotation of the basis functions, that is, one gets from one unit to the next one not via a simple translation, but by applying a combined symmetry (helix) operation. The four nucleotide base stacks and the sugar-phosphate (polySP) chain as well as the polycytidine (polyCSP) chain were taken in the same geometrical arrangement as in B-DNA. The coordinates of the heavy atoms were identical with those obtained by Arnott and co-workers from X-ray investigations [45]. The positions of the hydrogens were adjusted to these atoms by assuming standard bond angles and bond distances. The lattice sums were truncated after the second interacting neighbours using the method discussed previously. The tails of the electronic wavefunction also extended over the third and fourth neighbours, respectively, to obtain proper electrostatic balance. The smallest eigenvalues of the Fourier transformed overlap matrix, $S(k)$, had the order of magnitude of 10^{-1} in all of these calculations. In concordance with the error analysis, no numerical instabilities were observed for these systems at the HF level. In subsequent calculations, an extended 4-31G basis set was applied, and the corresponding eigenvalues were obtained around 10^{-3}, a region where error handling becomes even more important.

Table VIII shows the two most important energy bands for the four poly-base stacks. For comparison, the corresponding MO energy levels are also shown. The most important information, which we can obtain from these results, is the fact that the face-to-face interaction of the bases in the nucleotide stacks is relatively

TABLE VIII. Limits and widths of the valence and conduction bands of the four nucleotide base stacks (in eV).

Chain	Valence band				Conduction band			
	ε^{MO}	ε_{min}	ε_{max}	$\delta\varepsilon$	ε^{MO}	ε_{min}	ε_{max}	$\delta\varepsilon$
Poly(C)	−5.61	−5.51	−4.65	0.86	6.00	6.07	6.91	0.84
Poly(T)	−6.66	−6.48	−5.88	0.60	5.97	6.02	6.33	0.51
Poly(A)	−6.10	−6.04	−5.57	0.47	6.46	6.56	6.86	0.29
Poly(G)	−5.08	−5.16	−4.34	0.82	6.45	6.41	7.15	0.74

strong. The splitting of the individual MO levels by 0.5—0.9 eV leads to energy band widths which are very well comparable with those of molecular conductors. We expect, therefore, that electronic delocalization will play an important role in several physical properties of these polymers.

In the case of polySP and polyCSP, instead of a K^+ ion a proton has been attached to the PO_4^- group, thus keeping these chains neutral. Table IX again contains the characteristics of the valence band and conduction band for the polySP and polyCSP chains. To demonstrate the connection with the polyC energy bands, these latter values are repeated in the table. It should be mentioned here that all the valence and conduction bands of the four nucleotide base stacks originate from the HOMO and LEMO levels of the constituent molecules. Such a classification is not possible in the case of the polySP and polyCSP chains. With the help of Mulliken's population analysis, the amount of transferred charge from the sugar-phosphate chain to the cytosine chain has been computed using the results of the polyCSP superchain calculation. The result we found is a 0.187 e excess negative charge on each cytosine molecule.

TABLE IX. Limits and widths of the valence and conduction bands of poly(C), poly(SP) and of the two highest filled and lowest unfilled bands of the polycitidine chain, poly(CSP) (all energy values in eV).

	poly(C)			poly(SP)			poly(CSP)		
	ε_{min}	ε_{max}	$\delta\varepsilon$	ε_{min}	ε_{max}	$\delta\varepsilon$	ε_{min}	ε_{max}	$\delta\varepsilon$
Conduction				7.43	8.00	0.57	7.40	7.96	0.56
band	6.07	6.91	0.84				6.55	7.38	0.83
Valence	−5.51	−4.65	0.86				−5.19	−4.36	0.83
band				−6.44	−6.28	0.16	−6.79	−6.71	0.08

Looking at the tables, we can see that the valence and conduction bands of the stacked bases and of the polySP chain have widths of several tenths of an eV, indicating that there is a possibility for Bloch-type conduction in these systems if free charge carriers are generated in them. On the other hand, the gap in all cases is more than 10 eV. Even if we remember the well known fact that a Hartree—Fock calculation always leads to too large a gap for conduction (which will be substantially reduced if correlation corrections are taken into account in the following paragraphs), this certainly rules out the possibility of intrinsic semiconduction in DNA.

On the other hand, our calculation of the polyCSP superchain has resulted in a substantial charge transfer of 0.19 e per molecule pair from the sugar-phosphate unit to cytosine. Although, in a restricted Hartree—Fock superchain calculation due to the method one obtains only completely filled bands if there is considerable charge transfer from one chain to the other, this fact indicates the possibility of creating free charge carriers in the system. Although, according to this calculation, the valence band of polySP is rather narrow (0.16 eV), one should bear in mind that the presence of positive ions may increase the bandwidths to a significant degree. Therefore, if one would repeat the polySP and polyCSP band structure calculations, assuming the presence of K^+ ions around the PO_4^- groups and not putting a proton chemically bound to them, one would expect a rather large broadening of the valence band of polySP.

Finally, we would like to point out that the charge transfer from the sugar-phosphate (SP) unit to the cytosine (C) molecule cannot be explained by the naive HOMO (D)-LEMO (A) picture, because the valence band of polySP (see Table II) lies below the valence band of polyC by about 0.8 eV. The same is true if one compares the upper limits of the valence bands of the three other nucleotide base stacks (see Table I) with the upper limits of the valence band of polySP. Since the conduction properties of DNA strongly depend on the possibility of internal charge transfer in DNA or of interchain charge transfer between a polypeptide chain and DNA, we plan to perform more detailed investigations on this problem. It is also desirable to increase the size of the basis set beyond the minimal one for these studies. The first results obtained with an extended set will be reported for polyC in connection with correlation calculations in a subsequent paragraph.

4.4. EFFECT OF HYDRATION ON THE ENERGY BANDS OF DNA

The problem of the proper theoretical treatment of solvation effects in DNA has been solved to a great extent in the past years. In a number of publications, the interaction energy between water molecules and the nucleotide bases, base pairs, single, and double helices of DNA has been calculated (for a review see [48]). Recently, the complete solvent structure of a B-DNA double helix fragment with 12 base pairs and the corresponding sugar and phosphate units has also been determined [49]. In these Monte Carlo simulations, 447 water molecules were included, and their interaction energies and probability distributions (at a temperature of 300 K) were calculated.

The hydration shell acts on the electronic system of the stacked nucleotide bases as an external electric field. By a proper representation of the electron distribution of the water molecules through point charges (see below), one can calculate the matrix elements of this perturbing field in terms of the Bloch functions of the infinite polymer and include in this way its effect on the band structure. The water molecules forming the hydration shell of B-DNA can be divided into clusters by inspection of their geometrical positions obtained from Monte Carlo calculations. The clusters are chosen in such a way that the intra-cluster interaction should be sustantially larger than the intercluster one. As a next step minimal basis *ab initio* MO calculations were performed for each water cluster [46]. The resulting canonical Hartree—Fock orbitals were then transformed to localized ones using Boys' method. To find a suitable representation of the electron distribution of the water clusters by point charges, the electrostatic potential $V_\ell^N(\mathbf{r}_{i\ell})$ was first determined for each localized orbital ℓ of the Nth cluster at a given set of positions $\mathbf{r}_{i\ell}$. Each of these point sets is located on spheres with different radii around the center of charge of the given orbital. The radii cover the short, medium, and long range regions of the molecular potential field.

As a next step, the resulting potentials were fitted using point charge potentials by minimizing the deviation of the two electrostatic fields in the least-square sense:

$$F_\ell^N = \sum_{i=1}^{m_\ell^N} \left[\sum_{j=1}^{M^N} \frac{q_{j\ell}^N}{|\mathbf{r}_{i\ell}^N - \mathbf{r}_{j\ell}^N|} - V_\ell^N(\mathbf{r}_{i\ell}^N) \right]^2 = \min., \qquad \ell = 1, 2, \ldots, n_N^*. \tag{65}$$

M^N stands for the number of point charges representing one localized orbital; m_ℓ^N is the number of points for which the two potentials have been calculated, and n_N^* is the number of occupied localized orbitals in the Nth water cluster. $q_{j\ell}^N$ and $\mathbf{r}_{j\ell}^N$ stand for the charge and location of the resulting point charges, respectively, and are determined by solving the system of linear equations resulting from the minimum condition in Eq. (65). In the calculations reviewed here, the values $m_\ell^N = 4$ were used which give satisfactory accuracy according to previous experiences gained by the application of the above described procedure to the problem of intermolecular interactions.

The 37 water molecules surrounding one cytidine unit, which have been taken from the results of [49], were divided into five clusters on the basis of their geometrical positions. It should be pointed out that, in this first model, calculation of the effect of the water surrounding DNA on its band structure we used the water structure around a cytidine (cytosine + sugar + phosphate) unit to generate an effective potential acting on a cytosine stack. This procedure, of course, does not give a fully realistic description of the effect of the water structure on the electronic structure of DNA, but it gives, in a relatively simple model calculation an order of magnitude estimate of the effects to be expected.

In a more realistic calculation, the water structure around a cytidine unit together with the counterions will have to be considered, and the effect of its effective potential on the band structure of a cytidine polymer will have to be investigated. In further steps, the aperiodicity of DNA as well the role of the polypeptide chain in the deep grove of DNA will have to be taken into account. In the light of these requirements, this calculation (but also a more consistent one which would have used the water structure computed for a cytosine stack which, at the time of our calculation, was not available) can be considered only as a rough, order of magnitude estimate of the investigated effect. It does not describe completely the role of the first hydration shell around a cytosine stack and at the same time overestimates the effect of the water molecules situated further away by not introducing any screening to account for the sugar and phosphate molecules between the water and the cytosine stack.

The inspection of the water clusters around a cytidine unit shows, however, that most of the water molecules are situated in the immediate neighbourhood of the cytosine molecules (forming the larger part of their first hydration shell), and only part of the water molecules are further apart than they would be in the case of a cytosine stack. Therefore, the error is less than one would expect without looking into the details. After having determined the proper point-charge representations of the water clusters the potentials of the other water molecules situated around neighbouring DNA subunits could simply be calculated by transforming the $r_{j\ell}^N$ vectors with the symmetry operation of the helix. After having represented in this way the charge distribution of each water cluster, we let them interact. The modification of their charge distribution due to the interactions was also taken into account. We finally calculated, the matrix elements of the electronic point-charge potentials together with those of the corresponding nuclei in terms of the AO basis functions also used for the construction of the Bloch orbitals of the polymers. These matrix elements were then added to the one-electron part of the Fock matrix.

The energy bands, obtained with an STO-4G basis set with without and hydration shell, respectively, are given in Table X. The results show that the

TABLE X. The two highest filled and the two lowest unfilled energy bands of a cytosine stack calculated with and without hydration (all energy values in eV).

Without hydration			With hydration		
ε_{min}	ε_{max}	$\delta\varepsilon$	ε_{min}	ε_{max}	$\delta\varepsilon$
8.421	8.697	0.276	7.026	7.331	0.305
5.916	6.738	0.822	5.608	6.473	0.865
−5.739	−4.872	0.867	−6.592	−5.679	0.913
−7.762	−7.386	0.376	−8.529	−8.110	0.419

potential of the hydration shell somewhat stabilized the base stack. All bands move downwards, and the total energy is also lowered in the presence of the water molecules by 8 eV per elementary cell. It is interesting to note that the band shifts are quite different and vary between 0.3 and 1.4 eV. Their values depend on the structures of the corresponding wave function, i.e. on the fact as to which atoms contribute mostly to the Bloch function in question and how many water molecules are situated in their neighbourhood.

Though no dramatic changes can be observed in the band structure as a whole, it should be pointed out that the hydration together with positive ions (Na$^+$, K$^+$, etc.) could play a very important role in the determination of the conductive properties of DNA. Our preliminary calculations showed that the presence of K$^+$ ions causes band shifts of many eV-s in DNA. These ions do not of course, appear alone but always in a solvated form. It will therefore, be very important to treat them together with the hydration shell to obtain a more realistic description of the electronic structure of DNA.

4.5. CORRELATION CORRECTIONS TO THE ENERGY BANDS IN DNA

The first DNA-model studied from the point of view of correlation effects was the polyC stack [50]. The first quantity of interest from these calculations is the total energy per cytosine molecule, calculated as the sum of electronic and nuclear contributions. Its electronic part was obtained at two different levels, without correlation ($E = E_{HF}$) and including correlation effects ($E = E_{HF} + E_{corr.}$). To calculate $E_{corr.}$, only virtual excitations within the valence shell were taken into account, since processes involving core electrons are not relevant for the properties studied with the wavefunctions obtained. Applying a Gaussian 4-31G basis set, $E_{HF} = -392.036264$ a.u. was obtained for the infinite polymer. The valence shell correlation energy per monomer is -0.663012 a.u. (-18.5 mH per valence electron) in polyC, which is in concordance with our previous experiences showing that about one half of the estimated full correlation energy can be achieved with the given method and basis set.

Turning now to the calculation of polaron corrections to the energy bands, we first present in Table XI the values of the self-energy terms obtained at the extrema of the conduction and valence bands, respectively (in fact, these corrections were calculated for three points in each band and fitted by a trigonometric polynomial to obtain a fine mesh the BZ for the calculation of the Green-matrix for the exciton studies to be described below). As we can see from Table XI, the self-energies again follow the previously discussed general trends and shift together the conduction band to lower and the valence band to higher energies, respectively. It can also be observed that the shift is larger at the top than at the bottom

TABLE XI. Different physical quantities contributing to the formation of the quasiparticle energy-band gap in poly(C): one-particle energies $\varepsilon(HF)$, electron and hole self-energies, $\Sigma(e)$ and $\Sigma(h)$, respectively, and quasiparticle energies $\varepsilon(QP)$. Results obtained using a 4-31G atomic basis set and second-order Møller-Plesset perturbation theory. All quantities are given in eV.

Quantity	Valence band	Conduction band
$\varepsilon_{max}(HF)$	-8.203	3.084
$\varepsilon_{min}(HF)$	-9.039	2.296
$\Sigma_{band\,max.}(e)$	-0.574	-1.470
$\Sigma_{band\,min.}(e)$	-0.631	-1.461
$\Sigma_{band\,max.}(h)$	1.846	0.203
$\Sigma_{band\,min.}(h)$	2.007	0.254
$\varepsilon_{max}(QP)$	-6.931	1.817
$\varepsilon_{min}(QP)$	-7.663	1.089

of the conduction band, leading to an effective band narrowing. The HF conduction band is shifted in this way by 1.2 eV downwards and the valence band by 1.3 eV upwards, respectively, reducing the fundamental gap from 10.5 eV to 8 eV.

Let us look more closely at the individual levels themselves and compare them with photoelectron spectroscopical results. If we would increase the distance between the nucleotide bases in the polynucleotides, the energy bands would be reduced on single levels lying approximately in the middle of the band. The Wannier functions of the polymer would be reduced at the same time for simple molecular orbitals. These levels can reasonably be compared to the results of spectroscopical measurements performed on single bases. As also observed in previous *ab initio* calculations on nucleic acid bases [88—90] and polynucleotides [42—44], the π and lone-pair levels are always mixed in our case, too. For the three highest filled and three lowest unfilled bands, respectively, the ordering n, π, π, π^*, π^*, n^* was found here. If the previously described correlation corrections are also performed for the monomer, the highest filled (π) level is obtained at −7.523 eV. This value will have to be compared with the first vertical ionization potential (IP) of 8.7 eV obtained by gas-phase UV photoelectron spectroscopy for cytosine [91]. A considerable part of the remaining discrepance can be eliminated if vibrational corrections (the formation of optical phonon-polarons) are also taken into account [92].

5. Electronically Excited States of Biological Macromolecules

Biopolymers represent a special class of solids from the point of view of their optical properties. The traditional methods of treating bound electron-hole pairs are inapplicable to them: their monomeric units strongly interact; therefore, intracellular and intercellular excitations must be treated on the same footing. This clearly contradicts the Frenkel model, based on the assumption that the excited electron primarily shares a unit cell with the hole. On the other hand, the dielectric screening in these materials is certainly too weak to produce Wannier-Mott-type excitons with a large radius.

A theoretical framework to bridge the gap between the above two limits of the exciton picture was introduced by Takeuti [51]. It is based on Green's function formalism developed earlier by Lax [52] and by Koster and Slater [53] to treat lattice vibrations and impurity states, respectively. A similar method was also worked out by Altarelli and Bassani [54] which is based on a Fredholm-type solution of the integral equation determining the expansion coefficients of the exciton wave function.

To obtain quantitative results, Takeuti had to assume at that time simple parabolic bands, and he had to restrict himself to the diagonal part of the electron-hole interaction matrix. These restrictions are, however, not necessary today and excitonic levels of polymers with medium size elementary cells like polynucleotides can be calculated from first principles without any semiempirical approximations made in evaluating the matrix elements of Green's function and the electron-hole (e-h) interaction.

It seems to be especially recommendable to avoid the use of empirical parameters if typical correlation effects such as excitonic binding are investigated in extended systems. For semiempirical schemes contain parameters adjusted to describe the experimentally observed properties of relatively small molecules. The effect of correlation is, therefore, built *ab ovo* into these parameters, and their use in a perturbation theoretical scheme to calculate e-h attraction may lead to inconsistent results. On the other hand, we may also have situations in infinite systems not appearing for a few atoms, for which the parameters were optimized.

5.1. GREEN'S FUNCTION THEORY OF DELOCALIZED EXCITONS
IN BIOPOLYMERS

In deriving the electron-hole interaction we start from the Hamiltonian (in atomic units):

$$\hat{H} = \int d^3x \psi^\dagger(x) \left[-\frac{1}{2}\Delta + V_p(x) \right] \psi(x) +$$

$$+ \frac{1}{2} \int d^3x \int d^3x' \psi^\dagger(x)\psi^\dagger(x') \frac{1}{|x-x'|} \psi(x')\psi(x), \qquad (66)$$

where x represents the spatial and spin variables ($x = [\mathbf{r}, \sigma]$) and $V_p(x)$ is the periodic potential due to the ions. The field operators $\psi(x)$, $\psi^\dagger(x)$ obey the usual fermion commutation rules and will be expanded, using as a basis the complete orthonormal set of Wannier spin orbitals belonging to the valence bands (v) and conduction bands (c), respectively:

$$\psi(x) = \sum_\ell a_{\ell v} w_v(x - \ell) + \sum_\ell a_{\ell c} w_c(x - \ell), \qquad (67a)$$

$$\psi^\dagger(x) = \sum_\ell a_{\ell v}^\dagger w_v^*(x - \ell) + \sum_\ell a_{\ell c}^\dagger w_c^*(x - \ell), \qquad (67b)$$

where the notation $w_j(x - \ell) = w_j(\mathbf{r} - \mathbf{R}_\ell) \cdot \xi(\sigma)$ is used ($\mathbf{R}_\ell$ stands for the lattice vector of the cell in which the Wannier function is localized). Substituting Eqs. (66) and (67) into (65) and separating the one- and many-particle terms, we obtain

$$\hat{H} = \hat{h} + \hat{g}, \qquad (68)$$

$$\hat{h} = \sum_{\ell, m} a_{\ell c}^\dagger a_{mc} h_{\ell m}^c + \sum_{\ell, m} a_{\ell v}^\dagger a_{mv} h_{\ell m}^v, \qquad (69)$$

$$\hat{g} = \frac{1}{2} \sum_{\ell_1,\ldots,\ell_4} \sum_{j_1,\ldots,j_4} a^\dagger_{\ell_1 j_1} a^\dagger_{\ell_2 j_2} a_{\ell_3 j_3} a_{\ell_4 j_4} \times g \begin{pmatrix} \ell_1 \, \ell_2 \; \Big\| \; \ell_3 \, \ell_4 \\ j_1 \, j_2 \; \Big\| \; j_3 \, j_4 \end{pmatrix}, \quad j_i = v, c. \tag{70}$$

The monoelectronic and bielectronic integrals are defined here by

$$h^j_{\ell m} = \int d^3x w^*_j(x - \ell) \left[-\frac{1}{2} \Delta + V_p(x) \right] w_j(x - m), \quad j = v, c, \tag{71}$$

$$g \begin{pmatrix} \ell_1 \, \ell_2 \; \Big\| \; \ell_3 \, \ell_4 \\ j_1 \, j_2 \; \Big\| \; j_3 \, j_4 \end{pmatrix} = \int d^3x \int d^3x' w^*_{j_1}(x - \ell_1) w^*_{j_2}(x' - \ell_2) \frac{1}{|x - x'|} \times$$

$$\times w_{j_3}(x' - \ell_3) w_{j_4}(x - \ell_4). \tag{72}$$

To obtain more transparent expressions, we introduce at this point the particle-hole picture using the operators $a^\dagger_\ell \equiv a^\dagger_{\ell c}$ and $a_\ell \equiv a_{\ell c}$ for the creation and destruction of a particle in the conduction band and similarly $d^\dagger_\ell \equiv a_{\ell v}$ and $d_\ell \equiv a^\dagger_{\ell v}$ in the case of a hole in the valence band. In terms of these operators, we can construct the wavefunction of an exciton through the following steps.

(i)   First, we create an e-h pair from the completely filled valence band $\Phi = d_{\ell_1} \ldots d_{\ell_{N_c}} \Phi_0 (\Phi_0$ is the vacuum) with a separation $\mathbf{R}_s$:

$$\Psi_{\ell+s,\ell} = a^\dagger_{\ell+s} d^\dagger_\ell \Phi. \tag{73}$$

(ii) As the second step, stationary eigenstates with quasimomentum $\mathbf{K}$ are formed in the Bloch form

$$\Psi_{s,\mathbf{K}} = N_c^{-1/2} \sum_\ell \exp(i\mathbf{K} \cdot \mathbf{R}_\ell) a^\dagger_{\ell+s} d^\dagger_\ell \Phi. \tag{74}$$

(iii) Finally, the wave function of an exciton with momentum $\mathbf{K}$ is formed as a linear combination of symmetry-adapted e-h pair functions with different separation:

$$\Psi_\mathbf{K} = \sum_s \Omega_{s,\mathbf{K}} \Psi_{s,\mathbf{K}}. \tag{75}$$

The weight of the different charge-transfer components in the excitonic wave function an be determined by solving the Schrödinger equation

$$\hat{H}\Psi_\mathbf{K} = E_\mathbf{K}\Psi_\mathbf{K}, \tag{76}$$

also providing the exciton band structure $E_\mathbf{K}$. For this purpose we partition the Hamiltonian by substituting the particle-hole operators in Eqs. (68)–(70). After some algebra we obtain

$$\mathscr{H} = E_{HF} + \hat{H}_e + \hat{H}_h + \hat{H}_{e,h}, \tag{77}$$

where $E_{HF} = \langle \Phi_{HF} | \hat{H} | \Phi_{HF} \rangle$ is the energy of the completely filled valence band in the HF approximation and

$$\hat{H}_e = \sum_{\ell, m} a_\ell^\dagger a_m \left\{ h_{\ell m}^c + \sum_p \left[g \left(\begin{matrix} \ell \, p \\ c \, v \end{matrix} \middle\| \begin{matrix} p \, m \\ v \, c \end{matrix} \right) - g \left(\begin{matrix} p \, \ell \\ v \, c \end{matrix} \middle\| \begin{matrix} p \, m \\ v \, c \end{matrix} \right) \right] \right\}$$

$$= \sum_{\ell, m} a_\ell^\dagger a_m \langle w_c(x - \ell) | \hat{F} | w_c(x - m) \rangle, \tag{78}$$

$$\hat{H}_h = - \sum_{\ell, m} d_m^\dagger d_\ell \left\{ h_{\ell m}^v + \sum_p \left[g \left(\begin{matrix} \ell \, p \\ v \, v \end{matrix} \middle\| \begin{matrix} p \, m \\ v \, v \end{matrix} \right) - g \left(\begin{matrix} p \, \ell \\ v \, v \end{matrix} \middle\| \begin{matrix} p \, m \\ v \, v \end{matrix} \right) \right] \right\}$$

$$= - \sum_{\ell, m} d_m^\dagger d_\ell \langle w_v(x - \ell) | \hat{F} | w_v(x - m) \rangle, \tag{79}$$

$$\hat{H}_{e, h} = - \sum_{\ell_1, \ldots, \ell_4} a_{\ell_1}^\dagger a_{\ell_4} d_{\ell_3}^\dagger d_{\ell_2} \times \left[g \left(\begin{matrix} \ell_1 \, \ell_2 \\ c \, v \end{matrix} \middle\| \begin{matrix} \ell_3 \, \ell_4 \\ v \, c \end{matrix} \right) - \right.$$

$$\left. - g \left(\begin{matrix} \ell_2 \, \ell_1 \\ v \, c \end{matrix} \middle\| \begin{matrix} \ell_3 \, \ell_4 \\ v \, c \end{matrix} \right) \right]. \tag{80}$$

To obtain $\mathscr{H}$, we neglected in Eq. (77) those contributions from $\hat{H}$ which contain more than two electron or hole operators, respectively. Since the wave function Ψ_K contains only one e-h pair operator, the matrix elements of those terms are automatically zero. The Wannier functions are not eigenfunctions of the Fock operator $\hat{F}$; it is, therefore, worthwhile switching to a Bloch representation to calculate the matrix elements of $\hat{H}_e$ and $\hat{H}_h$ with the exciton wave functions $\Psi_{s, K}$. We substitute the ws from Eq. (35) into (78) and (79) and form the matrix elements. After some algebraic computations we obtain

$$\langle \Psi_{r, K} | \hat{H}_e | \Psi_{s, K} \rangle = N_c^{-1} \sum_k \exp[i k \cdot (R_r - R_s)] \varepsilon_{c, k}, \tag{81}$$

$$\langle \Psi_{r, K} | \hat{H}_h | \Psi_{s, K} \rangle = N_c^{-1} \sum_k \exp[i k \cdot (R_r - R_s)] \, (-\varepsilon_{v, k-K}). \tag{82}$$

On the right-hand side of Eq. (76) we need, furthermore, the matrix element of E_K, whch can be written in the form

$$\langle \Psi_{r, K} | E_K | \Psi_{s, K} \rangle = N_c^{-1} E_K \sum_k \exp[i k \cdot (R_r - R_s)]. \tag{83}$$

Now applying the resolvent operator method for the perturbation $\hat{H}_{e, h}$ and fixing the origin of the energy by setting $E_{HF} = 0$ in Eq. (77), we can write

$$\Psi_K = [E_K - (\hat{H}_e + \hat{H}_h)]^{-1} \hat{H}_{e,h} \Psi_K. \tag{84}$$

Substituting here Eq. (75), multiplying from the left hand side with $\Psi_{r,K}$ and integrating we obtain

$$\Omega_{r,K} = \sum_s \sum_t \langle \Psi_{r,K} | [E_K - (\hat{H}_e - \hat{H}_h)]^{-1} | \Psi_{s,K} \rangle \times$$

$$\times \langle \Psi_{s,K} | \hat{H}_{e,h} | \Psi_{t,K} \rangle \Omega_{t,K}. \tag{85}$$

The matrix elements of the e-h interaction are again easy to calculate using the expansions of $\Psi_{s,K}$ from Eq. (80) and we obtain for them in terms of the ws

$$\langle \Psi_{s,K} | \hat{H}_{e,h} | \Psi_{t,K} \rangle = - \sum_u \exp(-i\mathbf{K} \cdot \mathbf{R}_u) \times$$

$$\times \left[g \begin{pmatrix} s+u\,0 \,\Big\|\, u\,t \\ c\quad v \,\Big\|\, v\,c \end{pmatrix} - g \begin{pmatrix} 0\,s+u \,\Big\|\, u\,t \\ v\quad c \,\Big\|\, v\,c \end{pmatrix} \right] \equiv$$

$$\equiv V^{(v,\,c)}(\mathbf{R}_s, \mathbf{R}_t, \mathbf{K}). \tag{86}$$

Collecting all the calculated terms from Eqs. (81)—(83) and (86), and substituting them into Eq. (85), we arrive at a system of homogeneous linear equations for the determination of the unknown coefficients $\Omega_{t,K}^{(v,\,c)}$:

$$\Omega_{r,K}^{(v,\,c)} = \sum_s \sum_t G^{(v,\,c)}(\mathbf{R}_r, \mathbf{R}_s, E_K) V^{(v,\,c)}(\mathbf{R}_s, \mathbf{R}_t, \mathbf{K}) \Omega_{t,K}^{(v,\,c)}. \tag{87}$$

$G^{(v,\,c)}(\mathbf{R}_r, \mathbf{R}_s, E_K)$ stands for the matrix element of Green's function belonging to the periodic polymer. It can easily be calculated using the results of previous band structure calculations as

$$G^{(v,\,c)}(\mathbf{R}_r, \mathbf{R}_s, E_K) = N_c^{-1} \sum_k \frac{\exp[i\mathbf{k} \cdot (\mathbf{R}_r - \mathbf{R}_s)]}{E_K - (\varepsilon_{c,\,k} - \varepsilon_{v,\,k-K})}. \tag{88}$$

In summary, we can determine the exciton band structure E_K in the following steps.

(i)   From the HF energy bands $\varepsilon_{j,\,k}$ we calculate the matrix elements of Green's function, using Eq. (88) in the region of the forbidden gap where excitonic levels can be expected.

(ii)   We transform the Bloch functions into a set of optimally localized Wannier functions and calculate the matrix elements of the e-h interaction in the whole Brillouin zone (BZ) using Eq. (86).

(iii) Finally, we determine $E_\mathbf{K}$ by finding the zeros of the determinant $D = |\mathbf{G} \cdot \mathbf{V} - \mathbf{I}|$, where $\mathbf{I}$ is the unit matrix, and solve the system of equations of (87) to calculate the normalized values of $\Omega_{r, \mathbf{K}}$.

From the numerical point of view, step (i) is the simplest one and only amounts to a few percent of the total computational efforts. Since the HF bands of polynucleotides show a relatively simple shape, the only problem in calculating elements of G arises from the fact that for more distant neighbours the integrand in Eq. (88) has an oscillatory character due to the exponential factor. We have, however, made very good experiences using for its integration the Gauss-Legendre method which has also been applied in our polymer-orbital calculations to obtain the long-range parts of the density matrix.

The most time-consuming part of the calculations is the evaluation of the matrix elements $V^{(v, c)}(\mathbf{R}_s, \mathbf{R}_t, \mathbf{K})$ defined by Eq. (86). For the purpose of the construction of the Fock operators $\hat{F}$, as a first step in the HF calculation, the bielectronic integrals defined in Eq. (12) are calculated in terms of atomic basis functions (in our case Gaussians), which are used to construct the Bloch functions. The expansion of the ws in Eq. (35) means that the list of integrals over AO's will have to be transformed to obtain the corresponding one over the ws. This integral transformation can be performed for polymers with acceptable computer time only, if the Wannier functions are properly localized.

The zeros of the function $D(E_\mathbf{K})$ can finally be found by first localizing the regions of its sign change in a preliminary step and then iterating to each solution with a finer mesh. Since $D(E_\mathbf{K})$ turned out to be a rather slowly varying function, a simple interpolation scheme was enough to find its zeros in $8-10$ iteration steps with an accuracy of $|D| < 10^{-7}$. The energy values $E_\mathbf{K}$ contain at least five significant digits in eV in this case. The corresponding excitonic wavefunction components were obtained by a subsequent solution of the system of homogeneous linear equations. With the above conditions, their accuracy was better than four significant digits.

Since the band structure calculations, which form the starting point for the investigation of the exciton spectrum, were performed at the restricted HF level (putting, for the same spatial Bloch orbital, two electrons with opposite spin), the Wannier functions as obtained from Eq. (38) also conserve this closed-shell property. The e-h pair wavefunctions are, therefore, eigenfunctions of the component of the total spin in the direction of the axis of the quantization. Since the Hamiltonian applied does not contain spin-dependent terms, we can take linear combinations of the e-h determinants to form eigenfunctions of a definite spin multiplicity. Substituting the appropriate combinations for singlet and triplet states, respectively, we obtain after simple algebra for the elements of **V** in Eq. (86):

$$^{M}V^{(v,c)}(\mathbf{R}_s, \mathbf{R}_t, \mathbf{K}) = -\sum_u \exp(-i\mathbf{K} \cdot \mathbf{R}_u) \times$$

$$\times \left[\tilde{g} \begin{pmatrix} s+u\,0 & u\,t \\ c & v & v\,c \end{pmatrix} - 2\delta_M \tilde{g} \begin{pmatrix} 0\,s+u & u\,t \\ v & c & v\,c \end{pmatrix} \right]. \qquad (89)$$

The definition of the bielectronic four-center integrals $\tilde{g}$ is the same as that of g in Eq. (8) except that, instead of spinorbitals, only the spatial part of the Wannier functions is used here. For true singlet states $\delta_M = 1$ and for triplets $\delta_M = 0$. The lattice sum over $\mathbf{R}_u$ in Eq. (89) is not problematic. The first 'Coulomb-type' integrals contains exponentially decaying charge distributions on both electrons. The second one (appearing only for singlet excitons) is, from the point of view of the e-h pair, an exchange contribution but it can also be regarded as a Coulomb-type interaction between two neutral charge clouds. Its leading part is the dipole-dipole coupling; therefore, one does not expect serious problems with its summation.

It should be noted here, however, that the dominance of the exchange term, also called 'excitation transfer' interaction in long-range effects, is not *ab avo* clear since the first integral in Eq. (89) also contains far-reaching contributions when near diagonal ($\mathbf{R}_s \sim \mathbf{R}_t$) elements of **V** are calculated. On the other hand, the off-diagonal elements of the Green's function in Eq. (88) decay very fast so that the spatial extension of the exciton is determined by the long-range contributions in **V**. For the bottom of the singlet exciton bands the oscillator strengths have also been calculated using the expression

$$f^{(v,c)} = \frac{4}{3} E_{K=0} \left[\sum_t \Omega^{(v,c)}_{t,K=0} \langle \Lambda_{v,0} | \mathbf{r} | \Lambda_{c,t} \rangle \right]^2, \qquad (90)$$

where

$$\Lambda_{j,t} = \sum_p d^t_{pj} \chi_p(\mathbf{r} - \mathbf{R}_t - \mathbf{R}_p), \qquad (91)$$

is the tth unit contribution to the WF, localized in the reference cell [49] ($j = v, c$).

5.2. EXCITONIC ORIGIN OF THE FIRST UV ABSORPTION PEAK IN POLYENE

Let us summarize first the experimental observations related to the nature of the first absorption peak in *trans*-polyene. From the large number of experimental works investigating the optical spectrum of polyene, those which deal either with stretch-aligned films [36] or with crystal polymerized under shear conditions [55] are most interesting for our purposes. For these techniques yield highly aligned polymer fibers which is the necessary condition for the correct measurement of anisotropic optical properties. Therefore, their results can be reasonably compared with calculations referring to single polyene chains. There is a general consensus between different measurements that the first strong absorption band of undoped *trans*-polyene has its edge at 1.4—1.5 eV and peak at 1.9—2.0 eV. The absorption is very anisotropic, favouring the direction parallel to the polymer axis [55—57].

The mechanism of the absorption at the 2 eV peak is usually interpreted as a direct interband transition [55—57] with a tail down to 1.4—1.5 eV, probably due to structural disorder. It is viewed as arising from a transition from the 1-d peak in the density of states in the valence band to that in the conduction band [58], though several semiempirical calculations have also proposed reasonable alternatives to this model [59—62] starting from an excitonic Hamiltonian. The success of these theoretical models was mainly limited by the severe approximations introduced to make the problem computationally tractable. As a consequence, they arrived at either too large [61] or too low [62] excitation energies.

We have reexamined the problem of the 2 eV absorption peak in polyene on the basis of the exciton model by using an *ab initio* treatment for all interaction matrix elements and avoiding semiempirical integral approximations. A similar theoretical procedure recently proved to be quite useful in interpreting the visible optical properties of polydiacetylene (PDA) crystals [63], whose electronic structure is closely related to that of polyenes. Computations of the exciton spectrum of PDA and of other structurally similar organic polymers both at semiempirical [64, 65] and first principles [63, 66] levels have undoubtedly shown that the excited electron and the hole do not share the same monomeric cell in these systems. Due to strong interactions between the subunits, the bound electron-hole (e-h) pair is delocalized over several diacetylene units in PDA crystals, for instance. The contribution of the Frenkel exciton to the binding energy is not more than 30% on the average [63].

On the other hand, the shape and size of the experimentally observed excitonic electroreflectance response in PDA's could plausibly be accounted for [67, 68] if the field-induced energy shift due to large charge transfer components in the excitonic wave function was taken into account. The delocalization of the exciton within a radius of 25—30 Å plays a predominant role, both for the singlet and triplet excitation spectrum, not only in PDA's but also in other polymers. Therefore, it seems to be absolutely necessary to also include this effect when investigating the optical properties of polyene.

Turning now to the theoretical results, we have to recall at this point that there still is a quantity within our earlier defined methodological framework (second order perturbation theory for correlation effects, Toyozawa model for band self-energy corrections and exciton picture for excited states) which may introduce a major uncertainty into the evaluation of the results. Since both the Bloch and

Wannier functions are expanded as a linear combination of Gaussian atomic orbitals in our method, we have to pay great attention to the proper choice of these basis sets at the HF level as well as in correlation calculations. To be able to observe the dependence of band structure and excitonic properties on basis set extension, we performed the computations using three sets of gradually increasing size: a minimal-type (STO-3G), an extended one with split valence shell character (6-31G) and, finally, one with p-type polarization functions on hydrogens and d-functions on carbons (6-31G**). Table XII demonstrates the dramatic changes for different basis sets in the polyene band structure, especially if correlation corrections are also included. Not only the quasiparticle fundamental gap is, for instance, from 7.6 eV to 3 eV, but also the band widths change substantially. Since these quantities are basic ingredients of the corresponding Green's functions, the use of a proper basis set is inevitable in the calculation of optical spectra.

TABLE XII. Single (*HF*) and quasiparticle (*QP*) fundamental band gaps and $K = 0$ singlet exciton energy obtained using different atomic basis sets for alternating *trans*-polyene. The bond alternation was optimized for each basis set. All quantities are given in eV.

Quantity	STO-3G	6-31G	6-31G**
HF band gap	8.282	4.926	4.427
QP band gap	7.671	4.036	2.990
$K = 0$ singlet excitation	4.862	2.583	1.862

Since the polyene energy bands exhibit a simple dispersion, the numerical integration in Eq. (88) could easily be performed using the Gauss-Legendre method. The zeroes of the funciton $D(E_K)$ were found by first localizing the regions of its sign change within the gap and then iterating to each solution with a finer mash. We employed nine different values of K to obtain all details of the exciton bands needed in our present work on exciton-phonon interaction and energy transport in polyene. In agreement with our previous experiences on PDA polymers [63], charge transfer excitations until eight neighbours had to be taken into account to obtain convergence for the excitation energies.

We demonstrate the development of the lowest singlet and triplet exciton bands from the electron polaron bands in Figure 9. These results were obtained with our best quality basis set. For comparison, also the HF valence and conduction bands are shown. The dependence of the fundamental gaps and singlet exciton energies on the atomic basis set is shown in Table XII.

The exciton energies are measured from the top of the valence band. The lower edge of the spectrum at $K = 0$ is situated at 1.86 eV, which could quite reasonably be compared to the experimental peak at 2 eV. On the other hand, we did not yet introduce dielectric screening into the matrix elements of the e-h interaction.

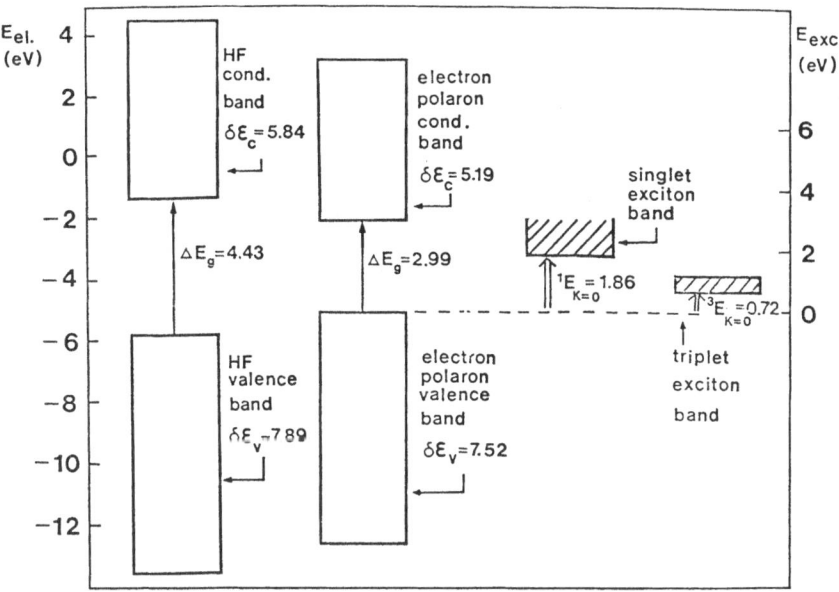

Fig. 9. HF and electron polaron π-electron valence and conduction bands and singlet and triplet exciton bands of *trans*-polyene obtained using the 6-31G** atomic basis set. The exciton energies are calculated from the top of the valence band.

Therefore, this agreement can only be fortuitous and, very probably, a consequence of the cancellation of two errors. At first, the quasiparticle gap of 3 eV is still too large in our calculations (by about 0.5 eV as compared to the experiment). The influence of dielectric polarization, on the other hand, can be estimated if we introduce screening into the matrix elements of the e-h interaction by the interpolation scheme of Hermanson and Phillips by writing

$$\varepsilon^{-1}(r) = \varepsilon_0^{-1} + \left\{ \frac{\varepsilon_0 - 1}{\varepsilon_0} \right\} \exp(-Qr), \qquad (92)$$

where Q^{-1} plays the role of a characteristic breakdown length for dielectric effects. It is related to the Thomas-Fermi wave number, and as a reasonable guess one can use for it the value $Q = a^{-1}$, where a is the lattice constant. More details on this procedure are given in [63]. Since the proper microscopic value of ε_0 is not known, we performed the calculations for $\varepsilon_0 = 2, 3, 4$. The realistic value should be somewhere around $\varepsilon_0 = 3$, meaning a reduction of the exiton-binding energy at $K = 0$ from 1.13 eV to about 0.6–0.5 eV according to Figure 10. The additional amount of e-h binding due to the lack of screening in our calculations, 0.5–0.6 eV according to this estimate is in its size comparable to the artificial increase of E_K due to a still somewhat large quasiparticle gap (by about 0.5 eV). The cancelation of these two opposite errors could explain the better than expected agreement

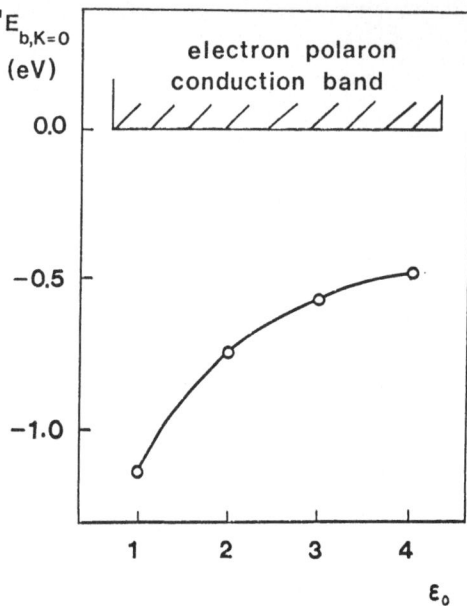

Fig. 10. $K = 0$ singlet exciton binding energy, $^1E_{b, K=0}$, as function of the dielectric constant, ε_0, in *trans*-polyene. Results obtained using the 6-31G** atomic basis set.

between experiment and theory at this level. Both of these errors deserve a more detailed examination in the future.

The second intense absorption in the polyene spectrum occurs at 8.5—9.0 eV [55—57], and it has been assigned to $\sigma \rightarrow \sigma^*$ transitions due to symmetry considerations [57]. On the basis of our data, we expect such transitions only at very much higher energies. In concordance with other previous suggestions [55], we propose that these peaks correspond to interchain $\sigma \rightarrow \pi^*$ or $\pi \rightarrow \sigma^*$ transitions, which we also plan to include as one of the next steps of these investigations.

5.3. SINGLET AND TRIPLET EXCITONS IN POLYPEPTIDES

Since our computational facilities limited the exciton calculations to the space of three electron and three hole bands (allowing for a maximal e-h separation of five peptide units), the previously presented corrections for the valence and conduction bands of the two polypeptide models were also applied to the next two highest filled bands and lowest unfilled ones, respectively, and the Green functions of the resulting nine band pairs were numerically computed. Since the procedure to calculate exciton levels is rather time-consuming, we could not yet explore all details of the exciton bands. The first three singlet and the first triplet bands could be calculated up to now at three points of the Brillouin zone, $K = 0$, $\pi/2a$ and π/a, respectively. Table XIII shows the obtained limits of the exciton bands for the

TABLE XIII. Singlet and triplet charge transfer exciton bands, $E(K)$, and the corresponding oscillator strengths, f, obtained for the α-helical polypeptide backbone. For comparison, the approximately calculated f-value of the corresponding monomer transition is also shown. Results were obtained using a 6-31G* atomic basis set and second-order Møller-Plesset perturbation theory. Energies are given in eV.

Quantity	1st singlet	2nd singlet	3rd singlet	1st triplet
$E(K=0)$	6.208	7.322	11.849	3.806
$E(K=\pi/a)$	6.602	7.830	11.588	3.918
δE	0.394	0.508	0.261	0.112
f(polymer)	0.246	0.162	0.020	—
f(monomer)	0.341	0.094	0.012	—

α-helical backbone together with the corresponding band widths and oscillator strengths, calculated at the bottom of the singlet bands.

To be able to judge the role of interpeptide interactions in building up the intensities of the individual transitions, we also calculated the f-value of the separated residues by substituting the renormalized zero cell contribution of the Wannier function into Eq. (88). As approximate transition energy of the monomer, the middlepoint energy in the corresponding exciton band was used. The corresponding quantities obtained for polyalanine are shown in Table XIV.

TABLE XIV. Singlet and triplet charge transfer exciton bands, $E(K)$, and the corresponding oscillator strengths, f, obtained for α-polyalanine. For comparison, the approximately calculated f-value of the corresponding monomer transition is also shown. Results were obtained using a 6-31G* atomic basis set and second-order Møller-Plesset perturbation theory. Energies are given in eV.

Quantity	1st singlet	2nd singlet	3rd singlet	1st triplet
$E(K=0)$	6.052	7.146	12.110	3.580
$E(K=\pi/a)$	6.398	7.628	11.849	3.688
δE	0.346	0.482	0.245	0.108
f(polymer)	0.228	0.157	0.022	—
f(monomer)	0.324	0.096	0.010	—

It is interesting to note in Tables XIII and XIV that the lowest singlet exciton for both polymers lost about 30% of its intensity as compared with the corresponding monomer transition. As expected from sum rules, the higher lying excited states compensate for this hypochromicity. The order of magnitude of the observed effect can reasonably be compared with experimental hypochromicity values reported for polyglutamate (36%) and polylysine (30%), respectively [69].

The importance of strong intermonomer interactions in building up the exciton is also reflected in the substantial delocalization of the excitonic wavefunction. As shown in Figure 11, charge transfer excitations for a distance of 4—5 residues contribute significantly to the calculated binding energy of the exciton. The Frenkel exciton ($N = 0$) possesses only about 70—80% of the total value. The lowest singlet exciton energy regions obtained of 6.0—6.2 and 7.1—7.3 eV can reasonably be related to the absorptions at wavelengths of 2060 and 1890 Å (6.0 and 6.5 eV), respectively, observed for polyglutamate [69, 70], and at 2050, 1900 and 1650 Å (6.0, 6.5 and 7.5 eV), respectively, observed for polyleucine [71].

Figure 12 summarizes the energy changes taking place in the e-h interaction

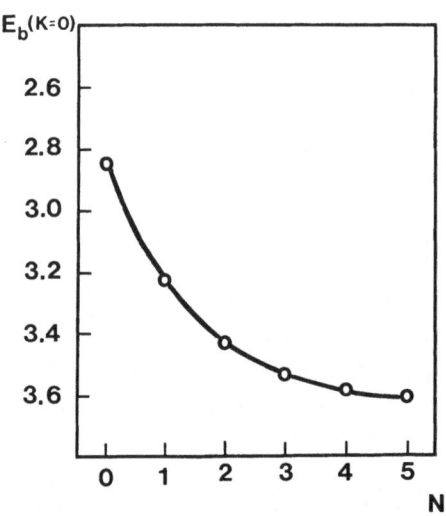

Fig. 11. Binding energy of the singlet exciton in the α-helical polypeptide backbone at $K = 0$ as function of the number of neighbouring peptide units for which the exciton is delocalized (in eV).

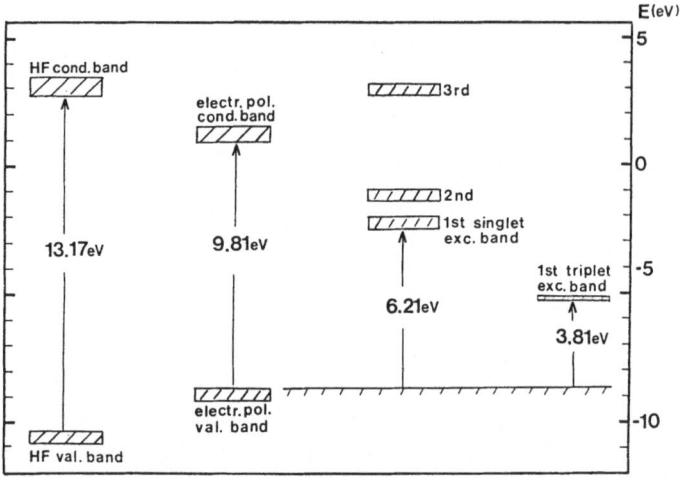

Fig. 12. Energy scheme of the formation of the charge transfer exciton in the α-helical polypeptide backbone in three steps: bare HF e-h pair, correlated free e-h pair, and singlet and triplet exciton (bound e-h pairs).

during the three major steps which lead to the formation of a delocalized exciton state in the polypeptide α-helix: (i) a free e-h pair is formed in the HF conduction and valence bands, respectively, upon investing an energy of 13.17 eV; (ii) the electron and the hole polarize the valence electron system creating a cloud of virtual exciton states, which accompany the particles during their motions, modify their self-energies (formation of polarons) and reduce the fundamental gap to 9.81 eV; (iii) the screened electron and hole polarons interact with each other and form singlet or triplet exciton states (depending on the mutual spin orientation).

The lowest energy transitions show hypochromism for both polyglycine and polyalanine, and the higher ones hyperchromism as compared to the spectrum of Nylon 6, where the random conformation prevents the cooperative excitation of the peptide units [71]. The observed differences in these spectra are, therefore, related to the previously described delocalization of the excitation energy for the regularly spaced peptide groups. The absence of long-range order in globular proteins, of course, sets a limit to this delocalization. On the other hand, at physiological temperatures, mostly due to the scattering on lattice vibrations, the travelling excitons have a finite free path anyway, comparable to or shorter than the length of identified helical segments in proteins. Therefore, the efficiency of energy transport via excitons in proteins is, in our opinion, primarily determined by the strength of the exciton-phonon interaction in these systems. The calculation of the phonon spectra from linear response theory and the determination of the exciton-phonon scattering matrix elements by perturbation theory are in progress in our laboratory.

5.4. CALCULATION OF THE UV SPECTRUM OF DNA

The precise knowledge of the excited electronic states in nucleic acids is of obvious importance for the proper description of photobiological and photo-chemical processes related to DNA. Besides their contribution to our under-standing of fundamental phenomena like energy migration, optical rotatory dispersion and hypochromism, studies of this kind also help to clarify the mechanism of ultraviolet (UV) radiation damage in DNA and its role in carcino-genesis as well as the charge transfer reactions between chemical carcinogens and DNA. On the other hand, the correct interpretation of experimental spectra plays a significant role in detecting biologically important structural transitions like denaturation-renaturation processes in nucleic acids. The optical absorption of highly polymerized nucleic acids is not a simple sum of the contributions coming from the constituent bases or base-pairs. In the ordered helical structure, the integrated intensity (oscillator strength) per chromophore is observed to decrease for the lowest energy transition as compared to partially or randomly oriented bases (hypochromic effect). In helical calf thymus DNA, with a first absorption maximum at 258 nm, for instance, the hypochromicity of the first absorption band is 37 per cent relative to the constituent mononucleotides [72]. The intensity lost for the long wavelength transition in the polymer is regained by bands at a shorter wavelength as required by general sum rules. This hypochromism can be thought to arise from the various intra- and interstrand interactions of the stacked nucleotide bases [72—75].

This hypothesis is also strongly supported by recent measurements of excimer fluorescence lifetimes performed under in vivo conditions for natural and synthetic DNA excited by synchrotron UV radiation [76]. The authors of that work not only proved the existence of an emission originating from a longer-lived state but also investigated the problem as to whether it is of excimer-type emitted from a delocalized exciton state of DNA (implying significant inter-base coupling) or whether it is a delayed monomer fluorescence. Arguments favouring excimer fluorescence are the long lifetime (relative to the monomer), the exponential decay and the absence of any nonlinear dependence on the exciting beam intensity (which should be observable if the excited but delayed monomer would arise from a triplet-triplet annihilation). The excimeric nature of the emission in DNA is supported by the presence of an unresolved band envelope red-shifted from the individual nucleotide bases. This feature has also been observed earlier at low temperatures [77—79].

The theoretical study of the hypochromic effect in polynucleotides has been the subject of several investigations based on models with less and less simplified assumptions [72—75, 80—87]. They reach from the classical oscillator theory to various semiempirical Hamiltonians used for the quantum-mechanical calculation of ground and excited states of nucleotide base monomers and dimers, respectively.

The results of our calculations strongly support the assignment of the observed emissions from longer-lived states in DNA [96] as originating from delocalized singlet exciton states. Both the significant exciton band width (0.6 eV) and the fact that a substantial part of the binding energy is missing if only Frenkel excitons are calculated point to the importance of charge transfer components in the excitation. In the case of the first singlet band, for instance, the binding energy of the Frenkel exciton is less than 80 per cent of the total value. At the same time, due to interbase interactions, the absorption of the polymer is somewhat red-shifted as compared with the separated monomers. For the first singlet band we can observe, for instance, a shift of 0.3 eV towards lower energies.

The most interesting result in Table XV is the significant reduction of the absorption intensity of the first singlet exciton band as compared again to the corresponding monomer transition. Since the quantum-mechanical expression for the intensities is an approximate one, the absolute values of the oscillator strengths are less important, but the reduction by about 26 per cent is certainly significant. As expected from the Kuhn-Thomas sum rule, it is partially compensated by the next transitions, which show a hyperchromic shift.

Accompanying the singlet exciton bands, we also obtained triplet ones which are relatively flat. The first triplet band starts at 2.97 eV ($K = 0$) and has a width

TABLE XV. Energies and intensities of the first three singlet charge transfer exciton bands of poly(C) and the corresponding monomer properties. Results were obtained using a 4-31G atomic basis set and second-order Møller-Plesset perturbation theory. Energies are given in eV.

Quantity	1st band	2nd band	3rd band
Exciton band min. $(K = 0)$	4.732	5.912	6.708
Exciton band max. $(K = \pi/a)$	5.384	6.184	7.173
Binding energy of the exciton	3.288	2.108	1.312
Monomer excitation energy	5.041	6.072	6.904
Oscillator strength of polymer	0.257	0.327	0.655
Oscillator strength of monomer	0.346	0.291	0.628
Hypochromicity (per cent)	26	−12	−4

of 0.22 eV. The singlet-triplet splitting is thus 1.76 eV at $K = 0$. Though the classification of the electronic transitions of the DNA macromolecule according to the π-type or lone-pair states of the monomers is only approximate, it is still possible to assign the exciton bands to monomer levels if the Wannier functions are contracted to molecular orbitals. As mentioned earlier, the π and lone-pair bands turn out to be always mixed. The first three singlet monomer excitation energies obtained at 5.04, 6.07 and 6.90 eV, respectively, will have to be compared with the experimentally observed values of 4.5, 5.2 and 6.1 eV [93]. A considerable part of the remaining discrepancy could probably be eliminated if vibrational corrections (the formation of optical phonon-polarons) were also taken into account in these calculations [92]. Since the level shift due to phonon-polaron binding has the same direction for the ground and excited state, the theoretical transition energies are somewhat more accurate than the individual levels.

From crystallographic studies on DNA it is well known that its structure is quite sensitive to the surrounding humidity. For all the calculations reported, the B conformation of DNA has been used [45] which is stable at high relative humidities. It is, therefore, of interest, to also investigate the dependence of the electronic properties also on hydration and ionic atmosphere. In a previous work [46], the influence of the hydration shell (as an external electric field) on the ground state properties of a polyC stack was calculated at a minimal basis level without correlation. Due to the HF approximation and the smaller basis set, these results cannot at the moment be compared directly with the above ones. However, they show the trend that in the presence of water the hole states are shifted by 0.8−0.9 eV to lower energies (in this way improving the agreement of the calculated IP's with experiment). The excitation energies themselves are, however, nearly unchanged in the presence of water. We expect more reliable information concerning this problem from similar calculations with larger basis sets and with the inclusion of correlation.

The hypochromism of native DNA shows significant dependence on the A-T content. Simple interpretations of this observation on the basis of additive contributions from A-T and G-C base pairs resulted in a linear relationship between A-T content and hypochromicity. Since these models did not take the strong interbase interactions into account, they were incapable of explaining important deviations from the linearity. Semiempirical time-dependent Hartree calculations of mixed DNA segments have shown, on the other hand [94—95], that the apparent near linear dependence is a direct consequence of the almost random nearest neighbour frequency distribution of these DNA's. On the other hand deviations from linearity were shown to be connected with non-random nearest neighbor frequencies.

As a first step in investigating these important effects using *ab initio* wave-functions, we calculated the exciton spectra of G-C and A-T homopolynucleotides with the help of minimal basis Wannier functions starting from HF electron and hole states. Though the excitation energies obtained are substantially larger in this case than the ones obtained using the previously discussed more sophisticated procedure, they show an increase in hypochromicity in poly(A-T) as compared to poly(G-C), a trend also observed experimentally. As the next step, we plan to investigate optical properties of native DNA's by calculating the matrix elements of Green's function and of the electron e-h interaction with the help of methods worked out for aperiodic solids.

6. Interaction of Polymers with an External Electromagnetic Field

Absorption of radiation means energy transfer from the external electromagnetic field to internal elementary excitations of the polymer. As a first step in this process, the photon field will have to be coupled to the electronic polarization waves of the solid, i.e. to the excitons. For the absorption to actually take place, the exciton-photon interaction should, however, not be stronger than the coupling between excitons and further lattice polarization modes (phonons) in the solid. Otherwise, the probability that photons will be re-emitted is too high and, instead of decaying into internal vibrational modes, the energy absorbed by the exciton will be returned to the radiation field.

The dispersion of exciton energy bands, even in polymers with strongly interacting units, is only a few eV, i.e. very weak on the scale of the photon dispersion. We, therefore, expect significant exciton-photon coupling only in the neighbourhood of the lower edge of the Brillouin zone (BZ). At wave vectors, however, where energy and momentum are comparable for both particles, the coupling will be so strong that neither can be regarded as an independent elementary excitation. Instead, exciton and photon merge into a single new particle, the *polariton*, whose further interaction with lattice vibrational modes represents the second step in the absorption process.

Experimentally, surface polaritons (SP's) have been observed and studied on a variety of materials. The microscopic mechanism leading to polariton formation may be very different for different types of solids. While optical phonon SP's are quite common on inorganic insulators, plasmon SP's have been detected on several metals. Both of them can occur on certain semiconductor surfaces, sometimes even strongly coupled in the same frequency region [96]. For organic solids, which are in their electronic structure very similar to biopolymers, SP's have been studied by attenuated total reflection measurements, most extensively in the case of anthracene [97], tetracyano-quinodimethane (TCNQ) [98, 99], the cationic cyanine dye crystal CTIP [100, 103] and for the PDA crystal poly-2,4-hexadiyn-1,6-diol bis(p-toluene sulfonate), PTS [100].

Results of measurements have mostly been analyzed in terms of phenomeno-logical models treating the solid as a continuum with complex refractive index [99, 100] and representing the dielectric function by a sum of Lorentz oscillators whose parameters (transition energies, half widths and oscillator strengths) were derived from a Kramers-Kronig analysis of reflectivity data [103]. Microscopic treatments for molecular crystals were proposed within the framework of the Frenkel exciton model [104–106], using semiempirical Hamiltonians to represent the matter field.

The method presented here extends previous work in two respects: First, we solve the eigenvalue problem of the interacting exciton-photon system going beyond the Frenkel approximation. For this purpose, we apply the Green's function formalism of the charge-transfer exciton theory supplemented by terms representing the radiation field. Second, we calculate all excitonic effects from first principles, using *ab initio* Bloch and Wannier functions for one-electron orbitals and including electronic correlation effects on the spectrum with the help of the electron polaron model and of Møller-Plesset perturbation theory.

6.1. FIRST PRINCIPLES FORMALISM OF EXCITON-PHOTON INTERACTION: POLARITON STATES IN POLYMERS

In the coupling region, where the electromagnetic field cannot be treated as a weak perturbation, its interaction with the electronic excitations of the solid has to be included in zeroth order. To calculate the energy spectrum of the new particles evolving from this interaction, we will have to diagonalize the Hamiltonian of the interacting electron-radiation field system:

$$\hat{H} = \hat{H}_{el} + \hat{H}_{ph}. \tag{93}$$

The electronic part, $\hat{H}_{el}$, is again expanded in terms of the field operators but now including the electromagnetic field as well:

$$\hat{H}_{el} = \int d^3x \psi^\dagger(x) \left[-\frac{1}{2} \left(\frac{1}{i} \nabla - \frac{1}{c} \mathbf{A} \right)^2 + V_p(x) \right] \psi(x) +$$

$$+ \frac{1}{2} \int d^3x \int d^3x' \psi^\dagger(x) \psi^\dagger(x') \frac{1}{|x - x'|} \psi(x') \psi(x). \tag{94}$$

A stands here for the vector potential of the electromagnetic field containing only transverse photons ($\nabla \mathbf{A} = 0$). It can be expanded in terms of the photon operators b and $b^\dagger$, obeying Bose statistics:

$$\mathbf{A} = \sum_{\mathbf{k}, j} \left(\frac{2\pi c^2}{V \omega_{\mathbf{k}, j}} \right)^{1/2} \{ \mathbf{e}_{\mathbf{k}, j} \exp[i\mathbf{k}\mathbf{r}] b_{\mathbf{k}, j} + \mathbf{e}_{\mathbf{k}, j} \exp[-i\mathbf{k}\mathbf{r}] b^\dagger_{\mathbf{k}, j'} \}, \qquad (95)$$

where $\omega_{\mathbf{k}, j}$ stands for the frequency of the photon with momentum $\mathbf{k}$ and polarization index j; $\mathbf{e}_{\mathbf{k}, j}$ is the corresponding polarization vector and V is the normalization volume. In terms of the b's, the Hamiltonian of the free photon field has the form

$$\hat{H}_{ph} = \sum_{\mathbf{k}, j} \omega_{\mathbf{k}, j} b^\dagger_{\mathbf{k}, j} b_{\mathbf{k}, j}. \qquad (96)$$

Excluding non-linear optical phenomena from the present calculations by the neglect of the $\mathbf{A}^2$-term in Eq. (94) and taking into account the use of Coulomb gauge, we may identify the term responsible for the electron-photon interaction:

$$\hat{H}_{el, ph} = \int d^3x \psi^\dagger(x) \left[-\frac{1}{ic} \mathbf{A}(x)\nabla \right] \psi(x). \qquad (97)$$

To be able to evaluate the matrix elements of this operator for a polymer, we again expand the fermion operators using as a basis the complete orthonormal set of Wannier spin orbitals and write the exciton wave function in the form

$$\Psi_{s, \mathbf{K}} = N_c^{-1/2} \sum_{\ell} \exp(i\mathbf{K} \cdot \mathbf{R}_\ell) a^\dagger_{\ell+s} d^\dagger_\ell \Phi_{HF} \equiv B^\dagger_{s, \mathbf{K}} \Phi_{HF}, \qquad (98)$$

defining the creation operator for an electron-hole (e-h) pair with separation s, $B^\dagger_{s, \mathbf{K}}$. The wave function of an exciton with momentum $\mathbf{K}$ is again formed as a linear combination of e-h functions with different separation:

$$\Psi_{\mathbf{K}} = \sum_{s} \Omega_{s, \mathbf{K}} \Psi_{s, \mathbf{K}} \equiv B^\dagger_{\mathbf{K}} \Phi_{HF}. \qquad (99)$$

The central problem of the present calculations is the evaluation of the exciton-photon interaction term. To proceed, we substitute expansions (95) and (67) for the electromagnetic and fermion fields, respectively, into Eq. (97). Four of the eight terms resulting from the threefold operator product $(d + a^\dagger)(b + b^\dagger)$ $(d^\dagger + a)$ contain two electron or two hole operators, respectively. They describe Compton-type scattering processes which are not relevant for our study. Two of the four remaining exciton-photon interaction terms describe scattering processes, in which an exciton and a photon are simultaneously created or annihilated. They violate energy conservation and would contribute only in higher orders of perturbation theory.

Before writing down the final form of the interaction Hamiltonian, we make two further, physically reasonable, approximations.

(i) As mentioned earlier, the momentum of the photon capable of efficiently interacting with the exciton is certainly much smaller than the size of the BZ, $k \ll \pi/a$, i.e. the wavelength of the absorbed photon is much larger than the lattice period, $\lambda_{ph} \ll a$. Consequently, terms of the type $\exp(i\mathbf{kr})$ in Eq. (94) can be substituted by $\exp(i\mathbf{kR}_\ell)$.

(ii) The electronic properties of most polymers are strongly anisotropic; absorption is usually most efficient along the chains containing delocalized electrons. It, therefore, seems to be reasonable to simplify our treatment by restricting the calculations to one photon branch polarized parallel to the polymer axes, $\mathbf{e}_{\mathbf{k},j} \rightarrow \mathbf{e}_{\mathbf{k}}$.

Finally, making use of the periodic symmetry, integrals of the type $\int w(\ell)w(\ell')$ can be handled as depending only on $|\ell' - \ell|$, and the interaction term is obtained in the form

$$\hat{H}_{exc,\,ph} = \sum_{\mathbf{k}} \sum_{s} \{ \gamma_{s,\,\mathbf{k}}^{c,\,v} B_{s,\,\mathbf{k}}^\dagger b_{\mathbf{k}} + \gamma_{s,\,\mathbf{k}}^{v,\,c} B_{s,\,\mathbf{k}} b_{\mathbf{k}}^\dagger \}. \tag{100}$$

The coupling constants $\gamma_{s,\,\mathbf{k}}^{j,\,j'}$ are defined by

$$\gamma_{s,\,\mathbf{k}}^{j,\,j'} = - \left(\frac{2\pi}{v_c \omega_{\mathbf{k}}} \right)^{1/2} \mathbf{e}_{\mathbf{k}} \int d^3 r w_j^*(\mathbf{r}) \frac{1}{i} \nabla w_{j'}(\mathbf{r} - \mathbf{R}_s), \tag{101}$$

where $v_c = V/N_c$ is the volume of the unit cell. Analogously to Eq. (95), the total energy of the exciton field can be written as

$$\hat{H}_{exc} = \sum_{\mathbf{k}} E_{\mathbf{k}} B_{\mathbf{k}}^\dagger B_{\mathbf{k}} = \sum_{\mathbf{k}} \sum_{s} \sum_{s'} \nu_{\mathbf{k}} \Omega_{s,\,\mathbf{k}}^* \Omega_{s',\,\mathbf{k}} B_{s,\,\mathbf{k}}^\dagger B_{s',\,\mathbf{k}}. \tag{102}$$

Now collecting the Hamiltonians from Eqs. (95), (100) and (101), the total energy of the interacting exciton-radiation system is obtained in the form

$$\hat{H} = \hat{H}_{exc} + \hat{H}_{ph} + \hat{H}_{exc,\,ph} \equiv \sum_{\mathbf{k}} \hat{h}_{\mathbf{k}}, \tag{103}$$

with

$$\hat{h}_{\mathbf{k}} = \sum_{s} \sum_{s'} \nu_{\mathbf{k}} \Omega_{s,\,\mathbf{k}}^* \Omega_{s',\,\mathbf{k}} B_{s,\,\mathbf{k}}^\dagger B_{s',\,\mathbf{k}} + \omega_{\mathbf{k}} b_{\mathbf{k}}^\dagger b_{\mathbf{k}} +$$

$$+ \sum_{s} \gamma_{s,\,\mathbf{k}}^{c,\,v} B_{s,\,\mathbf{k}}^\dagger b_{\mathbf{k}} + \sum_{s} \gamma_{s,\,\mathbf{k}}^{v,\,c} B_{s,\,\mathbf{k}} b_{\mathbf{k}}^\dagger. \tag{104}$$

To facilitate the diagonalization of $\hat{h}_k$, we build from the operator set $\{B, b\}$ the vector $\mathbf{C}$ defined by

$$C_i(\mathbf{k}) = B_{i-\ell-M,\mathbf{k}} \quad \text{for} \quad 1 \leqslant i \leqslant 2M+1,$$
$$C_i(\mathbf{k}) = b_{\mathbf{k}} \quad \text{for} \quad i = 2M+2, \tag{105}$$

where M denotes the radius of the exciton (i.e. charge transfer excitation components will be included from the reference cell, $M = 0$, to neighboring cells $-M$, $-M+1$, ... , $M-1$, M). Similarly, from the coefficients of the corresponding operators in Eq. (104) we construct the matrix $\mathbf{A}$:

$$A_{i,j}(\mathbf{k}) = \nu_{\mathbf{k}} \Omega_{i-1-M,\mathbf{k}} \cdot \Omega_{j-1-M,\mathbf{k}} \quad \text{for} \quad 1 \leqslant i, j \leqslant 2M+1.$$
$$A_{i,j}(\mathbf{k}) = \gamma^{c,v}_{i-1-M,\mathbf{k}} \quad \text{for} \quad 1 \leqslant i \leqslant 2M+1; \quad j = 2M+2,$$
$$A_{i,j}(\mathbf{k}) = \gamma^{v,c}_{j-1-M,\mathbf{k}} \quad \text{for} \quad 1 \leqslant j \leqslant 2M+1; \quad i = 2M+2, \tag{106}$$
$$A_{i,j}(\mathbf{k}) = \omega_{\mathbf{k}} \quad \text{for} \quad i, j = 2M+2.$$

It is easy to verify using a term-by-term comparison that, expressed with the help of these quantities, operator $\hat{h}$ takes the equivalent form

$$\hat{h}(\mathbf{k}) = \mathbf{C}^\dagger(\mathbf{k})\mathbf{A}(\mathbf{k})\mathbf{C}(\mathbf{k}). \tag{107}$$

In seeking the eigenvalues of $\hat{h}$, we first diagonalize $\mathbf{A}$:

$$\mathbf{A}(\mathbf{k}) \cdot \mathbf{U}(\mathbf{k}) = \mathbf{U}(\mathbf{k}) \cdot \Lambda(\mathbf{k}), \tag{108}$$

where $\Lambda(\mathbf{k})$ is a diagonal matrix and $\mathbf{U}(\mathbf{k})^\dagger = \mathbf{U}(\mathbf{k})^{-1}$. The transformation of our coefficient vector $\mathbf{C}$ with this unitary matrix,

$$\mathbf{D}(\mathbf{k}) = \mathbf{U}(\mathbf{k})\mathbf{C}(\mathbf{k}), \tag{109}$$

provides us with the eigenvector matrix of $\hat{h}$:

$$\begin{aligned}
\hat{h}(\mathbf{k}) &= \mathbf{D}(\mathbf{k})^\dagger \mathbf{U}(\mathbf{k})\mathbf{A}(\mathbf{k})\mathbf{U}(\mathbf{k})^\dagger \mathbf{D}(\mathbf{k}) \\
&= \mathbf{D}(\mathbf{k})^\dagger \Lambda(\mathbf{k})\mathbf{D}(\mathbf{k}) \\
&= \sum_{i=1}^{2M+2} \Lambda_{i,i}(\mathbf{k})\mathbf{D}_i(\mathbf{k})^\dagger \mathbf{D}_i(\mathbf{k}).
\end{aligned} \tag{110}$$

In this way we found new quasiparticles, the polaritons, corresponding to the eigenstates of the interacting exciton-photon field. Their energy spectrum is given throughout the BZ by $\Lambda_{i,j}(\mathbf{k})$.

As we can see from Eq. (104), four quantities are needed for the construction of the Hamiltonian of the coupled exciton-photon field, $\hat{h}_k$: (i) The frequency dispersion of the excitons, $\nu_{\mathbf{k}}$, (ii) the weight factors of e-h pairs with different separation in the total exciton wavefunction, $\Omega_{s,\mathbf{k}}$, (iii) the dielectric constant of the crystal, ε, entering the photon dispersion, $\omega_{\mathbf{k}}$, (iv) and the coupling constants $\gamma_{s,\mathbf{k}}^{c,v}$ and $\gamma_{s,\mathbf{k}}^{v,c}$, respectively. The exciton frequencies throughout the BZ can be obtained from Eq. (87).

For the dielectric constant of the crystal, ε, we could only use in these calculations the plausible value of 3. The exciton energies were calculated previously [63] as functions of ε, and they showed a slow variation of a few tenths of an eV around $\varepsilon = 3$. We expect, therefore, that the accuracy of this quantity will not substantially influence predictions concerning the polariton spectrum. Calculations, however, to elucidate the microscopic dielectric function, $\varepsilon(\mathbf{r})$, including local field effects will be needed in the future.

The most difficult part of these calculations is the evaluation of the exciton-photon coupling constants $\gamma_{s,\mathbf{k}}^{j,j'}$. As the first step in our exciton calculations, ab Initio Bloch functions were calculated and the Wannier orbitals were obtained by a variationally optimized Fourier transformation from the Bloch orbitals. It is convenient, therefore, to now return to the Bloch basis again by using the inverse of the transformation in Eq. (35) and substitute these expansions for the Wannier functions in Eq. (101). The resulting lattice sums are again easily evaluated making use of the translational symmetry, and the only remaining problem is the computation of integrals of the type $\int \phi_j \nabla \phi_{j'}$. For them, we applied the well known procedure of molecular physics used to calculate transition moments [107]. The approximation, involved in this step is the substitution of the exact Hamiltonian by the sum of the Fock operators. The computation of the remaining matrix elements of the coordinate operator is, finally, a standard step in our crystal orbital program.

The eigenvalues of the Hamiltonian $\hat{h}_{\mathbf{k}}$ were computed only at the lower edge of the Brillouin zone of the backbone of the polydiacetylene PTS where significant coupling can be expected. The alternating bond structure and its conjugated π-electron system make PTS into a very interesting model system, also for biopolymer studies. On the other hand, due to its very precisely known structure, it is much better accessible for detailed experimental investigations. In fact, from the point of view of optical properties, it is one of the best characterized polymer crystals. Figure 13 shows the two polariton bands obtained resulting from the lowest singlet exciton of PTS, in the quasi-momentum region of 0.005 to 0.015 π/a, where the eigenstates of the matter + radiation field significantly differ from those of the non-interacting fields. The second step of the absorption process, the interaction of these polariton states with bulk phonon modes, is presently under investigation in our laboratory. Future extension of this work will also include the corresponding surface states. Similar calculations are also in progress for polyene and for polypeptides. Finally, it should be mentioned that the theoretical method for the treatment of matter-field interaction, presented in this paragraph can easily be applied to the first principles investigation of ORD and CD spectra in biopolymers. Work along these lines is also in progress in our laboratory.

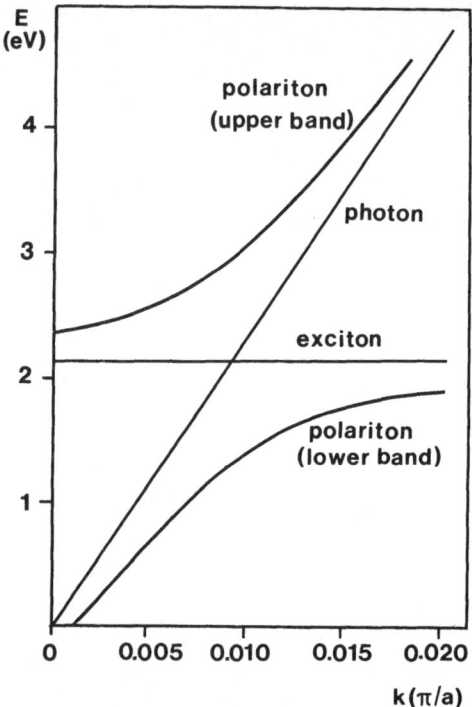

Fig. 13. Polariton dispersion resulting from the interaction of an incoming photon with the lowest-lying singlet exciton in the polydiacetylene model PTS.

References

1. Del Re, J. Ladik, and G. Biczo: *Phys. Rev.* **155**, 967 (1967).
2. J. M. André, L. Gouverneur, and G. Leroy: *Int. J. Quant. Chem.* **1**, 427 (1976); *Int. J. Quant. Chem.* **1**, 451.
3. F. E. Harris and H. J. Monkhorst: *Phys. Rev. Lett.* **23**, 1026 (1969); *Phys. Rev.* **B2**, 4400 (1970).
4. J. Ladik and S. Suhai, in H. Ratajczak and W. J. Orville-Thomas (Eds.): *Molecular Interactions*, J. Wiley and Sons, 1980, pp. 151.
5. I. I. Ukrainski: *Theor. Chim. Acta (Berlin)* **38**, 139 (1975).
6. C. Merkel: *Elektronische Eigenschaften von Molekülkristallen*, Thesis, München, 1977.
7. H. J. Monkhorst and M. Kertesz: *Phys. Rev.* **B23**, 3015 (1981).
8. J. M. André, J. L. Bredas, B. Themans, and L. Piela: *Int. J. Quant. Chem.* **23**, 1065 (1983).
9. P.-O. Löwdin: *Rev. Mod. Phys.* **39**, 259 (1967).
10. S. Suhai, P. S. Bagus, and J. Ladik: *Chem. Phys.* **68**, 467 (1982).
11. G. H. Wannier: *Phys. Rev.* **52**, 191 (1937).
12. E. I. Blount: *Solid State Phys.* **13**, 305 (1963).
13. N. H. March, W. H. Young, and S. Sampanthar: *The Many-Body Problem in Quantum Mechanics*, Cambridge University Press, Cambridge.
14. S. Suhai: *Int. J. Quant. Chem.* **23**, 1239 (1983).
15. S. Suhai: *J. Polym. Sci. Polym. Phys. Ed.* **21**, 1341 (1983).
16. S. Suhai: *Phys. Rev.* **B27**, 3506 (1983).
17. S. Suhai: *J. Chem. Phys.* **84**, 5071 (1986).

18. C. Møller and M. S. Plesset: *Phys. Rev.* **46**, 618 (1934).
19. Y. Toyozawa: *Progr. Theor. Phys.* **12**, 422, Kyoto (1954).
20. S. T. Pantelides, D. J. Mickish, and A. B. Kunz: *Phys. Rev.* **B10**, 2602 (1974).
21. B. Hudson and B. Kohler: *Ann. Rev. Phys. Chem.* **25**, 437 (1974).
22. S. Basu: *Advances in Quantum Chemistry*, P. O. Löwdin, Ed., Academic Press, New York, 1978, pp. 33.
23. C. K. Chiang *et al.*: *Phys. Rev. Lett.* **39**, 1098 (1977).
24. J. Paldus and J. Cizek: *Phys.Rev. A* **2**, 2268 (1970).
25. M. Takahashi, J. Paldus, and J. Cizek: *Int. J. Quantum Chem.* **24**, 707 (1983).
26. C. B. Duke *et al.*: *Chem. Phys. Lett.* **59**, 146 (1978).
27. A. A. Ovchinnikov, I. I. Ukrainsky, and F. Kventzel: *Usp. Fis. Nauk (USSR)* **108**, 81 (1972).
28. N. Tyutyulkov, I. Kanev, O. Castano, O. Polansky, and H. Barentzen: *Theor. Chim. Acta (Berlin)* **55**, 207 (1980).
29. J. W. Mintmire and C. T. White: *Bull. Am. Phys. Soc.* **27**, 313 (1982).
30. M. André and G. Leroy: *Chem. Phys. Lett.* **5**, 557 (1971).
31. M. Kertesz, J. Koller, and A. Azman: *Chem. Phys. Lett.* **5**, 557 (1971).
32. A. Karpfen and J. Petkov: *Solid State Comm.* **29**, 251 (1979).
33. S. Suhai: *J. Chem. Phys.* **73**, 3843 (1980).
34. S. Suhai: *Chem. Phys. Lett.* **96**, 619 (1983).
35. S. Suhai: *Phys. Rev. B* **27**, 3506 (1983).
36. R. Fincher *et al.*: *Phys. Rev. B* **20**, 1589 (1979).
37. P. S. Bagus, J. Pacansky, and W. Wahlgren: *J. Chem. Phys.* **67**, 619 (1977).
38. For a review see: B. Hudson and B. Kohler: *Ann. Rev. Phys. Chem.* **25**, 437 (1974).
39. S. Etemad, A. J. Heeger, L. Lanchlan, T.-C. Chung, and A. G. MacDiarmid: *Mol. Cryst. Liq. Cryst.* **77**, 43 (1981).
40. W. R. Salaneck, H. W. Gibson, E. W. Plummer, and B. H. Tonner: *Phys. Rev. Lett.* **49**, 801 (1982).
41. J. F. Yan, G. Vanderkooi, and H. Scheraga: *J. Chem. Phys.* **49**, 2713 (1968).
42. S. Suhai, C. Merkel, and J. Ladik: *Phys. Lett.* **61A**, 487 (1977).
43. J. Ladik, S. Suhai, P. Otto, and T. C. Collins: *Int. J. Quant. Chem.* **4**, 55 (1977).
44. J. Ladik and S. Suhai: *Int. J. Quant. Chem.* **7**, 181 (1980).
45. S. Arnott, S. D. Dover, and A. J. Wonacott: *Acta Cryst.* **B28**, 2198 (1964).
46. P. Otto, J. Ladik, G. Corongiu, S. Suhai, and W. Förner: *J. Chem. Phys.* **77**, 5026 (1982).
47. S. Suhai and B. Reiner: to be published.
48. E. Clementi: *Computational Aspects for Large Chemical Systems*, Lecture Notes in Chemistry, Springer, Berlin, 1980.
49. G. Corongiu and E. Clementi: *Biopolymers* **20**, 551 (1981).
50. S. Suhai: *Int. J. Quant. Chem. Quant. Biol. Symp.* **11**, 223 (1984).
51. Y. Takeuti: *Prog. Theor. Phys.* **18**, 421 (1957).
52. M. Lax: *Phys. Rev.* **94**, 1391 (1954).
53. G. F. Koster and J. C. Slater: *Phys. Rev.* **95**, 1167 (1954).
54. M. Altarelli and B. F. Bassani: *J. Phys. C* **4**, L328 (1971).
55. D. Baeriswyl, G. Harbeke, H. Kiess, E. Meier, and W. Meyer: *Physica* **117B & 118B**, 617 (1983).
56. S. Etemad, A. J. Heeger, L. Lauchlan, T.-C. Chung, and A. G. MacDiarmid: *Mol. Cryst. Liq. Cryst.* **77**, 43 (1981).
57. J. J. Ritsko, E. J. Mele, A. J. Heeger, A. G. MacDiarmid, and M. Ozaki: *Phys. Rev. Lett.* **44**, 1351 (1981).
58. J. J. Ritsko: *Phys. Rev. B* **26**, 2192 (1981).
59. J. A. Pople and S. H. Walmsley: *Trans. Faraday Soc.* **58**, 442 (1962).
60. D. Pugh: *Mol. Phys.* **26**, 1297 (1973).
61. N. A. Cade: *Chem. Phys. Lett.* **53**, 45 (1978).
62. M. Kertesz: *Chem. Phys.* **44**, 349 (1979).
63. S. Suhai: *Phys. Rev. B* **29**, 4570 (1984).
64. M. R. Philpott: *Chem. Phys. Lett.* **50**, 18 (1977).

65. D. R. Yarkony: *Chem. Phys.* **33**, 171 (1978).
66. S. Suhai: *Int. J. Quantum Chem. Quantum Biol. Symp.* **18**, 161 (1984).
67. L. Sebastian and G. Weiser: *Chem. Phys.* **62**, 447 (1981).
68. L. Sebastian and G. Weiser: *Phys. Rev. Lett.* **46**, 1156 (1981).
69. I. Tinoco, Jr., A. Halpera, and W. T. Simpson: (M. A. Stahmann (Ed.)) *Polyamino Acids, Polypeptides and Proteins*, Univ. of Wisconsin, Madison, 1962, pp. 147.
70. W. C. Johnson, Jr. and I. Tinoco, Jr.: *J. Am. Chem. Soc.* **94**, 4389 (1972).
71. S. Onari: *J. Phys. Soc. Jpn.* **29**, 528 (1970).
72. H. De Voe: *Ann. New York Acad. Sci.* **158**, 298 (1969).
73. I. Tinoco, Jr.: *J. Amer. Chem. Soc.* **82**, 4785 (1960).
74. W. Rhodes: *J. Amer. Chem. Soc.* **83**, 3609 (1961).
75. H. De Voe and I. Tinoco, Jr.: *J. Mol. Biol.* **4**, 5181 (1962).
76. J. P. Ballini, P. Vigny, and M. Daniels: *Biophys. Chem.* **18**, 61 (1983).
77. J. Eisinger, M. Guéron, R. G. Shulman, and T. Yamane: *Proc. Natl. Acad. Sci. USA* **55**, 1015 (1966).
78. K. Imakubo: *J. Phys. Soc. Japan* **24**, 143 (1968).
79. V. Kleinwachter: *Stud. Biophys.* **33**, 1 (1972).
80. J. Koutecky and J. Paldus: *Theor. Chim. Acta (Berlin)* **1**, 368 (1963).
81. T. A. Hoffmann and J. Ladik: *J. Theor. Biol.* **6**, 26 (1964).
82. J. Ladik and K. Sundaram: *J. Mol. Spectr.* **29**, 146 (1969).
83. K. Sundaram and J. Ladik: *Physiol. Chem. and Phys.* **4**, 483 (1972).
84. A. Pullman and B. Pullman: *Adv. Quant. Chem.* **4**, 267 (1970).
85. S. N. Volkov and V. I. Danilov: *FEBS Lett.* **65**, 8 (1976).
86. V. I. Danilov, V. I. Pechenaya, and N. V. Zheltorsky: *Int. J. Quant. Chem.* **17**, 307 (1980).
87. S. Suhai: *Int. J. Quant. Chem.: Quant. Biol. Symp.* **11**, 223 (1984).
88. E. Clementi, J.-M. Andre, M.-Cl. Andre, D. Klint, and D. Hahn: *Acta. Phys. Hung. Acad. Sci.* **27**, 493 (1969).
89. B. Mely and A. Pullman: *Theor. Chim. Acta* **13**, 278 (1969).
90. E. Clementi, J. Mehl, and W. Niessen: *J. Chem. Phys.* **54**, 508 (1971).
91. C. Yu, S. Peng, I. Akiyama, J. Liu, and P. R. LeBreton: *Chem. Soc.* **100**, 2303 (1978).
92. S. Suhai: *Biopolymers*, to be published.
93. L. B. Clark and I. Tinoco: *J. Amer. Chem. Soc.* **87**, 11 (1965).
94. E. Brown and E. S. Pysh: *J. Chem. Phys.* **56**, 31 (1972).
95. E. S. Pysh and J. L. Richards: *J. Chem. Phys.* **57**, 3680 (1972).
96. For reviews, see: A. Otto: *Optical Properties of Solids, New Developments*, edited by B. O. Seraphin, North-Holland, New York, 1976, pp. 677—724.
97. K. Tomioka, M. G. Sceats, and S. A. Rice: *J. Chem. Phys.* **66**, 2984 (1977).
98. M. G. Sceats, K. Tomioka, and S. A. Rice: *J. Chem. Phys.* **68**, 4486 (1977).
99. M. R. Philpott, P.M. Grant, K. Syassen, and J.-M. Turlet: *J. Chem. Phys.* **67**, 4229 (1977).
100. M. R. Philpott and J. D. Swalen: *J. Chem. Phys.* **69**, 2912 (1978).
101. M. Orrit, J. Bernard, J.-M. Turlet, and Ph. Kottis: *J. Chem. Phys.* **78**, 2847 (1983).
102. J. Bernard, M. Orrit, J.-M. Turlet, and Ph. Kottis: *J. Chem. Phys.* **78**, 2857 (1983).
103. G. Weiser, W. Fuchs, and H. J. Hesse: *Chem. Phys.* **52**, 183 (1980).
104. J. J. Hopfield: *Phys. Rev.* **112**, 1555 (1958).
105. V. M. Agranovitch: *Zh. Eksp. Teor. Fiz. Piz'ma Red.* **37**, 340 (1959).
106. M. Orrit, C. Aslangul, and Ph. Kottis: *Phys. Rev. B* **25**, 7263 (1982).
107. J. Lindenberg and Y. Öhrn: *Propagators in Quantum Chemistry*, Academic Press, London, 1973.

The Problem of Protonation in Rhodopsin and Model Schiff Bases

C. SANDORFY
Département de Chimie, Université de Montréal, Montréal, Québec, Canada, H3C 3J7

and

D. VOCELLE
Département de Chimie, Université du Québec à Montréal, Montréal, Québec, Canada, H3C 3P8

1. Introduction

This article is concerned with the protonation of non-aromatic Schiff bases (SB). Among these the most important in nature are without any doubt the chromophores of visual and bacterial Rh. The chromophore of visual pigments is the SB of 11-*cis*-retinal in which the oxygen of the aldehyde is replaced by a nitrogen belonging to an amino-acid residue, lysine, of the surrounding protein, called opsin (Figure 1a). So the SB is covalently bound to the protein. The chromophore is a conjugated system containing five C=C and one C=N double bonds. There is only a minor exception to this: some species have an additional C=C bond in their

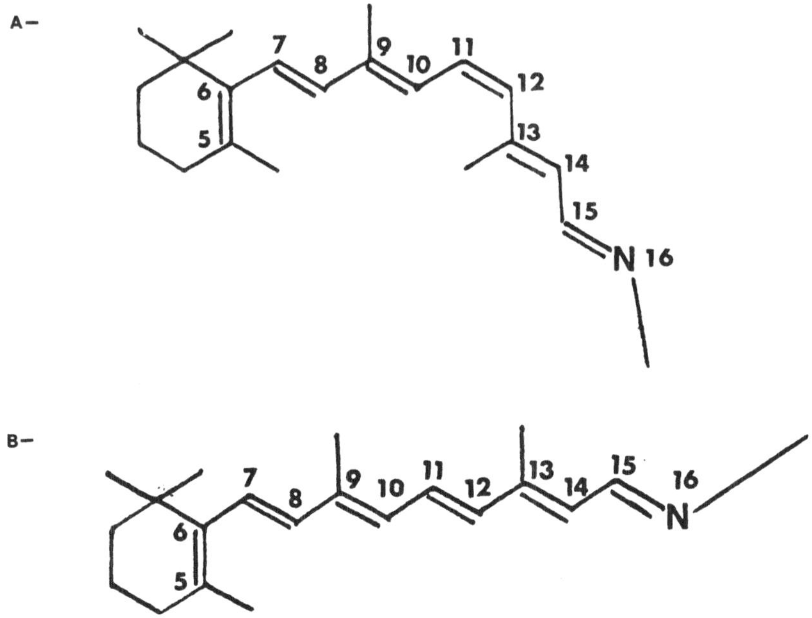

Fig. 1. The chromophores of visual and bacterial pigments: (a) 11-*cis*-retinal Schiff base of rhodopsin, (b) all-*trans*-retinal Schiff base of light-adapted bacteriorhodopsin.

Jean Maruani (ed.), Molecules in Physics, Chemistry, and Biology, Vol. IV, 195—211.
© 1989 *by Kluwer Academic Publishers.*

conjugated system (Porphyropsin). The chromophore of (light-adapted) bacterior-hodopsin (bR) which has been found in the purple membrane of *halobacterium halobium* is the SB of all-*trans*-retinal (Figure 1b). Other bacterial Rh became known in recent years. The function of bacterial Rh is to use light energy to push a proton or an other ion across a cell membrane against a chemical potential gradient. Membranes have an essential role to play in living organisms. The degree of permeability to a variety of ions controls many vital functions. In all likelihood a stimulus to the nervous system consists in changing the permeability of the neuronal membrane. Photon absorption by visual pigments also leads, eventually, to changing the permeability of the surrounding membrane to (sodium) ions thereby giving a signal to the central nervous system.

For all types of Rh photon absorption is followed by isomerization of the chromophore and concomitant changes in its environment. After five or six stages the visual chromophore of vertebrates is hydrolyzed down from the protein and is later restored chemically. For the visual chromophore of invertebrates and bacterial Rh, at the end of the photochemically induced cycle, the chromophore regains its original configuration and conformation.

There exists a nearly complete consensus to the effect that the SB chromophore is protonated and that both protonation and deprotonation occur along the photochemically induced cycle. The present review is devoted to the problem of protonation in Rh and is limited to the initial stage. It will try to answer the following questions.

(1) Why must protonation be invoked?
(2) What evidence do we have for the protonation of the retinylidene SB in Rh? Are they entirely 100% protonated?
(3) How can they be protonated under the given conditions and the by available acids?
(4) What is the nature and extent of involvement of the molecular environment of the chromophore?

In order to approach these problems we shall have to review the results of a number of authors including a series of works carried out on model systems in our laboratories. Most of the latter are based on nuclear magnetic resonance and Fourier transform infrared studies.

The scope of this paper is limited to the aspects of the protonation problems of Rh. The reader interested in other aspects of the structure and functioning of visual and bacterial pigments might consult some of the excellent recent reviews which were published in this field [1—10].

2. Why Invoke Protonation?

In human or bovine Rh the main intense absorption band is located in the middle of the visible spectrum. Its maximum is at 498 nm; in light adapted bR it is at

568 nm. However, for the retinylidene SB, in a test-tube in a neutral solvent, the maximum is only at 370 nm. Protonation needs to be invoked as a partial explanation for this bathochromic shift. Actually, protonated SB usually absorbs at frequencies lower than non-protonated SB. The reasons for this are not obvious but the following line of thoughts is probably reasonable. As is well known the stabilization that occurs when a proton binds to a nitrogen base is substantial; depending on solvent the difference in Hartree–Fock energy may amount to 100–200 kcal/mole. Then to obtain a bathochromic shift the excited state involved with the transition must come down even more. This is expected, since in the first excited (π, π^*) state the basicity of SB (and other nitrogen bases) is higher than in the ground state. This has been known since Förster's original experimental work [11] and Sandorfy's subsequent quantum chemical calculations [12, 13]. In the case of the retinylidene SB this increase in basicity was demonstrated by the excited state dipole moment measurements of Mathies and Stryer [14] and Corsetti and Kohler [15] and by the determination of the pK_a in the excited state by Becker [16]. However, the shift is a small difference between two large numbers and intuitive arguments and even the results of quantum chemical calculations, should be taken with caution.

Another reason for caution is the mixed character of the (π, π^*) excited state of the longer conjugated Schiff bases. This is easily explained on the example of *trans*-1,3-butadiene (see [17]). Under C_{2h} symmetry the ground state is totally symmetrical, A_g. The first excited state is B_u and the $^1A_g \rightarrow {}^1B_u$ transition is allowed, polarized in the direction containing the central C—C bond. The second and third (π, π^*) excited states have A_g symmetry to which transition from the ground state is forbidden. The intriguing fact is that, in a higher approximation, the two A_g excited states can mix; this brings one of the resulting states (A_g^-) down while the other (A_g^+) shifts upwards [18–20]. If the mixing is high, $^1A_g^-$ might be actually lower than ^1B_u. This does probably not happen in butadiene. It has been shown, however, by two photon spectroscopy that in long chain linear polyenes the lowest excited state is the A_g^- state [21–31]. The order of the states is solvent dependent. All this applies to all-*trans*-retinal and its SB [31]. Now, in the absence of a center of symmetry the A_g^--like and B_u-like states can actually mix to an extent which is also solvent dependent. Birge et al. [4, 30] demonstrated that protonation of the SB in retinal SB analogs produces a reversal so that ^1B_u becomes again the lowest. All this shows that when we try to understand the bathochromic shift due to protonation the mixing of the lowest excited states cannot be disregarded. This certainly applies to SB with four or more double bonds.

UV-VIS studies were carried out on model systems by Blatz and coworkers [31–33]. They used a SB of all-*trans*-retinal and a large number of strong acids (usually mineral). They found a most important effect: the distance between the azomethine nitrogen and the anion controls the λ_{max} of the cation in a non-polar solvent. Another way of viewing this effect is that the λ_{max} increases as the

N^+H---X^- H-bond weakens. Since in the pigments a bathochromic shift occurs in the primary step, one can venture that in bathorhodopsin and in vitamin K, the H-bond between the protonated Schiff base and its counter-ion has weakened considerably or perhaps completely severed. Based on this concept Honig and coworkers [34] have proposed for bathorhodopsin and intermediate K, a structure where the proton-bridge is completely broken and consequently, where a charge separation has been created by a geometric change in the chromophore. In this way a significant amount of energy is stored in the chromophore by charge separation alone in a low dielectric medium. In order for this model to be credible, at least two conditions should be known: what is the polarity of the milieu surrounding the chromophore and what kind of a proton-bridge is really present in the pigments? Rafferty and Shichi [35] and Hildebrandt and Stockburger [36] have shown that water molecules are present near the chromophore. Warshel and Barboy [37] have indicated that a low dielectric medium was unlikely for both Rh and bR. Bissonnette et al. [38—39] have used model SB and UV spectroscopy to show that weak acids having pK_as similar to glutamic or aspartic acids could only protonate partially a conjugated SB in a non-polar solvent. From these observations it might be concluded that the environment surrounding the chromophore must be polar in order to account for the water molecules found near it and if so, full protonation by weak acids could be expected. On the contrary, if the chromophore is in a non-polar environment, then the water molecules must be located at some distance from the chromophore and protonation would not be complete.

One might ask however: is protonation really required? Actually, the observed bathochromic shift could be obtained without protonation. For this we have to remember that whenever a nitrogen base is protonated the proton remains hydrogen bonded to the "original" proton donor. This entails a tautomeric equilibrium of the type:

$$—CH=N^+—H---:X^- \rightleftharpoons —CH=N:---H—X. \qquad (1)$$

In the case of Rh, X is probably an amino-acid like aspartic or glutamic acid, or tyrosine or arginine. Now, the bathochromic shift can be obtained by 'just a hydrogen bond' as on the right side of Eq. 1, provided that another polar group (or even more than one) is put near the SB chromophore. Leclercq's approximate quantum chemical calculations (Favrot et al. [40]) have shown that this is possible by placing a proton donor (X) in front of the nitrogen and a carboxylate ion close to the β-ionone ring at C_5, at about 3.1 Å or closer. The existence of these counter-ions must be assumed anyway: even full protonation would only account for a *part* of the observed bathochromic shift. What we mean saying at this point is that with two or more counter-ions the shift could be explained even without protonation. We do not imply that this is so in reality; all we intend to say is that the bathochromic shift is not in itself a proof for protonation. It gave, however, the starting impetus for an evolution of ideas which proved to be important for the understanding of the functioning of Rh.

3. What Evidence Do We Have that the Chromophore Is Protonated?

It would be natural to turn first to infrared spectroscopy to secure information on the state of protonation of the chromophore in Rh. Unfortunately, conditions are not favorable for this. In the pigments the retinylidene SB is surrounded by a protein, lipids, etc., full of functional groups which strongly absorb in the infrared. In particular, this prevents us from identifying the bands due to the N^+H stretching and bending and the $C{=}N$ stretching vibrations which are most directly affected by protonation. It is, however, possible to carry out infrared studies on model SB which are not subject to conditions prevailing in pigments.

Let us start with a simple non-conjugated SB, isobutylidene isopropylamine $(CH_3CHCH{=}NCH(CH_3)_2$, a model which was chosen by Favrot et al. [41]. The branched substituents stabilize this molecule to a large extent. In chloroform solution the free (non hydrogen bonded) $\nu(C{=}N)$ band is at 1664 cm^{-1}. In presence of an acid in excess, $(CF_3)_3COH$ with a pK_a value similar to that of acetic acid, this frequency goes up to 1704 cm^{-1} indicating protonation; this is corroborated by the low value of the broad OH band at 2530 cm^{-1}. The band at 1664 cm^{-1} reappears if the imine is in excess showing the delicate balance between protonation and non-protonation. The equilibrium indicated by these frequencies is clearly:

$$-CH{=}N\text{:---}HOR \rightleftarrows -CH{=}N^+H\text{---}{}^-OR. \tag{2}$$

So protonation is characterized by a high frequency shift of $\nu(C{=}N)$, about 40 cm^{-1} in this case and a low N^+H stretching frequency (a 'free' N^+H frequency would be around 3300—3200 cm^{-1} [42]). Other, similar, results were given in [41]; using picric acid, a strong acid, confirmed the above observations. The likely proton donor in Rh being a carboxyl group of an amino-acid Lussier et al. [43] examined, more recently, the extent of protonation of another non-conjugated SB by a series of acids, trichloro, dichloro, monochloro and monobromo acetic acids and propionic acid whose pK_a values are 0.66, 1.26, 2.85, 2.90 and 4.87, respectively. They found the $\nu(C{=}N^+)$ bands between 1708 and 1712 cm^{-1}. However, the $\nu(C{=}N)$ band of the non-protonated imine (1667 cm^{-1}) was also present in all cases. From the respective band areas it could be estimated that the percentage of protonation is, in the above order, about 90—95%, 80—90%, 75—55% and 10—20%. (Monochloro and monobromo acetic acids gave practically the same results). Moreover, the 2700—2000 cm^{-1} region of the spectra exhibits continuous absorption which is weak for the strongest and the weakest of the acids used but pronounced for the middle members of the series. This indicates that the proton is close to the midpoint of the N^+---H---O^- bridge. This type of system has been thoroughly studied by Zundel and his coworkers [44—46]. These infrared data confirm the existence of an equilibrium of the type represented by Eq. 1. It is important to recall in this respect that aspartic and glutamic acids have pK_a values equal to 3.86 and 4.25, respectively.

Lussier et al. [47—48] extended this investigation to SB conjugated with one, two or five C=C double bonds where the last one contains the same conjugated chain as the retinylidene SB. The basicity of the SB increases as the degree of conjugation increases. This is shown by the increasing percentage of protonation. For trans-retinal it is 100% for trichloroacetic acid, 80—90% for monochloro-acetic acid, 60—70% for 3-chloro-propionic acid and 10—20% for propionic acid. 3-chloropropionic acid has a pK_a value of 3.99, very close to that of aspartic acid. As expected, the $v(C=N)$ stretching frequencies go downwards with increasing conjugation: 1666, 1657, 1641 and 1619 cm^{-1} for no, one, two and five C=C bonds while $v(C=N^+)$ frequencies for the protonated imines are always higher: 1702, 1675, 1668 and 1646 cm^{-1} for protonation with HCl and 1712, 1685, 1679 and 1658 cm^{-1} for protonation with monochloroacetic acid.

Resonance Raman spectroscopy is an elegant way of extracting the spectrum of the chromophore from that of Rh. Since this technique magnifies, with a high degree of preference the bands due to vibrations which occur in the unit which is primarily affected by the electronic excitation, the chromophore's vibrations appear strongly and those of the protein are eliminated. The wavelengths of the widely used argon ion lasers are right in the desired spectral region. Even so the N$^+$H bands are lost since that group is too polar. However the C=N and C=C stretching bands and a number of C—C stretching and C—H deformation are seen to advantage. In both Rh and bR the $v(C=N^+)$ band is at 1655 cm^{-1} while the $v(C=N)$ band is at 1625 cm^{-1}. This parallels the above mentioned more recent infrared results obtained on model compounds. The Raman spectra of Rh have been the object of intensive research for several years. Instead of reviewing again the results we refer to the extensive works of Smith et al. [61] and Massig et al. [62].

More important than the $v(C=N)$ frequencies themselves is the shift of about 30 cm^{-1} which is observed upon deuteration [49]. It would hardly be expected if the proton was not covalently bound to the nitrogen. The effect on the C=N force constant is, of course, indirect.

In recent years infrared caught up with Raman in Rh research. Highly sophisti-cated differential infrared spectroscopic methods were developed in the labora-tories of Rothschild [50—51], Siebert [52—54] and Eisenstein [55—57]. While this is indeed a very interesting subject, it will be enough to state from the point of view of the present article, that the infrared data coincide with the Raman data as far as can be seen and they indicate protonation for all the Rh which were studied.

But why should the C=N stretching frequency increase upon protonation? Since protonation implies the formation of a new covalent bond on the nitrogen it would be reasonable to expect the opposite. This has been explained by Honig et *al.* [58] and by Aton *et al.* [59] on the basis of normal coordinate calculations. In the normal coordinate corresponding to the frequency of 1655 cm^{-1}, C=N motion is mixed with C=N$^+$H bending and other bending motions. The C=N$^+$H bend is thought to be at about 1350 cm^{-1} [60—62]. The mixing pushes ν(C=N$^+$) up and δ(N$^+$H) down. If the hydrogen is replaced by deuterium, the latter vibration, being a mainly N$^+$H motion moves to about 1000 cm^{-1}, the mixing is suppressed and ν(C=N$^+$) shifts to lower frequencies by about 30 cm^{-1}.

Does all this prove that the SB is protonated? Well, almost. Favrot *et al.* [41] found cases where the ν(C=N) frequencies of the protonated and non-protonated SB differ by only 7 or 9 cm^{-1}. It is a matter of unpredictable mixing which might depend on environmental effects or such other unknown factors as anharmonic coupling between the normal modes. Yet, it is fair to say that a great amount of accumulated Raman and infrared evidence favors protonation in both human and bovine Rh and bR.

Another strong argument for protonation comes from electronic spectroscopy. The fact of $^1B_u/^1A_g^-$ mixing was mentioned in the previous section. As was also mentioned, for the long chain polyenes the $^1A_g^-$-like state is at slightly lower energies than the ^1B_u-like state. This is the case for all-*trans*-retinal and its SB. Now, Birge *et al.* [63] determined the two photon spectra of the protonated SB of all-*trans*-retinal at room temperature and observed the reversal of the two states with the ^1B_u state being the lower. This reversal can be taken as a proof of protonation. Going a step forward, in their laboratories Birge and his coworkers were able to record two-photon spectra of the locked 11-*cis*-Rh prepared by Nakanishi and his coworkers [64]. A ring synthesized onto the central double bond prevents photoisomerization but this molecule fits the site of the chromophore in Rh with the same protein environment. The ^1B_u state was again found to be the one of lower energy that is, the first singlet excited state. As stated above the two states are actually mixed, so what is meant is the mainly ^1B_u-like state. Nevertheless, this experiment strongly indicates that the SB is protonated in Rh.

This does not mean that protonation of the SB is a prerequisite for photo-isomerization. Becker and Freedman [65] determined *cis-trans*-photoisomerization quantum yields for 11-*cis*-retinylidene butylamine, both non-protonated and protonated. The quantum yield for this SB varied from 0.01 in *n*-hexane to 0.34 in acetonitrile while for the protonated SB they found 0.24, practically independent of solvent. The role of protonation is, according to these authors, to promote mixing of the ^1B_u and ^1A_g-like excited states, thereby increasing the quantum yield of photoisomerization.

The existence of UV visual pigments should be noted in this context. Harosi and his coworkers [66] found cone pigments in several fishes which only absorb in the UV or at the limit of the UV and the visible. As an example, in the japanese dace there is one that has its main band at 405—415 nm and another which absorbs at 350—370 nm. These pigments 'bleach', that is they do have a photocycle just like the pigments that absorb well in the visible. Several other species are known which have UV absorbing pigments [67–70]. How are these UV absorbing Rh protonated? The low-lying absorption bands of human, bovine and other Rh led to the original suggestion that the SB chromophore is protonated in Rh. There is enough evidence to believe that generally it is; but maybe sometimes it is not.

NMR spectroscopy is a highly suited technique to investigate the protonation problem. It is a nondestructive method that probes the molecules in their ground state. This is an important asset since we are concerned with highly photosensitive pigments with high quantum yields. Three kinds of nuclei can be studied: ^{13}C and ^{15}N are used for the characterization of pigments while ^1H is highly informative in model studies. ^{13}C NMR can give useful informations on the question of protonation since this technique is sensitive to pH variations, to fields induced by charged groups and to steric interactions. ^{15}N NMR is even more suited since protonation of a Schiff base can lead to shifts of more than 100 ppm [72, 73]. These two techniques require that the chromophore be enriched in ^{13}C or that the protein be enriched in ^{15}N. Pigment NMR spectra are obtained in solutions (with the help of detergents) or in the solid state through lyophilization. Proton NMR is especially useful in model systems since protonation can be surveyed by the chemical shifts of the acidic proton or by its influence on adjoining hydrogens. The three techniques, when used together, are able to give a quite clear picture of protonation.

Works on the pigments (Rh and bR) have been done mainly by two groups: Mateescu et al. initiated the first studies [74–78] soon followed by Harbison et al. [79–83]. Mateescu et al. [71] found the following chemical shifts for Rh solubilized in octyl-β-glucoside: C_{15}, 165.9; C_{14}, 130.0; C_{13}, 168.1; C_9, 148.8 and C_6, 137.0 ppm. For bR, they found: C_{15}, 166.0; C_{14}, 130.0; C_{13}, 168.9; C_9, 148.5 and C_6, 135.6 ppm. In the solid state, they found for bR, C_{15}, 150; C_{13}, 143.7 ppm. This last result indicates that when lyophilisation is applied, depending on the conditions used, quite different results from those obtained in solution are obtained. Mateescu et al. used model compounds as reference points. These studies clearly showed that protonated Schiff bases had values of 121 ppm for C_{14} and 162.6 ppm for C_{13}. For the unprotonated Schiff base, the corresponding values were 130.0 and 145.5 ppm respectively. Compared to the pigment values, it would then appear that the chromophore is unprotonated as based on C_{14} results (pigment: 130; unprotonated Schiff base: 130 ppm) but protonated if C_{13} results are considered (pigment: 168; protonated Schiff base 162 ppm). In the solid-state, results indicate an unprotonated chromophore. For this last case, Mateescu et al. [71] made an important statement: the protonation-deprotonation process depends on the degree of hydration of the sample. It seems that the conditions used are such that lyophilization (which implies the removal of a large quantity of water) causes deprotonation as the water content falls dramatically. Considering the works of Rafferty and Shichi [84] and of Hildebrandt and Stockburger [85] on the presence of water around the chromophore, the study of pigment solid-state NMR could well indicate the involvement of water in the protonation or deprotonation of the chromophore. We have, for a certain number of years [86], advocated a proton transfer relay mechanism involving water. We consider this aspect of the utmost importance in solving the remaining problems in the photochemical transformation of Rh into bathorhodopsin (as well as of bR into K), and this point will be discussed more fully in Section 3 of this paper. In that respect, not only the red shift experienced by bathorhodopsin (or K) has not been fully explained, but the important problem of energy storage also remains to be elucidated. These problems have been reviewed by us recently [87].

In 1981, Yamaguchi et al. [88] studied ^{13}C-enriched bR (at carbons 15 and 14 of retinal) by NMR. They found for C_{15} and C_{14} chemical shifts of 160 and 118 ppm respectively. They also found peaks at 191 and 125 ppm which they ascribed to C_{15} and C_{14} of the free retinal. Since the peak at 125 ppm from TMS appears at about 130 ppm from TSP-d_4 (the internal standard used by Mateescu [71]), they argued that the peak appearing at 130 ppm (from TSP-d_4) is quite probably due to free retinal.

Using extensively solid-state NMR, Harbison et al. have demonstrated that in bR (dark-adapted), the chromophore is protonated. They arrived at this conclusion after comparing their C_{10}, C_{11} and C_{12} labeled chromophore with the chemical shifts of a model protonated Schiff base [80]. In a subsequent study, using this time 14-^{13}C retinyl-labeled bR, they observed a chemical shift of 122.0 ppm which favors a protonated Schiff base. Furthermore, by comparing chemical shift tensors, they arrived at a very important conclusion, dark-adapted bR contains 13-cis, 15-syn and all-trans, 15-anti-retinal derivatives [81]. This represents the first attempt to gain knowledge of the configuration of the C=N bond. We have also discussed the importance of the geometry of this bond in the mechanism of vision [86]; this work of Harbison et al. shows clearly that isomerization does occur in that group. Quite recently, the same group [82] has studied bR samples labeled at C_5, C_6, C_7, C_8, C_9, C_{13}, C_{15} and C_{18}; some were fully hydrated, others were lyophilized. A protonated chromophore was again observed. More importantly perhaps, they were able to show that the chromophore existed in a 6-s-trans-conformation rather than the 6-s-cis-conformation which is favored in solution. They also found indications that a negative-charge is present near C_5 while a positive charge might be present near C_7. These findings represent exciting new developments in the field of photosensitive pigments but the questions of how the charges are stabilized and how the energy is stored are not fully answered.

Another important question that needs an answer is to know with precision the relation that exists between the protonated Schiff base and its counter-ion. Blatz and coworkers [31—33] have shown that as the hydrogen bridge (N^+H- - -X^-) weakens, the λ_{max} of the Schiff base increases. Since in the primary step there is a bathochromic shift (for bovine Rh, the shift is 43 nm) it is tempting to ascribe it to a weakened hydrogen bridge between N^+—H and its counter-ion. Most studies have been done using models composed of a retinal Schiff base and strong mineral acids. It must be noted that strong acids offer poor relation to the most probable proton donor in the pigment which is in all likelihood, aspartic acid or glutamic acid (pK_as of 3.9 and 4.2 respectively) [89]. Since the protonating ability of weak acids toward a conjugated Schiff base was not known, we decided to investigate this question using ^{13}C and ^{15}N NMR and FTIR. In a non-polar solvent like chloroform, we found that acids with pK_as of 3—4 could only protonate the Schiff base partially. This was also true for methanol, even though the degree of protonation is greater in polar solvents. Using different carboxylic acids, we were able to show that with pK_as in the neighborhood of 4, only 75% protonation was possible in a medium with a dielectric constant of 32 [90]. In a non-polar solvent, the chemical shifts of C_{15} and C_{14} were between 154—158 ppm and 118—129 ppm respectively depending on the strength of the acids. In methanol, the figures are 159—162 ppm and 120—125 ppm. Remembering the data found by Mateescu et al. [71] and Harbison et al. [80], our study indicates rather clearly that complete protonation is only possible if water molecules or, perhaps, a proton relay network are present near the chromophore. These conclusions are the same as those reached by FTIR inspection of the same models [43, 47, 48]. These studies indicate that the role of water could be multiple: it is needed to insure a polar environment so that full protonation occurs even when the donor is a weak acid; it might also be needed in a proton relay and finally, it is needed to hydrolyse the Schiff base to all-*trans*-retinal.

^{15}N NMR substantiated all the findings found by ^{13}C NMR. On lyophilized samples, chemical shifts found by Mateescu et al. [71] differed from those found by Harbison et al. [80] and this difference was most probably due to differences in sample preparation. As stated before, one must be quite careful not to lyophilize too strongly the pigment since deprotonation could occur.

Finally, proton NMR is also quite useful in the investigation of the behavior of conjugated Schiff bases toward acids. Sharma and Roels [91] found that the chemical shift of the aldimine proton atom is slightly affected by acids. One would expect a strong deshielding since this proton is quite close to the positive nitrogen. This is not the case because on protonation the C=N bond loses some anisotropy which results in a shielding effect. Pattaroni and Lauterwein [92] and Rabiller and Danho [93] have studied in some detail the interaction of acids with Schiff bases. They were able to show that the acidic proton undergoes exchange with the donor: a fast equilibrium such as the one described before (see [Eq. 1]) has been proposed.

We were able to confirm the existence of such an equilibrium by observing the coupling behavior of the aldimine proton. This proton gives a doublet when the Schiff base is unprotonated and should become a doublet of doublets under protonation. This is indeed the case when strong mineral acids are used. For strong carboxylic acids (like trichloro or trifluoroacetic acids), only a doublet is obtained [92]. When excess acid is used or if the temperature is lowered to

−40 °C, a doublet of doublets appear [39]. This indicates clearly that an equilibrium such as the one described previously does exist.

In conclusion, NMR spectroscopy has shown quite convincingly that the chromophore is indeed protonated. It did also indicate that the milieu surrounding the chromophore must be highly polar and that water molecules are, in high probability, present near the retinal moiety and that their role is quite important in stabilizing the proton-bridge system.

The UV-visible, infrared, Raman and NMR spectroscopic results make us adhere to the suggestion that, in general, the chromophore of visual and bacterial Rh is protonated at its nitrogen atom. This does not close the argument, however. Two questions remain to be answered: are they protonated 100% and how can they be protonated in the given molecular environment?

4. How Can the Schiff Base be Protonated in Rh?

The results of Lussier et al. [43, 47—48], Zundel and coworkers [44—46] and Cossette et al. [90] have shown that, in a neutral medium non-conjugated and conjugated SB, including the retinylidene SB, can only be partially protonated by a carboxylic acid. From their works it can be estimated that a pK_a of about 2.0 would be needed to have the tautomeric equilibrium in Eq. 1, 100% in favor of protonation. Such acids are not present in either visual or bacterial pigments. Aspartic and glutamic acids, as said earlier, are likely candidates for being the proton donor. Clearly, some additional conditions must be fulfilled to have the SB 100% protonated or even for a large fraction of it to be protonated.

In this respect the work of Warshel and his coworkers represented an important step forward [37]. Contrary to previously held opinions, these authors pointed out that in Rh the chromophore must be in a polar environment in order to stabilize the zwitterion structure of the N^+—H- - -: O^- bridge. In this context, Blatz's 'levelling effect' should be recalled [31]. SB that are not or partly protonated by a given carboxylic acid in a neutral solvent, like carbon tetrachloride, will become protonated in a polar solvent, like methanol. Methanol helps in pushing the proton onto the SB.

A mechanism must exist in Rh for environmental stabilization of the zwitterion. This could be achieved in several ways. One possibility is using a proton relay network. An early suggestion for this was made by Bernard-Houplain et al. [94] in 1971. They studied the IR spectra of 2,6-di-tert-butylparacresol at low temperatures and found that up to concentrations of 1.0 M they do not form hydrogen bonds in neutral solvents. However, if some cyclohexanone was added to the solution, association OH bands appeared in the spectrum due to hydrogen bonds of the O—H- - -: O=C and - - -: O—H- - -: O—H- - -: O=C types. So the ketone group induced the formation of a hydrogen bonded chain. Similarly, protonation can be brought about by such a relay. The primary counter-ion which donates the proton to the SB nitrogen can be proton acceptor for another acid thereby increasing its own acidity. An early suggestion for such a mechanism was made by Khristoforov et al. [95] in 1974. Their system was retinal SB + p-cresol + p-cresol. In Rh it may consist of aspartic and glutamic acids, tyrosine and possibly other amino-acids. A variety of such systems were discussed by Denisov and Golubev [96], Rosenbuch [97], and Zundel and his coworkers [46].

Stabilization by through space dipole-dipole interactions with oriented polar groups of amino-acids was suggested by Warshel [98, 100]. Other models were proposed by Honig [99] and by Birge [64]. Birge et al. put forward a scheme of two counter-ions of which one has a negative charge and the other is neutral. The stabilization energy obtained by hydrogen bonding between the two counter-ions and the protonated SB amounted to about 3 kcal/mol.

Yet another idea (which the present writers favor) is based on the involvement of water in the stabilization mechanism. As said earlier there is a good amount of evidence for the presence of water molecules in the neighborhood of the chromophore in the pigments. Rafferty and Shichi [84] and Hildebrandt and Stockburger [85] observed that in a dry nitrogen atmosphere or in vacuo, pro-found changes are produced in the spectroscopic properties of the chromophore. The UV-visible absorption maximum of Rh undergoes a large hypsochromic shift from about 500 to about 390 nm suggesting that the SB ceased to be protonated. As to bR (all-*trans*) the maximum shifted from about 570 to 530 nm and the C=N stretching band in the Raman spectrum did not show any deuterium shift. Hildebrandt and Stockburger also observed that the width of the C=N band considerably narrowed when H_2O as a solvent was replaced by D_2O. This strongly indicates the interaction of the C=N group with water molecules. They suggested that the ion-pair is stabilized by water molecules.

Dupuis et al. [86] suggested, in order to explain Shichi's results that a water molecule is intercalated between the SB and the acid, so that a proton is transferred from the acid to water thereby forming a hydronium ion which then protonates the SB. Other water molecules would form a cluster around the ion-pair to stabilize it further. There is no direct proof for any of the above stabilizing mechanisms and the field is still open for research.

The mechanism of stabilization of the ion-pair appears to be even more complicated. Stoeckenius and his coworkers examined the effect of acidification and deionisation on bR [101−102]. When the pH is lowered to between 3 and 1 or the membrane is deionized on a cation exchange column, the purple color of bR changes to blue and the absorption maximum goes from 568 to 605 nm. It has been suggested [103] that these changes are due to the protonation of two counter-ions in the vicinity of the SB. The least one can say is that the protonation/nonprotonation equilibrium is very delicately regulated by anions and by metal cations which are probably hydrated. This may adapt the pigment to many different situations.

The whole problem is intimately connected with the nature of hydrogen bonding in ion pairs which in recent years has been the object of many investigations [104—113]. It is not enough to state that the SB is protonated. What would be needed is the shape of the potential surface governing the motions of the proton in the N^+---H---O^- bridge (which might include an intercalated water molecule). What is meant by protonation is the formation of a covalent N^+H bond, about 1 Å long; then the potential has a minimum close to the nitrogen. There may be another potential well close to the oxygen but the one close to the nitrogen would be the deeper. There is a fascinating other possibility, namely the existence of a double well potential with two nearly equal minima. In view of the presence of regions of continuous absorption in the infrared spectra of related model systems [46, 107, 114—118] this could easily be the case. Then all depends on the height of the barrier between the two wells. If the barrier is not too low and if both wells have a non-negligible population, the vibrational bands may correspond to one or the other well. These may or may not be found in the infrared since the bands corresponding to the protonation well (the one closer to the nitrogen) and to the non-protonation well (close to the oxygen) may have very different intensities. As shown by Leclercq et al. [119] the Raman lines might also correspond to the one or the other well. In a resonance Raman spectrum the relative intensities of the two will depend on the Franck-Condon factors with the electronic excited state from which the intensity is borrowed. It might happen that these factors are so different that the bands due to, say, a 90% populated well is not seen in the spectrum but those of the other 10% populated well are. While at present there is no evidence for it, this might be the case for Rh for some species.

If the barrier between the two minima is low, rapid proton transfer from one well to the other is possible thermally or by tunneling. Depending on the rapidity of the transfer this may broaden the NMR lines or even the vibrational bands.

That there are reasons to inquire further than merely stating that the SB is 'protonated' is clearly shown by recent experiments using differential infrared spectroscopy [54]. As we remember the frequency of $\nu(C{=}N)$ depends not only on the electronic structure or the force constant of the bond but also on vibrational coupling, mainly between the $C{=}N^+$, stretching and N^+H and CH bending motions. For this reason the identity of $\nu(C{=}N^+)$ for two Rh does not mean that the structure of the proton bridge is exactly the same in the two cases. Bagley et al. [120] deuterated either the one or the other or both hydrogens in the $HC{=}NH^+$ unit. When both hydrogens are replaced by deuterium the $\nu(C{=}N^+)$ is practically 'coupling-free'. The values they found are 1602 cm^{-1} for Rh, 1610 for bathorhodopsin (an intermediate in the photocycle of Rh) and 1614 for bR while the respective values for the non-deuterated analogs are 1655, 1655 and 1641 cm^{-1}. Now, the visible absorption maxima are at 498, 593 and 568 nm in the same order. While the visible data exhibit a greater effect of protonation in this order, the vibrational data indicate the opposite order.

5. What Difference Would It Make if the Schiff Base was Not Protonated?

Would vision be impossible without the protonation of the SB chromophore? Since non-protonated SB can also bleach (have photochemical cycles) the answer is probably no. We still might see but differently. What humans call 'visible' would not be the visible. For the many creatures, including humans, who have protonated SB chromophores, it would be shifted to higher frequencies.

The photochemical primary step would probably still contain *cis-trans*-isomerization but the proton, instead of moving away from the nitrogen after photon absorption would move onto the nitrogen:

$$\text{\Large$>$}C=N:\text{---}H\text{---}O\text{---} \xrightarrow{h\nu} \text{\Large$>$}C=N^+H\text{---}O^-\text{---}. \tag{3}$$

This might actually be the case for species having UV absorbing pigments as has been mentioned above.

Needless to say, the stages of the photochemical cycle would be different. More predictably the quantum yield would be different, probably much lower. The role played by protonation in mixing together the ^1B_u-like and $^1A_g^-$-like states would be lost. This is probably one of the main reasons why protonation is needed.

The extra stabilization provided by the hydrogen bond in the proton bridge would also be lost; unwanted thermal isomerization might sometimes occur.

Isomerization, protonation-deprotonation, positive and negative ions and water are all needed to keep functioning the astonishing system of vision. A great deal has been learned about this system in recent years; more remains to be discovered.

References

1. B. Honig: *Ann. Rev. Phys. Chem.* **29**, 31 (1978).
2. B. Honig: *Curr. Top. Membr. Transf.* **16**, 371 (1982).
3. M. Ottolenghi: *The Photochemistry of Rhodopsin* (Advances in Photochemistry, v. 12, Ed. J-N. Pitts, Jr., G. S. Hammond, K. Gollnick, and D. Grosjean), pp. 97—200. Wiley (1980).
4. R. R. Birge: *Ann. Rev. Biophys. Bioeng.* **10**, 315 (1981).
5. R. Uhl and E. W. Abrahamson: *Chem. Rev.* **81**, 291 (1981).
6. W. Stoeckenius and R. Bogomolni: *Ann. Rev. Biochem.* **51**, 587 (1982).
7. W. Stoeckenius, R. H. Lozier, and R. Bogomolni: *Biochim. Biophys. Acta* **505**, 215 (1979).
8. H. Shichi: *Biochemistry of Vision.* Academic Press (1983).
9. V. Balogh-Nair and K. Nakanishi: *New Comprehensive Biochemistry* (Stereochemistry, v. 3, Ed. C. Tamm), pp. 283. Elsevier Biomedical (1982).
10. D. S. Kliger: *Intl. J. Quant. Chem.* **16**, 809 (1979).
11. T. Förster: *Z. Elecktrochem.* **54**, 42, 531 (1950).
12. C. Sandorfy: *Comptes rendus Acad. Sci.* **230**, 861 (1950).
13. C. Sandorfy: *Can. J. Chem.* **13**, 439 (1953).
14. R. Mathies and L. Stryer: *Proc. Natl. Acad. Sci. (USA)* **73**, 2169 (1976).
15. J. P. Corsetti and B. E. Kohler: *J. Chem. Phys.* **67**, 5237 (1977).
16. A. M. Schaffer, T. Yamaoka, and R. S. Becker: *Photochem. Photobiol.* **21**, 297 (1975).
17. C. Sandorfy: *Electronic Spectra in Quantum Chemistry*, pp. 182—186. Prentice-Hall (1964).
18. J. Koutecky: *J. Chem. Phys.* **47**, 1501 (1967).

19. B. S. Hudson and B. E. Kohler: *Chem. Phys. Lett.* **14**, 299 (1972).
20. K. Schulten and M. Karplus: *Chem. Phys. Lett.* **14**, 305 (1972).
21. P. R. Monson and W. M. McClain: *J. Chem. Phys.* **53**, 29 (1970).
22. B. S. Hudson and B. E. Kohler: *Ann. Rev. Phys. Chem.* **25**, 437 (1974).
23. R. L. Christensen and B. E. Kohler: *J. Chem. Phys.* **63**, 1837 (1975).
24. R. R. Birge, K. Schulten, and M. Karplus: *Chem. Phys. Lett.* **31**, 451 (1975).
25. R. L. Swofford and W. M. McClain: *J. Chem. Phys.* **59**, 10 (1973).
26. G. R. Holtom and W. M. McClain: *Chem. Phys. Lett.* **44**, 436 (1976).
27. R. R. Birge, J. A. Bennett, B. M. Pierce, and T. M. Thomas: *J. Am. Chem. Soc.* **100**, 1533 (1978).
28. R. R. Birge, J. A. Bennett, L. M. Hubbard, H. L. Fang, B. M. Pierce, D. S. Kliger, and G. E. Leroi: *J. Am. Chem. Soc.* **104**, 2519 (1982).
29. R. R. Birge, D. F. Bocian, and L.M. Hubbard: *J. Am. Chem. Soc.* **104**, 1196 (1982).
30. R. R. Birge, L. P. Murray, B. M. Pierce, H. Akita, V. Balogh-Nair, L. A. Findsen, and K. Nakanishi: *Proc. Natl. Acad. Sci. (USA)* **82**, 4117 (1985).
31. P. E. Blatz, J. H. Mohler, and H. V. Navangul: *Biochemistry* **11**, 848 (1972).
32. P. E. Blatz and J. H. Mohler: *Biochemistry* **11**, 3240 (1972).
33. P. E. Blatz and J. H. Mohler: *Biochemistry* **14**, 2304 (1975).
34. B. Honig, T. Ebrey, R. H. Callender, U. Dinur, and M. Ottolenghi: *Proc. Natl. Acad. Sci. (USA)* **76**, 2503 (1979).
35. C. N. Rafferty and H. Shichi: *Photochem. Photobiol.* **33**, 229 (1981).
36. P. Hildebrandt and M. Stockburger: *Biochemistry* **23**, 5539, 1984.
37. A. Warshel and N. Barboy: *J. Am. Chem. Soc.* **104**, 1469 (1982).
38. M. Bissonnette and D. Vocelle: *Spectros. Int. J.* **2**, 120 (1983).
39. M. Bissonnette, H. Le Thanh, and D. Vocelle: *Can. J. Chem.* **62**, 1459 (1984).
40. J. Favrot, J. M. Leclercq, R. Roberge, C. Sandorfy, and D. Vocelle: *Photochem. Photobiol.* **29**, 99 (1979).
41. J. Favrot, D. Vocelle, and C. Sandorfy: *Photochem. Photobiol.* **30**, 417 (1979).
42. P. Chevalier and C. Sandorfy: *Can. J. Chem.* **38**, 2524 (1960).
43. L. S. Lussier, A. Dion, C. Sandorfy, H. Le Thanh, and D. Vocelle: *Photochem. Photobiol.* **44**, 629 (1986).
44. G. Zundel and J. Muehlinghaus: *Z. Naturforsch* **26b**, 546 (1971).
45. G. Zundel: in *The Hydrogen Bond* (Ed. P. Schuster, G. Zundel, and C. Sandorfy). Vol. II, pp. 683—766. North-Holland, Amsterdam (1976).
46. G. Zundel and H. Merz: in *Information and Energy Transduction in Biological Membranes.* Alan R. Liss, New-York, pp. 153—164 (1984).
47. L. S. Lussier, C. Sandorfy, H. Le Thanh, and D. Vocelle: *Photochem. Photobiol.* **45**, 801 (1987).
48. L. S. Lussier, C. Sandorfy, H. Le Thanh, and D. Vocelle: *J. Phys. Chem.* **91**, 2282 (1987).
49. A. R. Oseroff and R. H. Callender: *Biochemistry* **13**, 4243 (1974).
50. K. J. Rothschild and H. Marrero: *Proc. Natl. Acad. Sci. (USA)* **79**, 4045 (1982).
51. K. J. Rothschild, W. A. Cantore, and H. Marrero: *Science* **219**, 1333 (1983).
52. F. Siebert and W. Mäntele: *Biophys. Struct. Mech.* **6**, 147 (1980).
53. F. Siebert, W. Mäntele, and W. Kreutz: *Biophys. Struct. Mech.* **6**, 139 (1980).
54. F. Siebert and W. Mäntele: *Eur. J. Biochem.* **130**, 565 (1983).
55. K. Bagley, G. Dollinger, L. Eisenstein, A. K. Singh, and L. Zimanyi: *Proc. Natl. Acad. Sci. (USA)* **79**, 4972 (1982).
56. K. Bagley, G. Dollinger, L. Eisenstein, J. Vittitow, L. Zimanyi, T. G. Ebrey, and B. Nelson: *Biophys. J.* **41**, 337 (1983).
57. K. A. Bagley, V. Balogh-Nair, A. A. Croteau, G. Dollinger, T. G. Ebrey, L. Eisenstein, M. K. Hong, K. Nakanishi, and J. Vittitow: *Biochemistry* **24**, 6055 (1985).
58. B. Honig, T. Ebrey, R. H. Callender, V. Dinur, and M. Ottolenghi: *Proc. Natl. Acad. Sci. (USA)* **76**, 2503 (1979).
59. B. Aton, A. G. Doukas, D. Narva, R. H. Callender, U. Dinur, and B. Honig: *Biophys. J.* **29**, 79 (1980).
60. B. Curry, A. Broek, J. Lugtenburg, and R. Mathies: *J. Am. Chem. Soc.* **104**, 5274 (1982).

61. S. O. Smith, A. B. Myers, R. A. Mathies, J. A. Pardoen, C. Winkel, E. M. N. Van Den Berg, and J. Lugtenburg: *Biophys. J.* **47**, 653 (1985).
62. G. M. Massig, M. Stockburger, W. Gärtner, D. Oesterhelt, and P. Towner: *J. Raman Spectrosc.* **12**, 287 (1982).
63. R. R. Birge, B. M. Pierce, and L. P. Murray: in *Spectroscopy of Biological Molecules.* Reidel, p. 473 (1984).
64. R. R. Birge, L. P. Murray, B. M. Pierce, H. Akita, V. Balogh-Nair, L. A. Findsen, and K. Nakanishi: *Proc. Natl. Acad. Sci.* (*USA*) **82**, 4117 (1985).
65. R. S. Becker and K. Freedman: *J. Am. Chem. Soc.* **107**, 1477 (1985).
66. F. I. Harosi and Y. Hashimito: *Science* **222**, 1021 (1983).
67. J. A. Avery, J. K. Bowmaker, M. B. A. Djamgoz, and J. E. G. Downing: *J. Physiol.* (*London*) **334**, 23P (1983).
68. K. Hamdorf, J. Schwemer, and M. Gogala: *Nature* **231**, 458 (1971).
69. R. Paulsen and J. Schwemer: *Biochem. Biophys. Acta* **283**, 520 (1972).
70. F. I. Harosi: in *The Visual System.* Alan R. Liss, New-York, p. 41 (1985).
71. G. D. Mateescu, E. W. Abrahamson, J. W. Shriver, W. Copan, D. Muccio, M. Iqbal, and V. Waterhouse: *Spectroscopy of Biological Molecules* (Ed. C. Sandorfy and T. Theophanides), pp. 257—291. Reidel (1984).
72. D. Muccio, W. G. Copan, E. W. Abrahamson, and G. D. Mateescu: *Org. Magn. Res.* **22**, 121 (1984).
73. M. Allen and J. D. Roberts: *J. Org. Chem.* **45**, 130 (1980).
74. J. W. Shriver, E. V. Abrahamson, and G. D. Mateescu: *J. Am. Chem. Soc.* **98**, 2407 (1976).
75. J. W. Shriver, G. D. Mateescu, R. Fager, D. Torchia, and E. W. Abrahamson: *Nature* **270**, 271 (1977).
76. J. W. Shriver, G. D. Mateescu, and E. W. Abrahamson: *Biochemistry* **18**, 4785 (1979).
77. G. D. Mateescu, W. G. Copan, D. D. Muccio, D. V. Waterhous, and E. W. Abrahamson: *Proceedings of Int. Symp. on Synth. and Appl. of Isotopic. Labeled Compounds* (Ed. W. P. Duneau and A. Susan). Elsevier (1983).
78. J. W. Shriver, G. D. Mateescu, and E. W. Abrahamson: *Meth. in Enz.* **81**, 698 (1982).
79. G. S. Harbison, J. Herzfeld, and R. G. Griffin: *Biochemistry* **22**, 1 (1983).
80. G. S. Harbison, S. O. Smith, J. A. Pardoen, P. P. J. Mulder, J. Lugtenburg, J. Herzfeld, R. Mathies, and R. G. Griffin: *Biochemistry* **23**, 2662 (1984).
81. G. S. Harbison, S. O. Smith, J. A. Pardoen, C. Winkel, J. Lugtenburg, J. Herzfeld, R. Mathies, and R. G. Griffin: *Proc. Natl. Acad. Sci.* (*USA*) **81**, 1706 (1984).
82. G. S. Harbison, S. O. Smith, J. A. Pardoen, J. M. L. Courtin, J. Lugtenburg, J. Herzfeld, R. Mathies, and R. G. Griffin: *Biochemistry* **24**, 6955 (1985).
83. G. S. Harbison, P. P. J. Mulder, H. Pardoen, J. Lugtenburg, J. Herzfeld, and R. G. Griffin: *J. Am. Chem. Soc.* **107**, 4809 (1985).
84. C. N. Rafferty and H. Shichi: *Photochem. Photobiol.* **33**, 229 (1981).
85. P. Hildebrandt and M. Stockburger: *Biochemistry* **23**, 5539 (1984).
86. P. Dupuis, F. I. Harosi, C. Sandorfy, J. M. Leclercq, and D. Vocelle: *Rev. Can. Biol.* **39**, 247 (1980).
87. C. Sandorfy and D. Vocelle: *Can. J. Chem.* **64**, 2251 (1986).
88. A. Yamaguchi, T. Unemoto, and A. Ikegami: *Photochem. Photobiol.* **33**, 511 (1981).
89. E. A. Dratz and P. A. Hargrave: *Trends in Biochem. Sc.* **8**, 128 (1983).
90. D. Cossette and D. Vocelle: *Can. J. Chem.* **65**, 1576 (1987).
91. G. M. Sharma and O. A. Roels: *J. Org. Chem.* **38**, 3648 (1973).
92. C. Pattaroni and J. Lauterwein: *Helv. Chim. Acta* **64**, 1969 (1981).
93. C. Rabiller and D. Danho: *Helv. Chim. Acta* **67**, 1254 (1984).
94. M. C. Bernard-Houplain, C. Bourdéron, J. J. Péron, and C. Sandorfy: *Chem. Phys. Lett.* **11**, 149 (1971).
95. V. L. Khristoforov, E. N. Zvonkova, and R. P. Evstigneeva: *Zh. Obs. Khim.* **44**, 909 (1984).
96. G. S. Denisov and N. S. Golubev: *J. Mol. Struct.* **75**, 311 (1981).
97. J. P. Rosenbusch: *Bulletin de l'Institut Pasteur* **83**, 207 (1985).
98. A. Warshel: *Proc. Natl. Acad. Sci.* (*USA*) **75**, 5250 (1978).

99. B. Honig, T. Ebrey, R. H. Callender, V. Dinur, and M. Ottolenghi: *Proc. Natl. Acad. Sci. (USA)* **76**, 2503 (1979).
100. A. Warshel: *Proc. Natl. Acad. Sci. (USA)* **75**, 2558 (1978).
101. D. Oesterhelt and W. Stoeckenius: *Nature New. Biol.* **233**, 149 (1971).
102. P. C. Mowery, R. H. Lozier, S. Chae, Y. W. Tseng, M. Taylor, and W. Stoeckenius: *Biochemistry* **18**, 4100 (1979).
103. Y. Kimura, A. Ikegami, and W. Stoeckenius: *Photochem. Photobiol.* **40**, 641 (1985).
104. D. Hadzi and S. Bratos: in *The Hydrogen Bond* (Edited by P. Schuster, G. Zundel, and C. Sandorfy). Vol. XI, p. 565. North-Holland, Amsterdam (1976).
105. P. Schuster: *Ibidem.* Vol. I, p. 25.
106. J. Brickman: *Ibidem.* Vol. I, p. 217.
107. G. Zundel: *Ibidem.* Vol. II, p. 683.
108. L. Sobczyk, H. Engelhardt, and K. Bunzel: *Ibidem.* Vol. III, p. 934.
109. P. Schuster, P. Wolschann, and K. Tortschanoff: in *Molecular Biology, Biology, Biochemistry and Biophysics* (Edited by I. Pecht and R. Rigler. Vol. 24). Springer, Berlin, p. 107 (1977).
110. A. Novak: in *Structure and Bonding* (Vol. 28). Springer, Berlin, p. 177 (1974).
111. J. L. Wood: in *Spectroscopy and Structure of Molecular Complexes* (Edited by J. Yarwood). Plenum, New York, p. 303 (1973).
112. T. Zeegers-Huyskens and P. Huyskens: in *Molecular Interactions* (Edited by H. Ratajczak and W. J. Orville-Thomas. Vol. 2). Wiley, Chichester, p. 1 (1981).
113. P. Barczinski, Z. Dega-Szafran, and M. Szafran: *J. Chem. Soc. Perkin Trans. II*, 965 (1985).
114. E. G. Weidemann and G. Zundel: *Z. Naturforsch.* **25a**, 627 (1970).
115. R. Janoschek, E. G. Weidemann, H. Pfeiffer, and G. Zundel: *J. Am. Chem. Soc.* **94**, 2387 (1972).
116. R. Lindemann and G. Zundel: *J. Chem. Soc. Faraday Trans. II* **73**, 788 (1977).
117. B. Brzezinski and G. Zundel: *Chem. Phys. Lett.* **115**, 212 (1985).
118. H. Merz and G. Zundel: *Biochem. Biophys. Res. Commun.* **101**, 540 (1981).
119. J. M. Leclercq and C. Sandorfy: *Photochem. Photobiol.* **33**, 36 (1981).
120. K. Bagley, G. Dollinger, L. Eisenstein, M. Honig, J. Vittitow, and L. Zimanyi: in *Information and Energy Transduction in Biological Membranes* (Ed. C. L. Bolis, E. J. M. Helmreich, and H. Passow). Alan R. Liss, New-York, p. 27 (1984).

Molecular Neurobiology and Sociobiology

Modeling the Drug-Receptor Interaction in Quantum Pharmacology

J. S. GÓMEZ-JERIA
University of Chile, Faculty of Sciences, Casilla 653, Santiago, Chile.

1. Introduction

Quantum pharmacology (QP) is now a well-established branch of quantum chemistry. Its scope can be characterized by saying that QP studies the electronic and conformational properties of molecules possessing pharmacological activity, and seeks relationships between these properties and the drug's action mechanisms.

To study the electronic distribution and the spatial arrangement of nuclei, quantum pharmacology normally employs the now common methods of quantum chemistry used to obtain the wave function and the total energy of molecules: semiempirical [1—3] and *ab initio* [4—6] quantum-chemical methods.

Theoretical conformational studies can be of great help to the medicinal chemist due to the fact that they describe the whole conformational surface contrary to, for example, the NMR or the crystallographic studies that can reveal, respectively, only a section or one or several points of the surface.

It is also important to consider that a given molecule, in its crystalline form, may be in one or more conformations. The Quantum Pharmacologist can start from these conformations and explore only the surface available up to 7 Kcal mole^{-1}, without needing to analyze all the conformational space. In a family of molecules, it is therefore possible to create a conformational map of the common areas to select the most probable conformation(s) at the receptor level. This map can be improved by the inclusion of molecules with one or several degrees of restricted conformational freedom as shown in Figure 1 [7, 8].

From the electronic wave function, it is possible to calculate those electronic properties which depend on the electronic structure: atomic net charges, electronic densities at particular locations, electrophilic and nucleophilic superdelocalizabilities, dipole moments, etc. [9]. In the case of the total charge density, molecular orbital electronic density and molecular electrostatic potential, a visual picture is possible, permitting a fast analysis and comparison of the information presented [10—12]. In Figures 2 and 3 we present some of such examples.

Quantum pharmacology, through quantitative structure-activity relationships (QSAR), can provide a physical insight of the sequence of processes which form the basis of drug action: the pharmaceutical, pharmacokinetic and pharmacodynamic phases [13]. If we can establish a QSAR for a given family of molecules

Jean Maruani (ed.), Molecules in Physics, Chemistry, and Biology, Vol. IV, 215—231.
© *1989 by Kluwer Academic Publishers.*

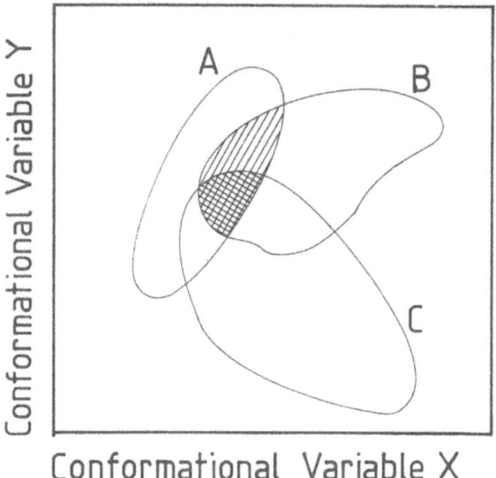

Fig. 1. Conformational map for molecules A, B, and C. The shaded areas show common conformational possibilities.

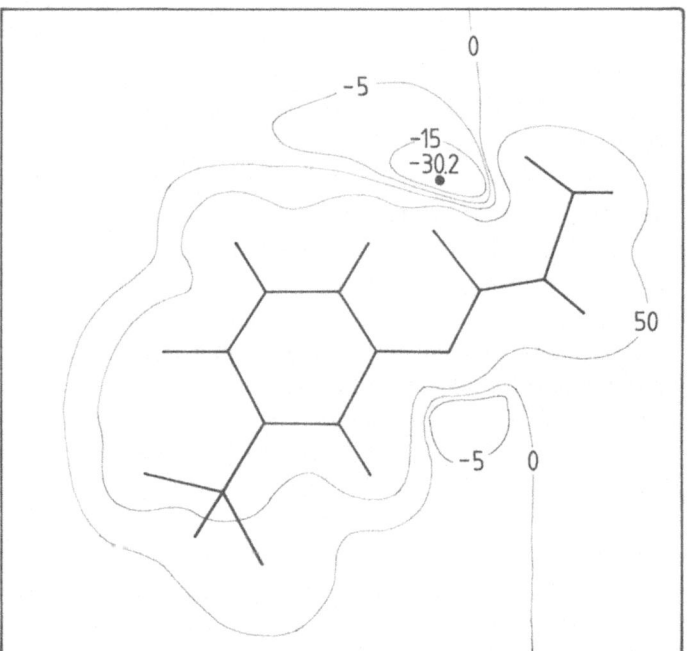

Fig. 2. Molecular electrostatic potential map for *m-t*-butyl-*N*-methylcarbamate. Values in kcal mol⁻¹.

possessing a similar effect, it is possible therefore to suggest new molecular structures with enhanced or diminished pharmacological activity. If we consider that, in general, it is necessary to synthetize, purify and test in animals and humans

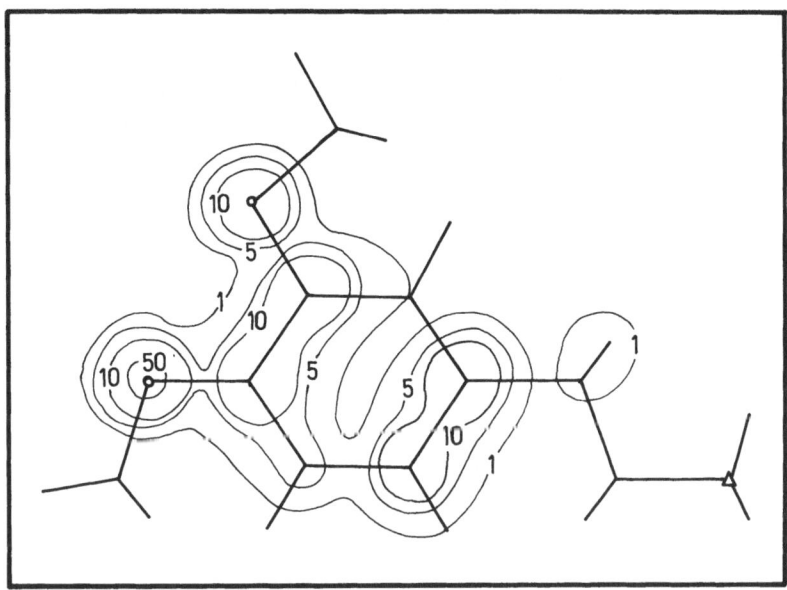

Fig. 3. Electronic density distribution map for the Highest Occupied Molecular Orbital of 3,4-dimethoxyphenylethylamine in the plane $z = 0.5$ Å. Values in 10^3 e.

several thousand molecules to get one useful for the market, we do not need to insist on the advantages (economical and time-saving) in using QP.

The usual way employed in QP to carry out a QSAR study is to use one of the two following approaches [14].

1. *Empirical methods*: they fit the experimental data and their main disadvantage is that they cannot provide a sure basis for a total theoretical understanding of the problem under study. The best examples of these methods are the Free-Wilson and the Hansch approaches [15–17].
2. *Model-based methods*: they start from a hypothesis of action and take into account the characteristics of the biological system. As the QSAR obtained were formally derived, it is expected that they will give a deep and physically correct insight of the underlying physics, and that they will be useful in predicting new active molecules.

With regard to the kind of parameters employed, the empirical methods may be divided into two groups.

1. Methods employing quantum-chemical reactivity indices (i.e., net charges, superdelocalizabilities, etc.).
2. Methods employing a mixture of quantum-chemical and classical (i.e., Hansch parameters, etc.) reactivity indices.

Hereafter, we shall center our attention on the model-based methods applied to the drug-receptor interaction step of the pharmacodynamic phase of drug action. This is for at least two reasons.

1. The drug-receptor interaction is well characterized by the equilibrium constant. This quantity has been measured in vitro with different methods [18, 19] for a very large quantity of molecules interacting with a variety of receptors: opiates [20], serotonergic [21], GABAergic [22], dopaminergic [23], etc.
2. The advantages of the model, mentioned above.

To carry out our purpose, we shall present an analysis of the drug-receptor interaction and present a model-based expression for the equilibrium constant. Then we shall give a brief sketch of the decomposition of the intermolecular interaction energy through perturbation theory and present two approaches to deal with the fact that we do not know the electronic-conformational structure of one of the partners. Finally, we present an example showing the clear advantages of formal Quantum Pharmacology over other approaches.

2. The Drug-Receptor Interaction

2.1. THE DRUG-RECEPTOR INTERACTION CAUSES

The drug-receptor interaction is caused by intermolecular forces. Table I shows their classification according to the distance between partners.

TABLE I. Classification of intermolecular forces according to the distance between partners.

Small distances (Zone A of Figure 4)	Intermediate distances (Zone B of Figure 4)	Large distances (Zone C of Figure 4)
QUASIMOLECULE (Exchange and Coulomb integrals)	1. Electrostatic 2. Exchange-polarization 3. Exchange-repulsion 4. Polarization 5. Exchange-dispersion 6. Dispersion	1. Electrostatic 2. Polarization 3. Dispersion

The receptor may be defined as a pattern A of forces of different origin, forming part of a biological system and having approximately the same structure as a certain pattern B of forces exhibited by the drug, in such a way that among the patterns A and B there is a complementary relationship [24].

Ariens [25] has proposed dividing the space around the receptor in the following three zones.

Zone I: in this space, the drug-receptor interaction occurs through inter-molecular forces.

Zone II: this covers the first zone, and is defined as the space in which only ionic forces are acting. This is where an accumulation, recognition and guiding of the drug molecule towards the receptor through long-range interactions occurs [25]. The recognition process can be associated with the matching of the molecular electrostatic potentials of the drug and the receptor (for examples, see Refs. 26 and 27), to produce a correct geometrical alignment.

Zone III: it consists of the remainder of the biophase in which there is not an influence of the receptor. Here thermal agitation will cause the passing of drug molecules from this Zone to Zone II. We shall center our attention on Zone I.

The magnitude characterizing the drug-receptor interaction is the equilibrium constant, that is normally measured in vitro. In this preparation, we can assume that the influence of both the pharmaceutical and pharmacokinetic processes is reduced to a minimum, so that the concentration of the drug in the biophase is equal or proportional to the one in the vicinity of the receptor [25]. Also, as the number of drug molecules in the bath is greater than the ones reaching the receptor, it is possible to use the concentration of drug as the dose added.

If we consider a state of thermodynamic equilibrium and a $1:1$ stoichiometry in the formation of the drug-receptor complex:

$$D + R \rightleftharpoons DR$$

where D is the drug molecule, R the receptor and DR is the drug-receptor complex (DR complex hereafter), the equilibrium constant is [28]:

$$K = (Q_{DR}/Q_D Q_R) \exp(-\Delta E/kT) \tag{1}$$

where ΔE is the difference between the ground-state energy of DR and the energies of the ground states of D and R: $\Delta E = E_{DR} - (E_D + E_R)$ and the Q's are the total partition functions measured from the ground state in solution [28].

We have shown that for the case where: (a) the receptor's mass is very much larger than the mass of the drug molecule, (b) the Boltzmann factors of the excited electronic states are negligible compared to those of the ground state, (c) the rotational and vibrational motions can be treated as independent and uncoupled, and (d) the temperature is 37 °C, we can approximate Eq. (1) as [28]:

$$\log(K) = a + b \log M_D + c \log(\sigma_D/(I_1 I_2 I_3)) + d\Delta E \tag{2}$$

where a, b, c and d are constants, σ_D is the drug molecule's symmetry number and $I_1 I_2 I_3$ is the product of its three moments of inertia about the three principal axes of rotation [28]. To determine the constants, we carry out a linear multiple regression analysis for a given family of drugs whose equilibrium constant has been measured in the same experimental conditions. The resulting equation will indicate which are the relevant structural indices accounting for the variation of $\log(K)$ in the family.

The interaction energy, ΔE, cannot be determined directly, either due to the size of the receptor or to the lack of knowledge of its molecular structure. Nevertheless, when we consider a drug-receptor interaction in which no covalent bonds are formed (i.e., a weak interaction), we can employ Perturbation Theory (PT) to evaluate ΔE.

Therefore, as a necessary step, we must model the interaction energy, maintaining the underlying physics in the interaction.

In the following, we shall give a brief sketch of the problems appearing in the application of Perturbation Theory to calculate ΔE.

2.2. THE INTERMOLECULAR INTERACTION ENERGY

Let us consider the interaction of a drug molecule D (with n_D electrons), with a receptor R (with n_R electrons). Our problem consists in solving the Schrödinger equation for the DR complex, for different intermolecular distances and relative orientations (r), remembering that our interest is mainly focused on the distances at which the DR complex is formed (Zone B of Figure 4).

At very large separations (Zone C in Figure 4), where the overlap of the charge distributions of the interacting molecules can be neglected, the straightforward application of the 'usual' Rayleigh-Schrödinger Perturbation Theory (RSPT) is valid. By 'usual' we mean that the unperturbed Hamiltonian, $\hat{H}^0$, is written as $\hat{H}^0 = \hat{H}^0_D + \hat{H}^0_R$, where $\hat{H}^0_D$ and $\hat{H}^0_R$ are, respectively, the Hamiltonian operators

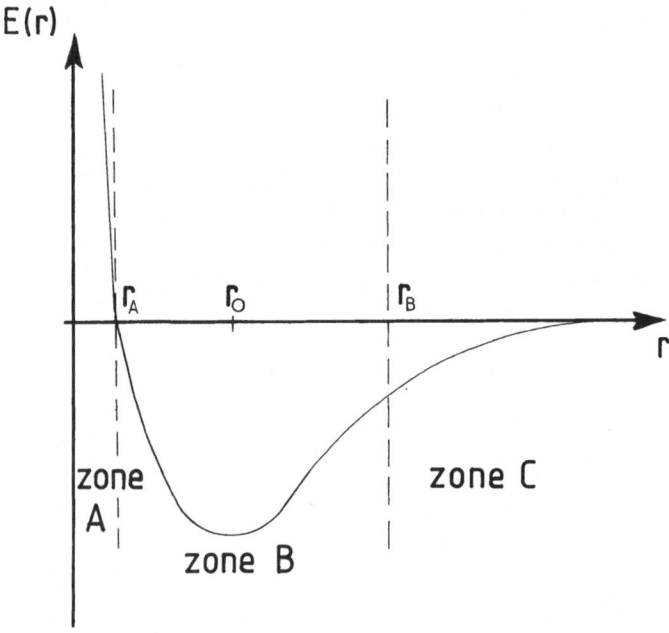

Fig. 4. Intermolecular potential. Zone A: small distances, Zone B: intermediate distances, Zone C: large distances.

of the drug and the receptor. This procedure, also called the 'polarization approximation', allots definite electrons to each partner. Up to second order, RSPT gives three contributions: the electrostatic $(E(el))$, polarization $(E(pol))$ and dispersion $(E(disp))$ energies (see Table II, first row). At large intermolecular separations we can employ the multipole expansion to express the electrostatic energy in terms of charge-charge, charge-dipole, dipole-dipole, . . . components [31].

TABLE II. Contributions to the intermolecular interaction energy for different wave functions.

Wave function	Perturbation contributions	
	First Order	Second Order
$\Psi_D \Psi_R + \cdots$	Electrostatic	Polarization, Dispersion
$A\Psi_D\Psi_R + \cdots$	Electrostatic Exchange-repulsion	Polarization, Dispersion, Exchange-dispersion, Exchange-polarization
$A\Psi_D\Psi_R + [A\Psi_D^+\Psi_R^- + A\Psi_D^-\Psi_R^+]^{(a)} + \cdots$	Electrostatic Exchange-repulsion	Polarization, Dispersion, Exchange-dispersion, Exchange-polarization, Charge transfer

Note: A is the antisymmetrizing operator.
[a] included to compensate for the incompleteness of the bases used for the separated molecules.

RSPT cannot be applied in its usual form to systems interacting very closely (Zone B of Figure 4). The reason can be briefly summarized as follows [32]. In the usual RSPT, $\hat{H}^0$ commutes with a group $G^0 = N \times S(n_D) \times S(n_R)$, where G^0 is the direct product of the symmetry group N of the nuclear configuration and of the groups $S(n_D)$ and $S(n_R)$ of permutations of the n_D and n_R electrons of the separate partners respectively.

On the other hand, the total spin-independent n-electron $(n = n_D + n_R)$ Hamiltonian operator, $\hat{H}$, commutes with a group G, which is the direct product of the symmetry group N and of the group $S(n)$ of permutations of all n electrons of the DR complex. In other words, $\hat{H}^0$ has a lower symmetry than the perturbed Hamiltonian.

This fact has provided arguments against the direct use of RSPT for short intermolecular distances. Among these, the very clear analysis of Claverie is the most important contribution and it can be summarized by saying that RSPT would give, instead of the physical interaction energy corresponding to the physical ground state of the DR complex, some 'mathematical' interaction energy corresponding to some 'mathematical' ground state [33, 34]. It is necessary to add that in the region of weakly interacting systems $(r = 7-8 \text{ Å}$ or more, Zone C of Figure 4), the physical curve is not different from the mathematical one. The advantages of such a procedure is that the interaction energy can be calculated directly as the sum of the first, second and higher order perturbation energies. This procedure can be applied to the study of the long-range interaction energy and of the best orientation (or recognition) of the drug by a receptor model.

At intermediate to small intermolecular distances, it becomes necessary to develop a perturbation formalism for the description at low orders of the effects of electron exchange between partners. A large number of exchange Perturbation Theories (or symmetry-adapted PT), have been suggested to treat this problem: the Murrel-Shaw-Musher-Amos [35, 36], the Eisenschitz-London-Hirschfelder-van der Avoid [37—39], the Hirschfelder-Silbey [40], etc., perturbation schemes.

This variety is due to the fact that there is a great deal of freedom available in the definition of the non-symmetric primitive function on which symmetry projection operators are applied to obtain one or more eigenvalues of $\hat{H}$ [41—43].

The second row of Table II shows, up to second-order, the different contributions to the interaction energy provided by symmetry-adapted PT. We avoided giving formulas, because of the diversity of exchange-perturbational schemes, but we can say that in order to model the short-range repulsive contributions it is necessary to introduce a proper term.

Therefore, we can conclude from this section that a correct modeling of the intermolecular interaction energy at short distances must take into account at least the electrostatic, polarization, dispersion and short-range repulsion ($E(\text{sr})$) energies.

2.3. THE CLAVERIE ET AL. (CLETAL) APPROACH

In a series of papers, Claverie et al. [44—53] have developed formulas to evaluate the interaction energy as the sum of long-range contributions (electrostatic, polarization and dispersion), and a short-range repulsive contribution. Later, a charge-transfer contribution was added [49]. In their simplest form, these formulas can be summarized as follows:

1. *Electrostatic Energy* [45—47]

The simplest way to account for the electrostatic energy is to represent it as:

$$E(\text{el}) = \sum_i \sum_j Q_i Q_j / R_{ij} \qquad (3)$$

where Q_i is the net charge of atom i, R_{ij} is the distance between atoms i and j, and the summation is over all the atoms of the partners. Now, considering that we are interested not in the values of the net charges per se, but in the charges whose variation can explain the variation of the equilibrium constant for a family of molecules interacting with the same receptor, $E(\text{el})$ may be represented by:

$$E(\text{el}) = \sum_i h_i Q_i \qquad (4)$$

where h_i is a constant and the summation is now only over the atoms of the drug (hereafter represented by i).

A more refined way to compute $E(\text{el})$ is to employ the multipole expansion up to the quadrupoles of the electron distribution if necessary [31, 48]. The same above scheme can be applied, i.e., to consider the receptor terms as constants in order to get formulas involving only the drug's multipole terms.

2. Polarization Energy [47]

The polarization energy is calculated as a sum of atom polarization contributions:

$$E(\text{pol}) = -(1/2) \sum_i \alpha_i (\varepsilon_i)^2 \tag{5}$$

where ε_i is the electric field created at atom i of the drug molecule by the receptor, and α_i is the mean polarizability of atom i. With the same criteria employed for $E(\text{el})$, we get:

$$E(\text{pol}) = -(1/2) \sum_i t_i \alpha_i \tag{6}$$

with t_i constant.

3. Dispersion and Short-Range Repulsion Energies [47]

These contributions can be evaluated by using the semiempirical Kitaigorodsky formula which involves atom-atom terms [54—56]:

$$E(\text{disp}) + E(\text{sr}) = E(\text{KIT}) = \sum_i \sum_j E(i, j) \tag{7}$$

where $E(i, j)$ represents each atom-atom contribution and is the sum of a dispersion and a repulsion term:

$$E(i, j) = k_i k_j [-(A/z) + (1 - Q_i/N_i^{\text{val}})(1 - Q_j/N_j^{\text{val}}) C \exp(-\alpha z)] \tag{8}$$

where $z = R_{ij}/R_{ij}^0$ with $R_{ij}^0 = [(2R_i^\omega)(2R_j^\omega)]^{1/2}$, R_i^ω being the van der Waals radius of atom i. The parameters α, A and C are kept independent of the atomic species. The values used are, for example, $A = 0.214$ kcal mol^{-1}, $C = 47.10^3$ kcal mol^{-1}, and $\alpha = 12.35$. For the van der Waals radii, we can use the values reported in [57] or others. The factors $(1 - Q_i/N_i^{\text{val}})$ represent the influence of the electronic populations on the repulsion, N_i^{val} being the number of valence electrons of atom i. k_i and k_j are factors allowing the variation of the minimum of $E(i, j)$ accordingly to the kind of interacting atoms ($k_H = k_C = 1$, $k_N = 1.18$, $k_0 = 1.36$) [45].

This formula is reduced in our case to:

$$E(\text{KIT}) = \sum_i \{-p_i k_i (R_i^\omega)^{1/2} + s_i k_i (1 - Q_i/N_i^{\text{val}}) \exp[-q_i (R_i^\omega)^{-1/2}]\} \tag{9}$$

where p_i, s_i and t_i are constants.

There are other ways to compute these terms, and most of them can be similarly adapted.

The adaptation of the CLETAL approach to our purposes shows that it is not difficult to calculate the terms involved in Eqs. 4, 6 and 9. Within this approach, the drug-receptor interaction energy is represented by:

$$\Delta E = \sum_i \{h_i Q_i + t_i \alpha_i - p_i k_i (R_i^\omega)^{1/2} +$$

$$+ s_i k_i (1 - Q_i/N_i^{val}) \exp[-q_i (R_i^\omega)^{-1/2}]\} \tag{10}$$

2.4. THE KLOPMAN-PERADEJORDI-GOMEZ (KPG) APPROACH

Klopman [58—61] has described a treatment, based on Perturbation Theory, in which allowance is made for ionic interactions. According to this method, the electronic energy change associated with the interaction of atoms i and j is:

$$\Delta E = \sum_p \left[Q_i Q_j/R_{ij} + (1/2)(\beta_{ij}^2) \sum_m \sum_{n'} D_{mi} D_{n'j}/(E_m - E_{n'}) - \right.$$

$$\left. - (1/2)(\beta_{ij}^2) \sum_{m'} \sum_n D_{m'i} D_{nj}/(E_{m'} - E_n) \right] \tag{11}$$

where Q_i is the net charge of atom i, $D_{mi} = \sum_\ell 2C_{i\ell}^2$ is the orbital charge of atom i in the Molecular Orbital (MO) m, $C_{i\ell}$ being the coefficients of the atomic orbitals (AO) of i, β_{ij} is the resonance integral; and E_m ($E_{m'}$) is the energy of the mth (m'th) occupied (virtual) MO of the drug, n and n' standing for the receptor. The value of β_{ij} is kept independent of the kind of AO because the drug-receptor complex does not involve covalent bonds. The summation on p is over all pairs of interacting atoms.

The first term of the right side of Eq. (11) represents the electrostatic interaction between two atoms having net charges Q_i and Q_j and is identical to the expression for the electrostatic energy in the CLETAL approach. The second and third terms introduce a partial electron transfer from MO m to MO n' and from MO n to MO m', respectively. It is clear that, for the last terms to be significant, a channel must be provided for the partial electron transfer. These terms do not appear in the CLETAL version above presented.

As we said before, the electronic-geometric structure of the receptor is not known in the great majority of cases, therefore we cannot directly evlauate Eq. (11).

Peradejordi et al. overcame this problem by replacing the MO energies of the receptor by constant values [62]. This permits transformation of Eq. (11) into the following:

$$\Delta E = \sum_i [f_i Q_i + g_i S_i^E + h_i S_i^N] \tag{12}$$

where f_i, g_i and h_i are constants, S_i^E and S_i^N are, respectively, the total atomic electrophilic and nucleophilic superdelocalizabilities of atom i [63, 64], defined as:

$$S_i^E = 2 \sum_m \sum_r C_{mr}^2 / E_m \tag{13}$$

where the summation on m is over the occupied MO's and the one on r is over the AO's coefficients of atom i contributing to one MO, and:

$$S_i^N = 2 \sum_{m'} \sum_r C_{m'r}^2 E_{m'} \tag{14}$$

where the summation on m' is now over the virtual MO's.

These reactivity indices can be directly interpreted. Within a given molecule, S_i^E represents the relative capacity to transfer electrons to an electron-deficient center and S_i^N represents the relative capacity to accept electrons. When we are comparing a family of drugs sharing a common skeleton, these indices can be compared for similar atoms within the family.

The Peradejordi *et al.* approximation can be justified by assuming that the drug-receptor interaction is charge controlled [59, 62], but if the processus is not, the problem of the evaluation of the orbital energies of the receptor remains.

The last approximation can be improved by considering two facts.

1. The receptor is, in general, a macromolecule composed by thousands of atoms and the Molecular Orbitals can be considered as forming part of bands. In this case we can replace the set of the receptor's MO energies by another set composed by average values corresponding to the arithmetic media of the band energy.
2. The Frontier Molecular Orbitals are probably the ones involved in weak interactions.

These considerations, coupled to the fact that we can employ a series expansion of the energy denominators, permit to arrive to the following expression for E [11]:

$$\Delta E = a + \sum_i [e_i Q_i + f_i S_i^E + s_i S_i^N] + \sum_i \sum_m [h_i(m)D_i(m) + j_i(m)S_i^E(m)] +$$

$$+ \sum_i \sum_{m'} [r_i(m')D_i(m') + t_i(m')S_i^N(m')] \tag{15}$$

where a, e, f, g, h, j, r and t are constants, $S_i^E(m)$, $S_i^N(m')$ and $D_i(m)$ are, respectively, the orbital electrophilic superdelocalizability of MO m at atom i, the orbital nucleophilic superdelocalizability of MO m' and the orbital electron density of MO m at the same atom. The summation on m includes a group of MOs close to the Highest Occupied Molecular Orbital (HOMO), and the HOMO itself. The summation on m' includes the Lowest Empty Molecular Orbital (LEMO) and a group of low-lying virtual MOs. The other terms are the same as those appearing in Eq. (12).

Up to now, there are no comparative studies of the CLETAL and KPG approaches. The CLETAL modeling has shown that it works very well for crystals and nucleotides [44—53], but no applications to Quantum Pharmacology are known. In the next section, we shall present a summary of the KPG approach applied to a pharmacological problem.

3. A Practical Example: the Serotonin Receptor Binding Affinity

Serotonin (5-hydroxytryptamine, 5-HT) is a neurotransmitter involved in a variety of actions: neuronal inhibition, smooth muscle contraction (rat stomach fundus) and relaxation (guinea pig ileum), tachycardia, hypotension, oedema, depolarization, etc. These actions are mediated through more than one 5-HT receptor [65].

The affinity for the 5-HT serotonin receptors of the rat stomach fundus has been measured by Glennon et al. for a very large quantity of indolealkylamines (Figure 5) and phenylalkylamines (Figure 6) ([66—74] and references therein). This system is of interest because there is experimental evidence showing that there is a linear correlation between the potencies of some indolealkylamines on the rat fundus and their capacity to inhibit Lysergic Acid Diethylamide (LSD) binding to brain membranes [75—76]. Therefore, the conclusions reached in the fundus preparation might also hold for one of the brain receptors.

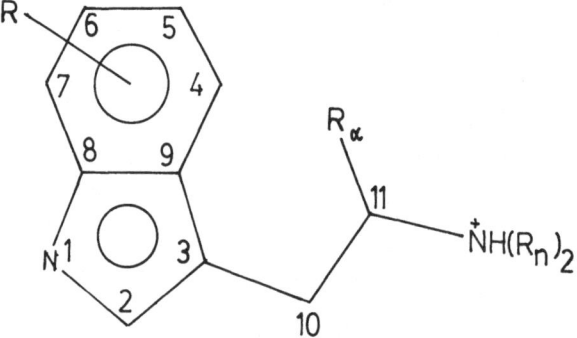

Fig. 5. General formula for indolealkylamines.

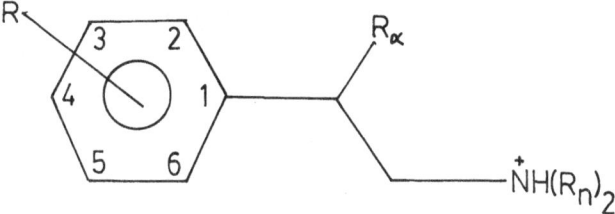

Fig. 6. General formula for phenylalkylamines.

QSAR studies on these molecules have been reviewed by Gupta *et al.* [77]. They concluded that none of the equations obtained were highly significant either because a small number of data points were employed, or because the data had been obtained using methods involving too many approximations [77]. To their criticism, we may add that the previous studies have normally examined a few individual parameters through the use of totally empirical methods.

Our first QSAR study with the KPG approach relating indolealkylamine serotonin receptor binding affinity (pA_2) to electronic structure indices obtained one equation explaining the variation of the pA_2 for 24 molecules, an equation that was the best one to that date [78]. Nevertheless, our study considered an excessive diversity in the structures of the molecules. For this reason, we carried out two more studies, one for 5-substituted tryptamines [79] and the other for 7-substituted tryptamines [80]. The results are presented in Tables III and IV

TABLE III. Experimental and calculated pA_2 for 5-substituted tryptamines.

Molecule	R_5	R_N	Experimental pA_2[a]	Calculated pA_2[b]	
1	Me	H	6.86	6.56	6.70
2	OH	Me	7.41	7.39	7.33
3	OMe	Me	7.08	7.40	7.35
4	Me	Me	6.52	6.29	6.29
5	OCOMe	Me	7.71	7.40	7.38
6	COMe	Me	5.86	6.01	5.98
7	OCOEt	Me	7.27	7.40	7.38
8	OCO-N-prop	Me	7.32	7.40	7.39
9	SMe	Me	6.84	6.65	6.59
10	OMe	Me, Et	6.85	6.91	7.03
11	OMe	Et	6.94	6.91	6.91
12	NH_2	Me	7.08	7.21	7.19
13	NH_2	H	7.53	7.38	7.53
14	OCO-*t*-but	Me	7.42	7.40	7.39
15	$OCOCH(Me)_2$	Me	7.40	7.40	7.39
16	H	H	6.27	6.24	6.38
17	H	Me	6.00	6.18	6.11
18	H	H, Me	5.97	6.16	6.11
19	H	Et	5.79	5.76	5.74

a. Refs. [66–75]; b. Ref. [79].

TABLE IV. Experimental and calculated pA_2 for 7-substituted tryptamines.

Molecule	R_7	R_α	R_N	Experimental pA_2[a]	Calculated pA_2[b]	
1	H	H	H	6.27	5.91	6.22
2	H	H	Me	6.00	6.14	6.14
3	Me	H	Me	6.29	6.05	6.19
4	OMe	H	Me	5.33	5.36	5.20
5	OH	H	Me	4.88	5.02	5.12
6[c]	H	H	Me	6.04	6.22	5.83
7[d]	H	H	Me	6.02	6.19	6.31
8[e]	H	H	Me	6.03	5.93	5.88
9	H	H	H, Me	5.97	6.02	6.01
10	Br	H	Me	6.51	6.49	6.37
11[f]	H	H	Me	5.68	5.55	5.64
12[g]	H	Me	H	5.49	5.49	5.49
13[h]	H	Me	H	6.46	6.44	6.55
14	H	H	Et	5.79	5.88	5.84
15	Et	H	Me	6.31	6.33	6.24

a. Refs. [66—75]; b. Ref. [80]; c. With a Me group at pos. 2; d. With a Me group on the indole N; e. With a S atom instead of the indole NH; f. With a CH_2 group instead of the indole NH; g. (S) $(+)$ isomer; h. (R) $(-)$ isomer.

respectively. Their inspection reveals that the predicted pA_2's are in excellent agreement with the experimental ones.

The results for the 7-substituted tryptamines were very striking because they suggested that the 7-substituent influenced the pA_2 through its steric effect (related to its size or to its influence in rising or diminishing the molecule's probability to reach the correct geometrical alignment permitting the DR interaction), and its effect on the total atomic electrophilic superdelocalizability of atom 7 ([80] see also Figure 5). Older theories suggested that it was the hydrophobicity of the 7-substituent the magnitude influencing the pA_2 [81].

On the other hand, a QSAR study employing the KPG model for the case of indolealkylamines [11] showed that the equations were very similar to the ones for indolealkylamines.

If indolealkylamines and phenylalkylamines interact with the rat stomach fundus serotonergic receptor by analogous mechanisms, our studies suggest that a phenylalkylamine carrying a small hydrophylic group at position 4 (see Figure 6), must have a high pA_2 value if its electrophilic superdelocalizability at C-4 is high enough. Moreover, if this compound is able to pass the blood-brain barrier it must be hallucinogenic at low doses. This is the case of the 1-(2,5-dimethoxy-4-nitrophenyl)-2-aminopropane (DON), having a pA_2 of 7.07 for the racemic mixture [82]. Our preliminary pharmacological studies showed that DON is a very potent hallucinogen [83].

This last result, obtained from a model-based study of the experimental evidence, clearly indicates that these methods are much better than the empirical ones.

I would like to conclude this section by saying that with the computing facilities available now, the employ of the CLETAL KPG or any other model-based approach to design new drugs or to study the physics of the DR interaction must replace the older methods. Naturally, there are other problems that are still not modeled in a formal way (the pharmaceutical and pharmacokinetic steps of drug action), but we have the tools to do it.

Acknowledgements

This paper is dedicated to Professor Raymond Daudel on the occasion of his 66th birthday. Thanks are due to Dr. Federico Peradejordi (Madrid), who first called my attention to Quantum Pharmacology, and to Dr. Jean Maruani (Paris), for his kind hospitality during my visit to the Laboratoire de Chimie Physique, where this article was finally structured.

Departamento de Investigacion y Bibliotecas (DIB), University of Chile, Fondo Nacional de Ciencia, Dr. Camilo Quezada, Dean of the Faculty of Sciences, CONICYT and ICSU are gratefully acknowledged for financial support. I would also like to thank Mr. Ruben Madrid, Director of SECI, University of Chile, for providing computer time.

References

1. J. A. Pople and D. L. Beveridge: *Approximate Molecular Orbital Theory*, McGraw-Hill (1970).
2. G. A Segal, Ed.: *Semiempirical Methods of Electronic Structure Calculations*, Part A, Plenum (1977).
3. J. N. Murrel and H. J. Harget: *Semi-empirical Self-consistent-field Molecular Orbital Theory*, Wiley (1976).
4. W. J. Hehre, L. Radom, P. v. R. Schleyer, and J. A. Pople: *Ab initio Molecular Orbital Theory*, Wiley (1986).
5. D. B. Cook: *Ab initio Valence Calculations in Chemistry*, Butterworths (1974).
6. H. F. Schaefer III, Ed.: *Methods of Electronic Structure Theory*, Plenum (1977).
7. D. G. Lister, J. N. MacDonald, and N. L. Owen: *Internal Rotation and Inversion*, Academic Press (1978).
8. B. Pullman, Ed.: *Quantum Mechanics of Molecular Conformations*, Wiley (1976).
9. E. Steiner: *The Determination and Interpretation of Molecular Wave Functions*, Cambridge U. Press (1976).
10. P. Politzer and D. G. Truhlar, Eds.: *Chemical Applications of Atomic and Molecular Electrostatic Potentials*, Plenum (1981).
11. J. S. Gómez-Jeria and D. Morales-Lagos: 'The mode of binding of Phenylalkylamines to the Serotonergic Receptor' (*QSAR in Design of Bioactive Compounds*, Ed. M. Kuchar), pp. 143–175. J. R. Prous (1984).
12. A. Hinchliffe and J. C. Dobson: *Chem. Soc. Rev.* **5**, 79 (1976).
13. E. J. Ariens, A. J. Beld, J. F. Rodriguez de Miranda, and A. M. Simonis: *The Pharmacon Receptor-Effector Concept* (The Receptors, v. 1, Ed. R. D. O'Brien), pp. 33–91. Plenum (1980).
14. Y. C. Martin: *Quantitative Drug Design*, Marcel Dekker (1978).

15. S. M. Free and J. W. Wilson: *J. Med. Chem.* **7**, 395 (1964).
16. T. Fujita and T. Ban: *J. Med. Chem.* **14**, 148 (1971).
17. C. Hansch and T. Fujita: *J. Am. Chem. Soc.* **86**, 1616 (1964).
18. F. G. van den Brink: 'General Theory of Drug-Receptor Interactions' (*Kinetics of Drug Action*, Ed. J. M. van Rossum), pp. 169—254. Springer-Verlag (1977).
19. J. M. Boeynaems and J. E. Dumont: *Outlines of Receptor Theory*, Elsevier/North-Holland (1980).
20. E. J. Simon: *Annals N.Y. Acad. Sci.* **463**, 31 (1986).
21. R. A. Glennon: 'Involvement of serotonin in the action of hallucinogenic drugs' (*Neuropharmacology of Serotonin*, Ed. A. R. Green), pp. 253—280. Oxford U. Press (1985).
22. J. B. Penney and H. S. Pan: 'Quantitative Autoradiography of GABA and Benzodiazepine binding in studies of mammalian and human basal ganglia function' (*Quantitative Receptor Autoradiography*, Ed. C. A. Boast), pp. 29—52. Alan R. Liss, New York (1986).
23. A. Barnett, H. Ahn, W. Billard, E. H. Gold, J. D. Kohle, D. Glock, and L. I. Goldberg: *Eur. J. Pharmacol.* **128**, 249 (1986).
24. F. W. Schueler: *Chemobiodynamics and Drug Action*, McGraw-Hill (1960).
25. E. J. Ariens, A. M. Simonis, and J. M. van Rossum 'Drug-receptor interaction: Interaction of one or more drugs with one receptor system' (*Molecular Pharmacology*, v. 1, Ed. E. J. Ariens), pp. 119—286. Academic Press (1964).
26. J. J. Kaufman, P. C. Hariharan, H. E. Popkie, and C. Petrongolo: *Ann. N.Y. Acad. Sci.* **367**, 452 (1981).
27. J. S. Gómez-Jeria, D. Morales-Lagos, and J. I. Rodriguez-Gatica: *Acta Sud Amer. Quim.* **4**, 1 (1984).
28. J. S. Gómez-Jeria: *Int. J. Quant. Chem.* **19**, 1969 (1983).
29. P. Arrighini: *Intermolecular Forces and their Evaluation by Perturbation Theory*, Springer-Verlag (1981).
30. J. O. Hirschfelder: *Chem. Phys. Lett.* **1**, 325 (1967).
31. R. Rein: 'On Physical Properties and Interactions of Polyatomic Molecules, with Applications to Molecular Recognition in Biology' (*Advances in Quantum Chemistry*, v. 7, Ed. P. O. Löwdin), pp. 335—396. Academic Press (1973).
32. W. Kolos: 'Some Problems of the Theory of Intermolecular Interactions' (*The World of Quantum Chemistry*, Eds. Я. Daudel and B. Pullman), pp. 31—42. D. Reidel Publ. Co. (1974).
33. P. Claverie: *Int. J. Quant. Chem.* **5**, 273 (1971).
34. P. Claverie: 'Elaboration of approximate formulas for the interactions between large molecules: applications in Organic Chemistry' (*Intermolecular Interactions: from Diatomics to Biopolymers*, Ed. B. Pullman), pp. 69—305. John Wiley (1978).
35. J. N. Murrel and G. Shaw: *J. Chem. Phys.* **46**, 1768 (1967).
36. J. I. Musher and A. T. Amos: *Phys. Rev.* **164**, 31 (1967).
37. R. Eisenschitz and F. London: *Z. Phys.* **60**, 491 (1930).
38. J. O. Hirschfelder: *Chem. Phys. Lett.* **1**, 363 (1967).
39. A. van der Avoid: *J. Chem. Phys.* **47**, 3649 (1967).
40. J. O. Hirschfelder and R. Silbey: *J. Chem. Phys.* **45**, 2188 (1966).
41. W. H. Adams: *Phys. Rev. Lett.* **32**, 1093 (1974).
42. D. M. Chipman: *J. Chem. Phys.* **66**, 1830 (1977).
43. W. H. Adams and E. E. Polymeropoulos: *Phys. Rev.* **A17**, 11, 18, 24 (1978).
44. P. Claverie: *Stud. Bioph.* (*Berl.*) **24/25**, 161 (1970).
45. J. Caillet and P. Claverie: *Biopolymers* **13**, 601 (1974).
46. J. Caillet and P. Claverie: *Acta Cryst.* **A31**, 448 (1975).
47. J. Caillet, P. Claverie, and B. Pullman: *Acta Cryst.* **B32**, 2740 (1976).
48. N. Gresh, P. Claverie, and A. Pullman: *Int. J. Quant. Chem.* **S13**, 243 (1979).
49. N. Gresh, P. Claverie, and A. Pullman: *Int. J. Quant. Chem.* **22**, 199 (1982).
50. N. Gresh, P. Claverie, and A. Pullman: *Int. J. Quant. Chem.* **29**, 101 (1986).
51. P. Claverie: 'Intermolecular Interactions and Solvent Effects: Simplified Theoretical Methods' (*Quantum Theory of Chemical Reactions*, v. 3, Eds. R. Daudel, A. Pullman, L. Salem, and A. Veillard), pp. 151—175. D. Reidel Publ. Co. (1982).

52. M. J. Huron and P. Claverie: *J. Phys. Chem.* **76**, 2123 (1972).
53. M. J. Huron and P. Claverie: *J. Phys. Chem.* **78**, 1862 (1974).
54. A. I. Kitaigorodski: *Tetrahedron* **14**, 230 (1961).
55. A. I. Kitaigorodski and K. W. Mirskaya: *Sov. Phys. — Crystallogr.* **9**, 137 (1965).
56. A. I. Kitaigorodski, K. W. Mirskaya, and V. V. Nachitel: *Sov. Phys. — Crystallogr.* **14**, 769 (1970).
57. A. Bondi: *J. Phys. Chem.* **68**, 441 (1964).
58. R. F. Hudson and G. Klopman: *Tetrahedron Lett.* **17**, 1105 (1967).
59. G. Klopman and R. F. Hudson: *Theor. Chim. Acta* **8**, 165 (1967).
60. G. Klopman: *J. Am. Chem. Soc.* **90**, 223 (1968).
61. G. Klopman: 'The Generalized Perturbation Theory of Chemical Reactivity and its Applications' (*Chemical Reactivity and Reaction Paths*, Ed. G. Klopman), pp. 55—165. John Wiley (1974).
62. F. Peradejordi, A. N. Martin, and A. Cammarata: *J. Pharm. Sci.* **60**, 576 (1971).
63. K. Fukui, T. Yonezawa, and C. Nagata: *Bull. Chem. Soc. Jap.* **27**, 423 (1954).
64. K. Fukui, T. Yonezawa, and C. Nagata: *J. Chem. Phys.* **27**, 1247 (1957).
65. P. B. Bradley, G. Engel, W. Feniuk, J. R. Frozard, P. P. A. Humphrey, D. N. Middlemiss, E. J. Mylecharane, B. P. Richardson, and P. R. Saxena: *Neuropharmacol.* **25**, 563 (1986).
66. R. A. Glennon and P. K. Gessner: *Res. Commun. Chem. Pathol. Pharmacol.* **18**, 453 (1977).
67. R. A. Glennon, S. M. Liebowitz, and E. C. Mack: *J. Med. Chem.* **21**, 822 (1978).
68. R. A. Glennon and P. K. Gessner: *J. Med. Chem.* **22**, 428 (1979).
69. R. A. Glennon, S. M. Liebowitz, and G. M. Anderson: *J. Med. Chem.* **23**, 294 (1980).
70. R. A. Glennon: *Life. Sci.* **29**, 861 (1981).
71. R. A. Glennon, R. Young, F. Benington, and R. D. Morin: *J. Med. Chem.* **25**, 1163 (1982).
72. R. A. Glennon, J. M. Jacyno, R. Young, J. D. McKenney, and D. Nelson: *J. Med. Chem.* **27**, 41 (1985).
73. R. A. Glennon, E. Schuber, and J. M. Jacyno: *J. Med. Chem.* **23**, 1222 (1980).
74. R. A. Glennon: *Life. Sci.* **24**, 1487 (1979).
75. R. A. Glennon: *Res. Commun. Psychol. Psychiat. Behav.* **4**, 333 (1979).
76. J. P. Green, C. L. Johnson, H. Weinstein, S. Kang, and D. Chou: 'Molecular determinants for interaction with the LSD receptor: biological studies and Quantum Chemical analysis' (*The Psychopharmacology of Hallucinogens*, Eds. R. S. Willette and R. C. Stillman), pp. 28—60. Pergamon Press (1978).
77. S. Gupta, P. Singh, and M. Bindal: *Chem. Rev.* **83**, 633 (1983).
78. J. S. Gómez-Jeria and D. Morales-Lagos: *J. Pharm. Sci.* **73**, 1725 (1984).
79. J. S. Gómez-Jeria, D. Morales-Lagos, J. I. Rodriguez-Gatica, and J. C. Saavedra-Aguilar: *Int. J. Quant. Chem.* **28**, 421 (1985).
80. J. S. Gómez-Jeria, D. Morales-Lagos, B. K. Cassels, and J. C. Saavedra-Aguilar: *QSAR Phys., Chem. and Biol.* **5**, 153 (1986).
81. C. L. Johnson and J. P. Green: *Int. J. Quant. Chem.* **QBS1**, 159 (1977).
82. R. A. Glennon, R. Young, F. Benington, and R. D. Morin: *J. Med. Chem.* **25**, 1163 (1982).
83. J. S. Gómez-Jeria and B. K. Cassels: *Proc. 16th Chilean Chemistry Meeting.* Osorno, Chile, p. 126 (1985).

Raman and Infrared Study of Acetylcholine and Postsynaptic Membranes

DIMITRINA ASLANIAN
Laboratoire de Physique et Spectroscopie des Solides associé au CNRS,
Université Pierre et Marie Curie, 4, place Jussieu, 75252 Paris Cedex 05, France.

1. Introduction

Raman and infrared (IR) spectroscopies provide largely complementary information on the structure and interaction of organic and biological molecules [1—6].

Raman spectroscopy followed the discovery of the inelastic scattering process in 1928 by the Indian physicist C. V. Raman his eye as detector, the sun as source and a telescope as receiver. Raman was awarded the Nobel prize in 1930 and the new phenomenon has been named the Raman effect. The spectroscopy based on this effect proved to be an exceptional tool for the investigation of molecules. Ten years after its discovery far more than 2000 papers reporting the Raman spectra of more than 4000 compounds had been published. In this period Raman spectroscopy has also been used to study biological systems. Thus, Edsall and coworkers [7—13] employed Raman analysis to characterize the ionic properties of amino acids and dipeptides using a mercury arc as a light source and photographic recording. However, rapid advances in IR instrumentation after the second world-war made IR spectroscopy an easier and quicker method. Thus, in spite of the efforts of many researchers, Raman spectroscopy remained primarily a tool of academic interest. The development of lasers as powerful monochromatic excitation sources gave rise to a remarkable revival of Raman spectroscopy. This, together with other instrumental and sampling improvements in the period of 1960—1970, favoured a more extensive use of the method and its application to molecules and macromolecular systems.

This review reflects the author's interest in the application of Raman and IR spectroscopy to the study of chemical transmission in the nervous system. We will focus on how both spectroscopic methods have extended our knowledge on the conformation of two principal elements involved in the chemical transmission of nerve impulsions. The first element, acetylcholine (Ach), is the most important chemical transmitter of the peripheral nervous system. The second principal element contributing to the chemical transmission of the nerve impulsions is the postsynaptic membrane.

In this paper, the principles of Raman and IR spectroscopy and their complementarity will be outlined first. Then, the biological characteristics and the preparation of samples will be discussed and this will be followed by comments on

Jean Maruani (ed.), Molecules in Physics, Chemistry, and Biology, Vol. IV, 233—280.
© 1989 *by Kluwer Academic Publishers.*

the recording and assignments of spectra. In the last part will be presented the experimental results obtained in the author's laboratory. Thus, Raman and IR data on the conformation of Ach together with data on the conformation of some Ach analogues (nicotine, muscarine, β-methylacetylcholine) gave very useful correlative information on the conformational possibilities of these molecules in transition from solid state to aqueous solution. With this information in hand as well as that obtained from Raman and IR study of the non-enzymatic hydrolysis of Ach, the first experimental model for the interaction of Ach with a surface has been proposed using a third vibrational spectroscopic method i.e., inelastic electron tunneling spectroscopy, IETS. Also, original information on the conformation of the proteins and lipids which compose the postsynaptic membranes has been presented.

2. Principles of the Methods

Both Raman and IR spectroscopy give data which describe molecular vibrations determined by the small oscillations (10^{10} to 10^{11} Hz) of the individual atoms in a molecule around their mean position [4—6]. However, the means by which the information is obtained is quite different in each case, because these spectroscopies rely upon fundamentally different physical processes, namely scattering and absorption.

The Raman effect is a molecular light scattering phenomenon in which a change in the frequency of light occurs. When a monochromatic laser beam with frequency ν_0 (usually 3×10^{14} to 7×10^{14} Hz) interacts with a molecule, the electric component of light exerts some force on all electrons of the molecule and tends to displace them from their average positions around the positively charged nuclei. The molecule develops an induced frequency and the induced dipole moment vibrates at frequency ν_0 with an amplitude proportional to the polarizability of the molecule. As a result the molecule emits Rayleigh radiation of frequency ν_0 (elastic scattering).

The polarizability of the molecule depends on its size, shape and orientation and may be modified as a result of change of these parameters [14]. Thus, Raman scattering arises when the polarizability of the electrons in the electric field is modified as a result of a change in the shape of the molecule due to the atomic nuclear vibrations. If the vibration of a given molecule involves displacement of the nuclei which changes the polarizability, then one observes a shift in the frequency of the scattered light corresponding to the frequency of the nuclear motions (inelastic scattering). As a result of these processes, the light scattered from the molecules contains not only the frequency of the excitation light (ν_0) but also the sum and difference of the molecular vibrational frequencies ($\nu_R = \nu_0 \pm \nu_v$). The frequency shifts $\nu_0 - \nu_v$ and $\nu_0 + \nu_v$ are called Stokes and anti-Stokes vibrations, respectively. They are associated with quantized changes in vibrational energies of the individual molecules as well as with quantized changes in intermolecular motions in the condensed state (lattice frequencies in crystals).

An energy level diagram for Raman scattering is shown in Figure 1. The energy levels lie far below those required to excite the first electronic level. When the molecule is excited from its fundamental vibrational level $n = 0$ to vibrational state $n = 1$ Stokes vibrations with frequency $\nu_R = \nu_0 - \nu_v$, arise. Anti-Stokes vibrations appear when the molecule, initially in the $n = 1$ state, scatters radiations of frequencies $\nu_0 + \nu_v$ and reverses to the $n = 0$ state. As the population of the molecules at $n = 0$ is higher than that at $n = 1$ level, the Stokes vibrations are always more intense than the anti-Stokes ones. In Raman studies, the Stokes vibrations are usually recorded.

If the energy of the incident light (ΔEn_0) is identical to the energy differences (ΔE) between two vibrational levels, for instance $\Delta E = E_{n=1} - E_{n=0}$, then the molecule can absorb light. This kind of absorption takes place in IR spectroscopy (Figure 1). Infrared spectroscopy is a molecular light absorption phenomenon which is due to the periodical variations of the permanent electric dipole moment of the molecules. The atoms of a molecule vibrate at frequencies in the IR region (10^{10} to 10^{11} Hz) which are identical to the Raman vibrational frequencies (Figure

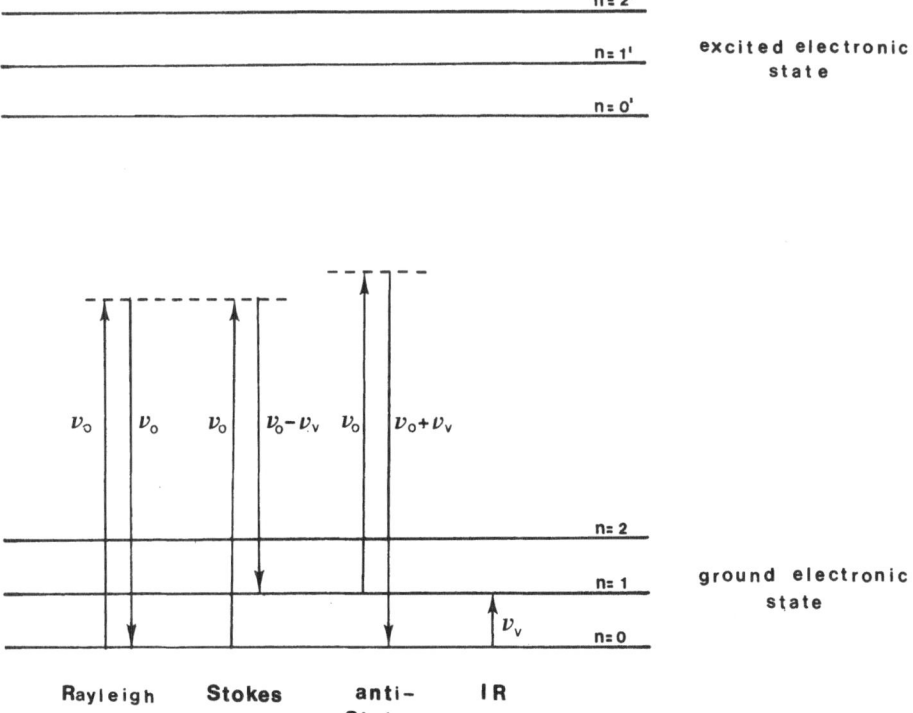

Fig. 1. Energy level diagram for light scattering.

1). However, due to the fundamentally different physical processes in both spectroscopies, many molecular vibrations that yield weak Raman signals show intense IR signals and conversely. In Raman spectroscopy, the intensity of vibrations depends upon the degree of modulation of the polarizability of the scattering species. Thus, symmetric vibrations of symmetric species and groups which contain polarizable atoms tend to scatter strongly (covalent bonds). In the case of IR spectroscopy, the absorption intensity relies upon a change in the dipole moment of the absorbing species. Since asymmetric species have large dipole moments than more symmetric species, strong IR features arise from polarized groups and asymmetric vibrations of symmetric groups.

Therefore, through combined utilization of both methods, it is possible to obtain complementary information regarding the identification and the characterization of specific chemical groups in a molecule and to elucidate their interactions with other nearby bonds.

3. Biological Characteristics of the Samples

The samples whose conformation is the subject of discussion in this article, are two indispensable elements for the chemical transmission of nerve impulses in the peripheral nervous system: the transmitter molecule, acetylcholine (Ach) and the postsynaptic membrane. Binding of Ach to the intrinsic channel protein embedded in the postsynaptic membrane, the Ach-receptor protein, assures the transmission of signals from one neuron to another or to a muscle.

In most neurons three regions may be distinguished: the cell body which contains the nucleus, a number of short branches (dendrites) and a single relatively straight process — the axon. The latter, transmits cell responses with the help of a terminal button — the synapse (Figure 2).

Chemical synapses can be conveniently divided into two components: (1) the presynaptic component, which is secretory, and (2) the postsynaptic component, which is receptive and transductive. A synaptic cleft (some 20 nm across) separates them. An essential feature of the presynaptic terminal is the accumulation of vesicles which contain the chemical transmitter (Figure 2).

The chemical transmitter Ach is an organic molecule $[CH_3OCOCH_2CH_2N^+-(CH_3)_3]$ which was the first specific chemical substance demonstrated as being released in response to nerve stimulation [15]. Ach is the transmitter used by motor neurons of the spinal cord and therefore, at all nerve-skeletal muscle junctions in vertebrates. In the vegetative nervous system, it is the transmitter for all preganglionic neurons and for the parasympathetic postganglionic neurons. Ach could be found also in many synapses throughout the brain.

The biochemistry of Ach is well understood and has been elucidated in 1952 by Nachmansohn and coworkers [16].

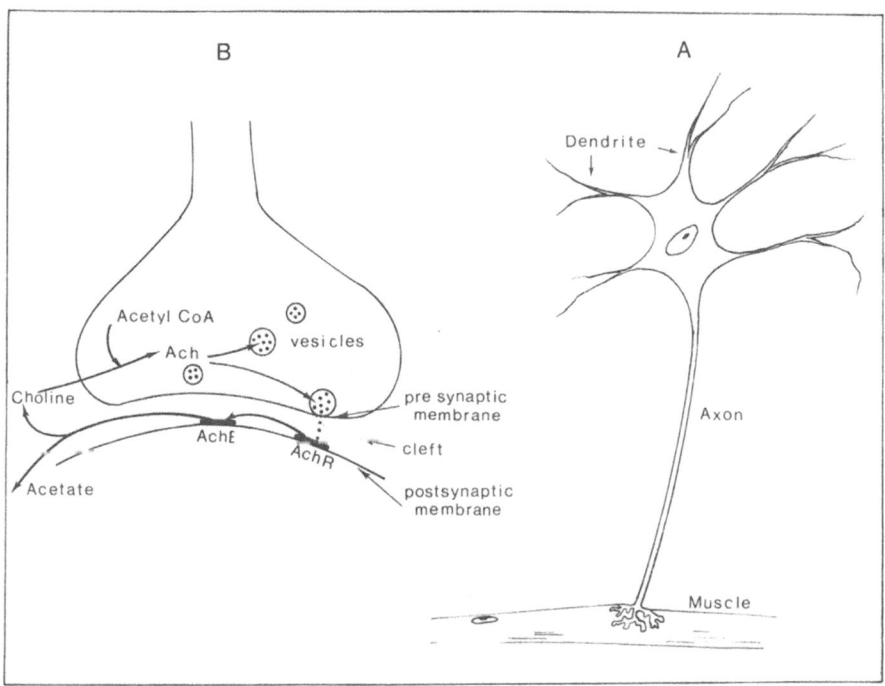

Fig. 2. Diagram of a motor neuron (A) and a synapse (B).

The synthesis of Ach in neurons is catalyzed by the enzyme choline acetyltransferase (CAT) which was discovered in the early 1940s [17, 18]. The catalysis consists of a transfer of acetate from acetyl CoA to choline [19]:

$$CH_3\overset{\overset{O}{\|}}{C}-CoA + (CH_3)N^+CH_2CH_2OH \longrightarrow CH_3\overset{\overset{O}{\|}}{C}-O-CH_2CH_2N^+(CH_3)_3 + CoA$$

| acetyl-coenzyme A | + | choline | (CAT) | acetylcholine | + | coenzyme A |

The enzyme choline acetyltransferase (CAT) can be found only in the nervous system. Cholinergic neurons (those that liberate Ach from their terminals) have a high level of this enzyme. The nervous tissue cannot synthesize choline which is delivered to neurons through the blood stream [20, 21]. The active transport of choline across the neuronal membrane is assured by a sodium dependent system [22, 23].

According to the vesicles hypothesis [24, 25] the transmission of nerve impulsions is accomplished by the liberation of Ach in packets (quantum release) into the synaptic cleft. When impulses reach the presynaptic terminal, some of the vesicules fuse with the membrane at specific release sites and liberate their contents by exocytosis into the synaptic cleft. It seems that the vesicular mechanism is not the only one the cell uses [26—28]. Proof of the vesicular hypothesis has been difficult to establish. It relies mostly on the ability to explain a rapid and large release of Ach, and on the lack of viable alternatives.

The diffusion of Ach across the 20 nm synaptic cleft is followed by its binding to specific receptors on the postsynaptic membrane and to AchE enzyme which hydrolyzes it (Figure 2).

There are two known types of receptors of Ach, the nicotinic and the muscarinic. The nicotinic receptors are found at neuromuscular junctions of the skeletal muscle in the autonomic ganglia and in the electroplax of the *Electric eel* or *Torpedo marmorata* fish. Pharmacologically, they are characterized by the property that they are stimulated by nicotine and blocked by curare. The muscarinic receptors are found in visceral autonomic cells, in smooth muscles and in brain, and are stimulated by muscarine and blocked by atropine. Both, of course, are stimulated by Ach [29—31].

It is a matter of considerable interest to elucidate:

1. the molecular parameters that enable the Ach molecule to interact with both groups of receptors;
2. the receptor organization and localization, which depend on the current understanding of the postsynaptic membrane structure and function.

The importance of knowledge on the conformation of Ach has stimulated a great amount of experimental and theoretical work. Starting from 1960, numerous attempts have been made to study the conformation of Ach molecule alone and in comparison with related substances of cholinergic systems. They were reviewed recently [32].

On the other hand, the pioneering work of Nachmansohn, Karlin and Changeux [33—35] stimulated many other researchers to study the structure and dynamics of the Ach-receptor as well as the correlation between the Ach-receptor structure and its gating properties. Actually, the best characterized receptor is the nicotinic Ach-receptor from the electric organ of electric fish.

The electric organ from electric fish *Torpedo marmorata* or *Torpedo californica* is a remarkable system for the study of synaptic transmission because its ontogeny resembles the development of muscle. In *Torpedo*, the embryonic fish contains muscle cells which develop into an oriented stack of polar cells creating the electric organs. Schematically the electric organ may be viewed as a muscle which has lost its contractibility but preserved its excitability. Thus, on the molecular level, it consists of cells that are counterparts of muscle cells.

Torpedo is commonly found in shallow coastal waters. The European species, *Torpedo marmorata*, is 40 cm long when adult and weighs about 1500 g. Each fish possesses two electric organs, one on each side of its head. These organs consist of some 400 prisms which are stacks of thin flat cells. These cells are called electroplax and generate the electric discharge which may be as large as 250 V and

0.5 A [36]. Electroplax are highly asymmetric cells and synapses are found only along the ventral side of these cells. Thus, only one face receives nerve terminals and is excitable whereas the other one is specialized for active transport. It was estimated that there are more than 5×10^{14} nerve terminals in each electric organ [28]. In addition, all these synapses function with the same chemical transmitter, acetylcholine [35]. The electric organ, therefore, contains an immense collection of identical synapses. Such richness and homogeneity make out of it a unique study material for the chemistry of synaptic transmission. Thus, using the tissue of electric organ as well as dissected single electroplax, numerous biochemical investigations, physiological measurements and pharmacological assays have been developed [35].

We shall mention here only two very important biochemical experiments connected with the samples that were subject to investigations by vibrational spectroscopy.

1. Purification of membrane fragments rich in Ach-receptor protein which still responded to Ach by a change of permeability to cations [37, 38].
2. Purification of homogeneous Ach receptor molecule [39—43].

The isolation and purification of membrane fragments rich in Ach-receptor as well as that of the receptor alone, has been successful using the radio-labelled snake venom α-bungarotoxin ([35] and ref. therein). This toxin is typically a high neurotoxic polypeptide of molecular weight between 7000 and 8000 D which binds irreversibly the Ach-receptor [44].

The isolated membranes form closed vesicles with diameters ranging from 0.1 to 0.5 μm [45]. Studied in electron microscopy by the freeze-etching technique, these membranes disclose an original arrangement of the rosettes in double rows which has been identified with the Ach-receptor protein molecule [45—46]. In most purified membranes the receptor represents up to 30—40% of the total protein content in a state in which its most characteristic functional properties are preserved. The surface density of the molecule has been found to be 8000—12 000/μm^2. Thus, the postsynaptic membrane emerges as an extremely specialized edifice with concentrated receptor molecules.

The electron microscopic observations, X-ray diffraction studies and use of analytical probes such as proteases or anti-Ach-receptor-specific antibodies [47—50] revealed an asymmetric distribution of the protein perpendicular to the lipid bilayer.

The Ach-receptor, a glycoprotein molecule with a molecular weight about 250 000 D [38, 51], appears as a cylindrical pentameric complex which extends 55 Å into the synaptic side of the membrane in a funnel-shaped form and 15 Å on the cytoplasmic side [49] (Figure 3). The complex is formed by four different subunits which are indicated as α, β, γ and δ with the stoichiometry 2α, β, γ, δ [52—54].

Using a variety of affinity labels and neurotoxins, the Ach-receptor binding sites

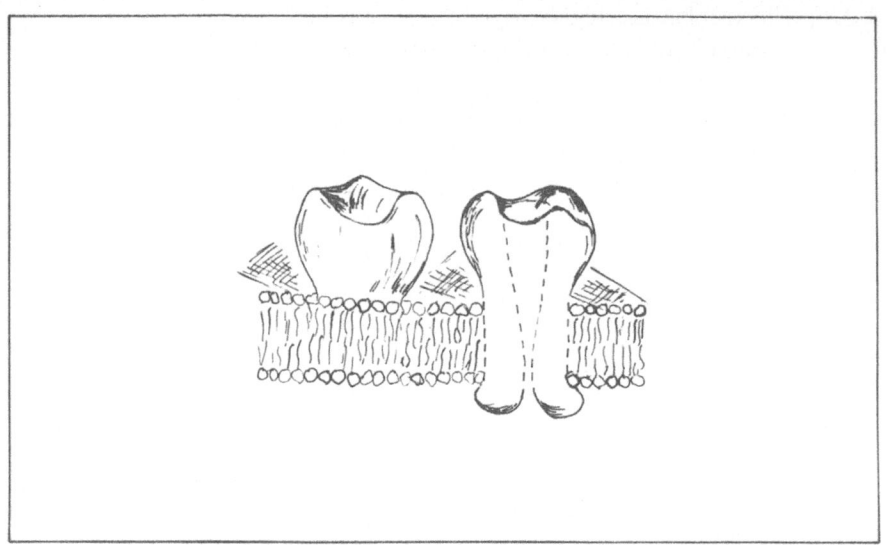

Fig. 3. A structural model of the acetylcholine receptor protein (according to Kistler and Stroud [47]), see text for details.

have been located on the two α-subunits [55]. Topographically, the subunits are distributed around a central channel in such a way that both α-subunits are adjacent to the β-subunit on one side and separated by the γ- and δ-subunits on the other side [48]. In the funnel shape part of the Ach-receptor molecule the channel is about 40 Å in diameter and narrows to about 7 Å at the level of the membrane surface [48, 49]. This value is close to the maximal diameter derived from electrophysiological measurements of conductivity of various cations with different sizes.

The amino acid sequences of the four Ach-receptor subunits have been determined in *Torpedo californica* [56, 57]. The α-sequences differ only by five or six residues from those obtained in the case of *Torpedo marmorata* [58, 59] and the subunit sequences differ only by one residue [60]. The four subunits are highly homologous and probably evolved from the same protein. Each subunit threads through the membrane several times, with the hydrophobic N-terminal portion on the synaptic cleft facing the membrane. How the separate subunits are brought together and assembled into the complete receptor complex within the post-synaptic membrane is not yet understood.

Another important protein of the postsynaptic membrane is a peripheral, non-receptor protein of apparent molecular weight of 43 000 daltons, referred as 43 kDa protein. This peripheral protein is localized on the cytoplasmic side of the membrane and can be extracted by treatment at alkaline pH [61, 62]. The extraction has essentially no effect on the functional properties of the membrane. However, important alterations such as increases in rotation [63], translation [64,

65], mobility and sensitivity to proteolysis and heat denaturation of Ach-receptor [66] have been reported. It seems that the 43 kDa protein may contribute to the immobilization of the receptor protein via a direct interaction.

The lipid composition of Ach-receptor-rich membrane has been analyzed by Popot *et al.* [67]. These authors found a rather high cholesterol/phospholipid weight ratio (0.40). A somewhat lower value (0.30) for the same ratio has also been reported [68]. The major part of total phospholipid (95%) in the Ach-receptor-rich membrane lipids is composed of three main categories of molecules: ethanolamine phosphoglycerides, choline phosphoglycerides and serine phosphoglycerides. Their fatty acids are characterized by the abundance of long chains of poly-unsaturated residues. Thus, the phospholipid composition, in general, seems to be comparable to the known phospholipid composition found in other cell membranes. On the contrary, the postsynaptic membrane distinguishes itself from other known cell membranes by a very high protein-to-lipid ratio which can reach values as high as 1.5−2.3 (w/w) [67].

4. Sample Preparation

Optical homogeneity and purity of samples are very important elements for successful spectroscopic studies. Especially in Raman spectroscopy, a great problem is the high luminescence background of samples, which can often obscure completely the weaker Raman signal. The luminescence stems from two possible sources: first, from natural chromophores which form a part of the sample under study and second, from impurities introduced during the preparation. The last source is the more frequent one and can be eliminated by purification of the samples. Therefore, special care should be given to assure spectroscopically pure samples. Organic molecules samples can be purified by multiple recrystallization, distillation and sublimation. For biological samples, the biochemical separations by simple or sophisticated methods, can improve considerably the sample purity.

4.1. ACETYLCHOLINE AND ANALOGUES

Acetylcholine chloride, bromide and iodide (AchCl, AchBr, AchI), Acetyl-β-methylcholine chloride and bromide (β-MeAchCl, β-MeAchBr) and Muscarine chloride (MuCl) come from Sigma Chemical Company and were used without further purification.

Nicotine (Nic) (Sigma) was purified by distillation in vacuum at 8°C. Nicotine dichlorhydrate (Nic2HCl) (Sigma) was purified by sublimation in vacuum at 110°C. Nicotine monochlorhydrate (NicHCl) was prepared by dissolving Nic in distilled water and titrated by HCl. After solvent evaporation the solid was vacuum dried. Deuterated NicHCl (NicDCl) was obtained from NicHCl in an exchange reaction with D_2O.

4.2. NATIVE POSTSYNAPTIC MEMBRANES

Ach-receptor-rich membranes from freshly dissected electric organ of *Torpedo marmorata* are routinely obtained by successive centrifugations and subsequent separation on density gradient [69]. Their specific activity (~ 2000 mM α-toxin binding sites per g protein) was determined according to Cohen *et al.* [70]. The total protein concentration was estimated by the method of Lowry [71] using bovine serum albumin as standard. Samples were used immediately after preparation or stored in liquid nitrogen and thawed at room temperature before use.

4.3. ALKALINE EXTRACTION OF THE 43 kDa PROTEIN

Alkaline extraction of 43 kDa and other non-receptor proteins was performed at 4°C using the method of Neubig *et al.* [61]. The alkaline treated membranes were pelleted, neutralized by resuspension of the pellet in a large volume of buffer (10 mM Tris-HCl, pH 7.5). After an additional centrifugation the samples were resuspended in a final concentration of about 20—30 mM toxin sites. The soluble, alkaline extracted material was neutralized upon controlled addition of HCl resulting in the aggregation of essentially the 43 kDa protein.

4.4. DETERGENT EXTRACTION OF THE ACH-RECEPTOR PROTEIN

The Ach-receptor protein was extracted from the membrane bilayer using the detergent dimethyldodecylamine oxide (DDAO). Alkaline treated membranes (about 10 mg protein/ml) were first supplemented with 0.6% mercapto-ethanol and 3% DDAO, shaken about 1 min and centrifugated at 100 000g for 20 min to remove the insoluble material. Further purification was achieved upon centrifugation of the soluble material on 5—20% sucrose gradient supplemented with 0.5% DDAO, at 40 000 rpm for 16 h in a Beckman SW 41 rotor. The fractions corresponding to the 9S 'light' form of Ach-receptor were dialysed for 5 h against the buffer (10 mM Tris-HCl, pH 7.5) supplemented with 0.5% DDAO to remove sucrose, and finally concentrated about 10-times in a 5 ml Diaflo cell using Amicon XM 30 membrane. Final protein concentration was about 20—25 mg protein/ml,and SDS gel analysis discloses almost exclusively the 4 polypeptide chains of the Ach-receptor. α-toxin binding site concentration was determined with 3H-labeled Naja nigricollis α-toxin as in [61].

5. Technical Procedures

5.1. RAMAN SPECTRA

Our Raman spectra were recorded using Coderg double (PHO and PH1) and triple (T-800) monochromators and Ar⁺ or Kr⁺ laser excitation sources, usually tuned at 488 nm and 514.5 nm.

Spectra of solid Ach, β-MeAch and Mu were recorded using a conical cavity realized by pressing the powder which is placed in a horizontal tube with a diameter of 3 mm. The advantage of this method resides in a multiple internal light diffusion. For the solid nicotins (NicHCl, Nic2HCl) a spinning cell (~ 2500 r/min) has been used in order to avoid sample decomposition under the laser beam. Slits were 2 cm^{-1} to 4 cm^{-1}. For aqueous solution samples cells of 1 cm^3 and 0.3 cm^3 with multipassage of the laser beam have been used. In these cases slits were 4 cm^{-1} to 6 cm^{-1}. Laser power at sample level varied between 90 and 600 mW.

Raman spectra of postsynaptic membranes were recorded using capillary tubes (1 mm internal diameter) held in vertical position in the middle of a brass holder. The volume of samples was usually 10 ml (20—30 mg protein/ml). Temperature was controlled by a thermocouple inserted into the sample holder adjacent to the capillary. Slits were 6 cm^{-1} to 8 cm^{-1} and laser power at the sample level was about 100—300 mW.

Spectra of all organic molecules (Ach and analogues) have been studied by classical recording techniques.

The biological samples are much more complex and most of them have been studied by a recording technique which involves computer assistance. Thus, the signal counter and the spectrometer motor were interfaced with an Apple IIe microcomputer. Points were collected at every 0.5 cm^{-1} in the range 400—1800 cm^{-1} and 2500—3120 cm^{-1}. The number of scans varied between 8 and 30 depending on the signal to noise ratio. Subtraction of the buffer and water spectrum, and the fluorescence baseline correction were made according to Williams [72, 73]. The smoothing of spectra was achieved by Fourier transforms.

5.2. INFRARED SPECTRA

Infrared spectra were recorded with a Perkin-Elmer 180 spectrometer. Faces of CaF$_2$, KRS-5 and CsI have been used.

6. Vibrational Frequency Assignment

The interpretation of vibrational spectra of molecules in terms of their chemical properties is the key for the use of Raman and IR spectroscopy. This is possible when one can relate the position of the peaks in a Raman spectrum or of the bands in a IR spectrum to the molecular conformation.

6.1. ASSIGNMENT OF ACH SPECTRA

As it has been noted above, the nature of both Raman and IR spectra is determined by small oscillations of the individual atoms in a molecule around their mean position.

For simple molecules, which have sufficient elements of symmetry, the combined study of Raman and IR spectra leads, by use of mathematical treatment, to a very precise knowledge of the relative position of their atoms in space.

For more complex molecules, a rigorous mathematical treatment cannot be carried out because these molecules do not normally have sufficient symmetry. Although the Ach molecule, which possesses 26 atoms is small by the standards of organic chemistry, it is a very complex one for rigorous mathematical treatment of the spectra.

According to the theory of molecular vibrations, a non-linear molecule containing N atoms would possess $3N$-6 fundamental modes of vibrations. Therefore, if the molecule has few elements of symmetry there could be indeed, $3N - 6$ different fundamental vibrations all of which might be 'allowed' both in the Raman and IR spectra.

Thus, the molecule of Ach will have 72 fundamental vibrations and with the additional overtone bands and combinations bands it could exhibit a vibrational spectrum of great complexity. However, the comparative studies made of spectra of large numbers of molecules have demonstrated that in reality the spectra are not as complicated as the above reasoning implicates. The most important factor which contributes to a simplification of the vibration spectra of large molecules is the concept of specific *group frequencies*. As vibrational analysis was extended to an increasing number of molecules, it became apparent that certain groups of atoms give rise to vibrations at or near the same frequency irrespective of the particular molecule in which the group occurs. Such a vibration can be used to characterize the group which vibrates.

The vibrations associated with covalently linked atoms were classified in two types: one which involves a periodic extension and contraction of the bond is known as the *stretching vibration*. The second which involves a periodic bending of the bond is known as *bending vibrations* (this kind of vibrations can be rocking or twisting, in plan or out of plan).

From a comparative study of the vibrational spectra of many large molecules and compounds, a great number of such characteristic group vibrations has been established [74]. Fairly well defined frequency ranges can be assigned to these different types of group vibrations. However, the precise position of the vibrations within these ranges is often modified by the nature of neighbouring molecular structure, the physical state of compound and other structural factors. This makes the assignments of the vibrational spectra to be a complicated task.

In our work, the frequency assignments of Ach were based on the solid state and aqueous solution Raman and IR spectra of AchI [75]. In order to distinguish between the vibrations associated with the choline part of the molecule and those corresponding to the acetyl part, the solid state and aqueous solution spectra of ChI alone have been used. Useful for the vibrational assignments of the acetyl part were the spectra of ethyl acetate [76], N,N-dimethylacetate [77], tetramethylammonium [78], tetramethylamine [79]; ethyl ether [80] and n-paraffins [80, 81].

The assignments of the most important fundamental frequencies of the three Ach compounds (AchCl, AchBr, AchI) in solid state and in aqueous solution are listed in Table I.

Table I. Solid state and aqueous solution R and IR spectra of AchCl, AchBr, AchI and ChI. In aqueous solution all of the three compounds give the Ach$^+$.

Solid state								Aqueous solution		Assignments
AchCl		AchBr		AchI		ChI		Ac$^+$		
R	IR	R	IR	R	IR	R	IR	R	IR	

Choline part

AchCl R	AchCl IR	AchBr R	AchBr IR	AchI R	AchI IR	ChI R	ChI IR	Ach$^+$ R	Ach$^+$ IR	Assignments
418w		425w	423w	422w	423w	420w	410w			δ(O—C—C—N)
452vs	452m	452vs	454s	450vs	543m	449s	450w	450vw		choline skeleton
462s	462w	472s	478w	485m	482m	468sh	468w			
539vw	539m	547w	534w	535m	532vw	531m	531w	525vw		
721vs		723vs		720vs		713vs		723vs		δ(CH$_2$)$_2$
875vs	876s	871m	870s	863m	864m	859s	860m	875s	874m	ν_s(C—N)
938m	935vw	916m	916m	913m	913m	896s	895m		960s	δ(CH$_3$)N r
955s	960vs	949s	952vs	953vs	955vs	956vs	956vs	952s	951s	ν_s(C—O)
1050sh	1057vs	1053w	1053vs	1055vw	1052vs	1054w	1053m	1055w	1055s	ν(C—C) + δ(CH$_3$) r
1107vw	1103m	1082vw	1077s	1139vw	1072vs	1079w	1129m	1143w	1089m	
1144m			1135s		1136m	1133w			1135w	
1229w	1233vs	1223w	1230vs	1221w	1223vs	1231vs	1231vw	1240vw	1248vs	ν_a(C—O)
1286vw	1288sh	1280w	1286m	1277w				1271vw		δ(CH$_2$) t or ω
	1309w	1307w	1308s	1304w	1303vs	1334m				ν_a(C—N)
1366w	1363vs	1370vw	1367m	1362vw	1366vs		1367sh			δ(CH$_2$) ω or t
		1413m	1406vw	1409s	1410m	1409s	1409m			δ(CH$_2$)N
1447sh	1448sh	1452sh	1450m	1445s	1448vw	1454sh	1454sh	1450vs	1445sh	δ(CH$_3$)
1460s	1456m	1460s		1459sh		1458vs				
1475sh	1472m	1475sh	1478m	1475sh	1477s		1477m	1475sh	1478s	δ(CH$_2$) sciss
									1489s	

Table I (Continued)

	Solid state								Aqueous solution		Assignments
	AchCl		AchBr		AchI		ChI		Ach+		
	R	IR	R	IR	R	IR	R	IR	R	IR	
Choline part											
	2860w		2880w		2867vw		2885m	2881w			$\nu_s(CH_2)$
	2920s	2920m	2925s		2920m		2930s	2925vw			$\nu_s(CH_3)N$
	2954m	2944s	2960vs	2952w	2946s		2954s	2948vw	2945sh	2973w	$\nu_a(CH_2)$
								2989vs	2978sh		
	3000sh	3001sh	3004sh	3008m	3005vs	3002vs	3006m				
	3011vs	3011vs	3010s	3012s	3011vs		3012vs				$\nu_a(CH_3)N$
	3021sh	3020sh	3020vs					3023m			
	3030sh		3045w	3040sh	3030vw	3029w	3027s		3045w	3040vw	
Acetyl part											
	602vw	601w	612m	608s	608w	607m					$\delta(C{-}C{-}O)$
	646vs	644m	622s	649s	644vs	644m			645vs		acetyl skeleton
	846m	846m	825vs	824s	828vs	827m			834vs		$\delta(CH_3)C$
		1386s	1378vw	1380m		1381vs					
	1735m	1733vs	1745m	1746vs	1740m	1737vs			1735vs	1734vs	$\nu(C{=}O)$

w — weak; m — medium; s — strong; sh — shoulder; v — very; r — rocking; ω — wedging; t — twisting; sciss — scissoring.

6.2. ASSIGNMENT OF POSTSYNAPTIC MEMBRANE SPECTRA

Assignment of the postsynaptic membrane spectra was facilitated by the numerous studies which have been made in the last 15 years on biological membranes using Raman spectroscopy [5, 6]. IR spectroscopy is rather difficult to use for biological membrane studies since water, a natural biological medium, has a great capacity of absorbing infrared radiations. On the contrary, the Raman spectrum of water is not very strong and can be eliminated by subtraction methods. Moreover, as we had already pointed out, Raman spectroscopy needs only a very small quantity of the sample.

The principal compounds of biological membranes are proteins and lipids. Thus, the application of Raman spectroscopy to biomembrane studies is based on the vibrational assignments of proteins and lipids.

6.2.1. *Assignment of Protein Vibrations*

Proteins are large molecules which usually contain between 500 and 3000 atoms. As a result, they have between 1500 and 9000 normal vibrations ($3N$). Six of these will be quasivibrations of zero frequency corresponding to three translations and three rotations.

One interesting feature of proteins is that they are usually linear peptide chains in which the same peptide group is repeated by means of rotation and translation along a single chain. In this respect, proteins may be regarded as a deformed one-dimensional crystal with different masses at each side. Most of the protein's normal vibrations occur at frequencies in the range of normal molecular group frequencies ($200-3000$ cm^{-1}). Thus, the concept of group frequencies such as the peptide group or a CH_2 group, is applicable to proteins.

The observed Raman spectrum of a protein consists of contributions from peptide backbone vibrations (main chain vibrations) and various amino acid vibrations (side chain vibrations).

6.2.1.1. *Backbone (main chain) vibrations*

The frequencies of Amide I and Amide III vibrations are the most used ones for characterization of the protein backbone conformation. The latter depends on the determination of the characteristic frequencies for helical, β-sheet and random coil protein conformations. This is commonly accomplished by using polypeptide models and proteins of known conformation.

The Amide I band arises from the coupled C=O stretching vibrations of the peptide bond. The theoretical basis to understand the Amide modes has been provided by Miyazawa [82]. His theory has been modified by Krimm and Abe [83] for much less regular hetero-polypeptide chains of proteins.

In polypeptides, the α-conformation normally appears in the region $1645-1660$ cm^{-1}; β-sheet conformation is in the region $1665-1680$ cm^{-1} and an unordered (random coil) conformation is in the region $1660-1670$ cm^{-1} [84, 85].

The Amide III band has a large contribution from (N—H) in plane-bending vibrations. The α-conformation usually gives very high Amide III bands at 1265 cm^{-1}, β-sheet at 1230—1240 cm^{-1} and random coil at 1240—1260 cm^{-1} [86—91]. The Amide III frequency is about three times more sensitive to conformational changes than Amide I.

Because of the wide variations which were found for the Amide I and Amide IIIα and the random coil frequencies in different proteins (glucagon [92], insulin [87, 93], keratin [94], α-casein [95]), assessments of protein secondary structure from these regions required a careful analysis. Progress has been made by the development of three methods to quantify the various amount of secondary structure using the Raman spectrum [72, 73, 96, 97]. In our current work, we apply the method of Williams [72, 73].

6.2.1.2. Side chain vibrations

Proteins are complicated copolymers containing amino acid residues which have aromatic, aliphatic and reactive functional groups as side chains. Several of these side chain vibrations scatter strongly in the Raman spectrum. Fortunately, these vibrations do not appear with strong intensity in the Amide I and Amide III regions.

Raman vibrations due to side chains are themselves of interest because of their sensitivity to the local environment of the side chain. Some of the amino acid residues such as tryptophane and tyrosine have strong conformational sensitive Raman vibrations [98—101]. Yu et al. [102] reported that the doublet 853—828 cm^{-1} in the Raman spectra of gly-1-tyrosine markedly changes its relative intensity depending on the local environment. They correlated the intensity ratio of these two vibrations with tyrosine residues in two different environments: normal or 'exposed' and 'buried' within the protein in hydrophobic regions. It was shown [103] that the doublet arises from Fermi resonance between a ring-breathing vibration and the overtone of an out-of-plane ring bending vibration. The intensity ratio is sensitive to the nature of hydrogen bonding or state of ionization of the phenolic OH group. If the OH group is strongly bound to a negative acceptor the intensity ratio is near 0.3; if the OH forms moderately strong H-bonds to H_2O, it is near 1.25, and if the OH participates as a strong H-bond acceptor, it is near 2.5 [103].

Among the tryptophane vibrations in the Raman spectra of proteins the one at 1360 cm^{-1} has been indicated to be sensitive to the environment [91, 92, 104, 105]. Presence or absence of the 1360 cm^{-1} band in the spectrum of tryptophane containing proteins suggests 'buried' or 'exposed' tryptophanes, respectively. When present, the 1360 cm^{-1} vibration appears as the peak of higher frequency of the doublet at 1360—1340 cm^{-1}. Recently Harada et al. [106] analyzed Raman spectra of tryptophane and related compounds under various conditions and arrived to the conclusion that the high sensitivity of the doublet is due to the Fermi resonance between one skeletal stretching fundamental vibration and one or two combinations of the out-of-plane vibrations of the indole ring. Another trypto-phane vibration found at 880 cm^{-1} has also been indicated to be useful as an additional probe for the tryptophane environment [107].

The disulfide group (S—S) which appears between 510—540 cm^{-1} is also con-formation sensitive. Although there exist different interpretations of this frequency region [108, 109] it is a very useful monitor of disulfide conformation and has been of value in studies of peptide hormones and neuro-toxines [110—114].

Aliphatic (C—H) stretching vibrations of amino acid side chains are observed as intense bands in the 2800—3100 cm^{-1} region near 2900, 2940 and 2970 cm^{-1}. They are assigned respectively to the methylene symmetric ($\nu_s CH_2$), methylene asymmetric ($\nu_a CH_2$) and methyl asymmetric ($\nu_a CH_3$) vibrations [115].

6.2.2. Assignment of Lipid Vibrations

The lipidic constituents of biomembranes are distinguished by their amphiphilic character as manifested by a strong hydrophilic headgroup and a long hydro-phobic tail. In water, amphiphils dominantly form bilayers with the headgroups on the external surface and a layer of juxtaposed hydrocarbon chains between them.

Phospholipids are the major class of lipid found in membranes. Phospholipids are usually derived from glycerol and consist of a glycerol backbone, two fatty acid chains and a phosphorylated alcohol. The most frequently occurring alcohols are serine, choline, ethanolamine, glycerol and inositol.

Two varieties of phospholipid preparation have long served as simple model of the biomembrane systems. Phospholipid dispersions are prepared in excess water by mild agitation above the lipid phase transition. Dispersions are large aggregates (typically 1—4 μm diameter) and consist of multiple bilayers (liposomes). Pro-longed ultrasonication followed by chromatographic or centrifugal separation results in the small (radius 125 Å) unilamellar vesicles [116]. Both unilamellar vesicles and liposomes were very useful model systems. Thus, it was observed, that they undergo a characteristic phase transition. In the gel phase the hydrocarbon chains are essentially in all-*trans* conformation about (C—C) bonds. In the liquid crystalline phase, the hydrocarbon chains possess some *gauche* isomers. Raman spectroscopy was very useful for studying the structural characteristics of phospho-lipid molecules [117—124]. In the vibrational assignments of lipid Raman spectra, the main base was the quite extensive literature available on the vibrational spectroscopy of aliphatic hydrocarbons. Two vibrational regions are usually used, the skeletal optical region and the stretching region.

6.2.2.1. Skeletal optical region (1000—1150 cm^{-1})

The vibrations of this region situated at 1060, 1100, 1130 cm^{-1}, are particularly sensitive to the conformational state of hydrocarbons [117, 118]. Both bands at 1060 and 1130 cm^{-1} were assigned to the vibrations of all-*trans* chain segments and the vibration at 1100 cm^{-1} is considered as resulting from structures contain-ing *gauche* rotations [119—128]. When a phosphate group is present, a distinct band appears at 1080 cm^{-1} which has been assigned to the PO$_2^-$ symmetric stretching mode [129]. Detailed studies on the assignments and the conformational dependence of the lipid skeletal optical vibrations have been carried out in a number of laboratories [130, 131]. It has been shown that the skeletal optical region is sensitive to *trans-gauche* isomerization of the hydrocarbon chains in phospholipids.

6.2.2.2. (C—H) stretching region (2800—3100 cm^{-1})

This region is determined by the methylene vibrations due to the relative abundance of the methylene groups in the hydrocarbon lipid chains. Two characteristic strong lipid vibrations occur in this region near 2850 and 2880 cm^{-1} which are assigned as symmetric and asymmetric stretching methylene vibrations (v_sCH$_2$, v_aCH$_2$), respectively. It has been shown that the intensity ratio I_{2880}/I_{2850} can serve as a measure of the lipid chain mobility: that is crystal → liquid crystal transition can be studied [124, 132]. The use of this ratio is possible because the peak at 2850 cm^{-1} is temperature independent whereas the 2880 cm^{-1} band changes its intensity with thermally induced changes in the lipid chain conformation. The temperature sensitivity of 2880 cm^{-1} vibration has been explained by a Fermi resonance interaction between the v_sCH$_2$ fundamental vibration and the δCH$_2$ deformation overtones [127, 133]. The δCH$_2$ bending fundamental vibrations situated at 1440—1460 cm^{-1} are sensitive to the chain configuration and, therefore, their modifications amplified by the resonance interactions in 2800—3000 cm^{-1} range, could be more easily detectable in the latter region. The resonance band is broad with a strong maximum near 2890 cm^{-1}. Thus, the shape of the 2880 cm^{-1} vibration in the Raman spectrum is determined by the overlap of the peaks of the v_aCH$_2$ fundamentals and v_sCH$_2$ Fermi-resonance-enhanced vibrations. Indeed, the latter vibrations cause an intensity maximum at this frequency range. The intensity decreases as a function of temperature in the course of lipid transition.

7. Experimental Results

7.1. ACETYLCHOLINE

7.1.1. Conformation of Acetylcholine

In the period of 1960—1970 several papers appeared using IR spectroscopy to study the Ach molecule. Conclusions on the conformation of Ach dissolved in ethanol has been made by investigating only the IR vibration of the carbonyl group (C═O). Thus, from the higher frequency of the (C═O) group in Ach than in ethyl acetate, Fellman and Fujita [134] concluded that the Ach molecule has a cyclic structure determined by the electrostatic attraction between the quaternary nitrogen and the polarized carbonyl group. Shortly after, the same authors [135] reexamined their proposal and argued against a cyclic conformation in favor of the inductive influence of the nitrogen atom on the carbonyl group. Canepa and Mooney [136] applying the method of the attenuated total reflexion, observed the carbonyl stretching frequency for Ach in water and ethanol. Since the frequency is dependent on the nature of the solvent these authors underlined that the cyclic conformation suggested by Fellman and Fujita [134] is not in agreement with the experimental evidence. In subsequent papers Fellman and Fujita [137, 138] proposed that the conformation of Ach in crystal and in water involves an interaction between nitrogen and the unshared electrons of ester oxygen.

It became clear that partial application of IR spectroscopy could not provide the information sufficient enough for establishing the precise conformation in that case. On the other hand, the rapid development of Raman spectroscopy in the early 1970s due to the application of laser beams, has greatly enhanced the importance of Raman spectra. The simultaneous application of both vibrational spectroscopic methods appeared to be even more important in studying the molecular conformation, because they mutually validate and complement their respective results.

The first paper on the simultaneous application of Raman and IR spectroscopies to study the conformation of Ach in solid state and aqueous solution appeared in 1974 [75]. The solid state and aqueous solution Raman and IR spectra of AchCl, AchBr and AchI are shown in Figure 4. In solution the three compounds give positively charged ions (Ach$^+$) which are the biologically active forms of the Ach molecule. In Table I are listed the most important fundamental frequencies and their assignment.

The comparison between the AchCl, AchBr and AchI solid state Raman and IR spectra indicated that:

1. the conformation of the three compounds is different, but the AchI one is quite close to that of AchBr;
2. the conformation of the methyl groups bound to the nitrogen as well as the conformation of the methylene groups appear to be different, whereas the conformation of the methyl group bound to the carbon atom appear similar;
3. the frequencies and intensities of both bonds, ν(C=O) and ν(C—O), are very intense in the IR spectra and do not practically vary from one compound to another.

At the time when the spectroscopic investigations were developed, only X-ray studies on the crystal structure of AchCl and AchBr were available [139, 140]. The estimated values for the length of both (C=O) and (C—O) bonds were as follows: 1.35 and 1.32 Å for AchBr [139] and 1.18 and 1.38 Å for AchCl [140]. Thus, one can expect important variations in frequency, especially for the ν(C=O) vibration which is not overlapped by another one. However, the spectral results do not indicate substantial differences [75]. On the basis of more precise X-ray reinvestigation of the AchBr crystal structure [141], a correction of the length of (C=O) bond to 1.19 Å and of (C—O) bond to 1.36 Å has been made for this

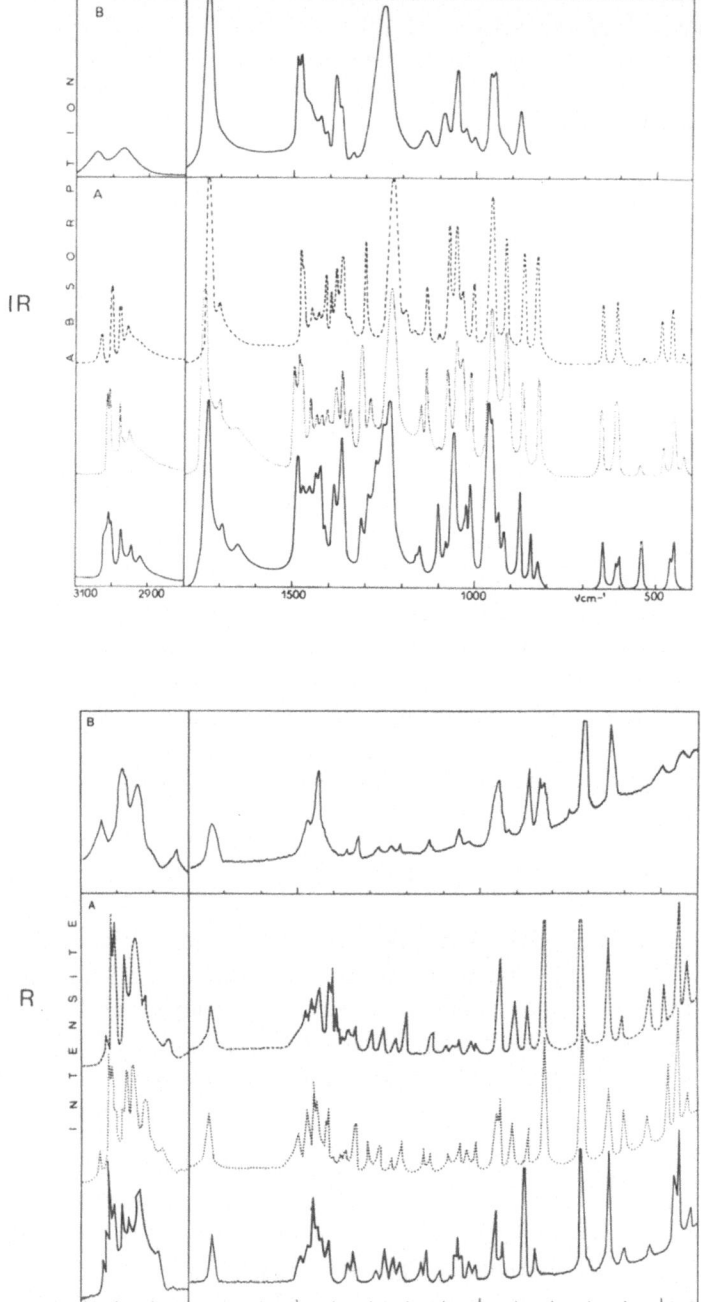

Fig. 4. Raman (R) and infrared (IR) spectra of AchCl (—), AchBr (...) and AchI (---); A —
crystal state, B — aqueous solution (Aslanian *et al.* [75]).

compound. Also a study of AchI crystal structure has been carried out [142] which indicated respectively 1.20 Å and 1.37 Å for the length of (C=O) and (C—O) bonds. Thus, the spectroscopic results are in excellent agreement with current X-ray data.

The transition of Ach from solid state to aqueous solution has been characterized by the following phenomena: the conformation of the acetyl part does not practically change while the spectral differences in the choline part are considerable — certain vibrations disappear whereas for others their frequency increases or their intensity changes. The vibrations which disappear belong to the choline part of Ach (Table I). This can be explained by an internal rotation of the molecule. The Ach molecule is a flexible molecule whose flexibility is due to the possibility of rotation about $(O_1—C_5)$ and $(C_5—C_4)$ bonds [143, 144].

The modification of the conformation of the choline part during the transition from solid state to aqueous solution, as shown in the Raman and IR spectra (Table I) indicates a rotation around the $(O_1—C_5)$ bond. The considerable increase in frequency of the stretching vibrations of all methyl groups could be explained by modification fo the molecular interactions in aqueous medium [75].

7.1.2. Non-Enzymatic Hydrolysis of Ach

Raman spectroscopy found successful application in the study of modifications of Ach molecules during the course of non-enzymatic hydrolysis which takes place in a basic medium [145]. The time evolution of the Raman spectra (Figure 5) has been related to the course of the reaction. As shown in the Figure 5, in the course of hydrolysis at pH 10, the Ach vibrations progressively disappear while the frequencies attributed to the final products of the reaction (choline and acetate ion) progressively appear and increase in intensity. The relative intensity of six Raman vibrations as a function of time are shown in Figure 6. The vibrational intensities being proportional to the concentration of the species which they stem from, one can note $A = I_0/I_t$ for the vibrations with decreasing intensity and $A = I_\infty/(I_\infty - I_t)$ for vibrations with increasing intensity (I_0, I_t and I_∞ are the intensities of the vibrations, respectively, at initial time, at the instant of measurement and at infinite time where the reaction can be considered as being over). Plotting log A versus time results in the graphs of Figure 7 which shows linear plots for pH 10 and pH 11 values.

The velocity of the reaction is $v = k(Ach)^a(OH^-)^b$. When pH is kept constant it is $v = k'(Ach)^a$ with $k'(OH^-)^b$.

The linear plots in Figure 7 lead also to the conclusion that the reaction is

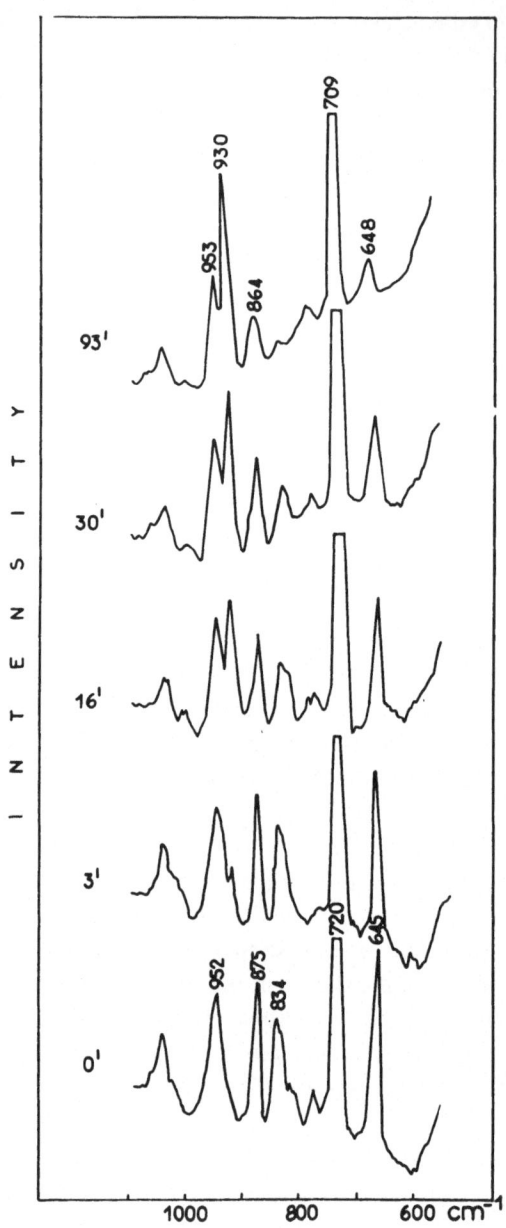

Fig. 5. Evolution of Ach Raman spectra at pH 10 during the course of hydrolysis ('*Non-enzymatic hydrolysis of Ach*', Lautié *et al.*, *J. Raman Spectroscopy* [145], Copyright 1978. Reprinted by permission of John Wiley & Sons).

indeed of the first order with respect to Ach ($a = 1$). From the slopes of these plots one can obtains k' which is 0.035 min^{-1} at pH 10 and 0.30 min^{-1} at pH 11. The ratio of the constant k' at pH 10 and pH 11 gives a value for b, which is close

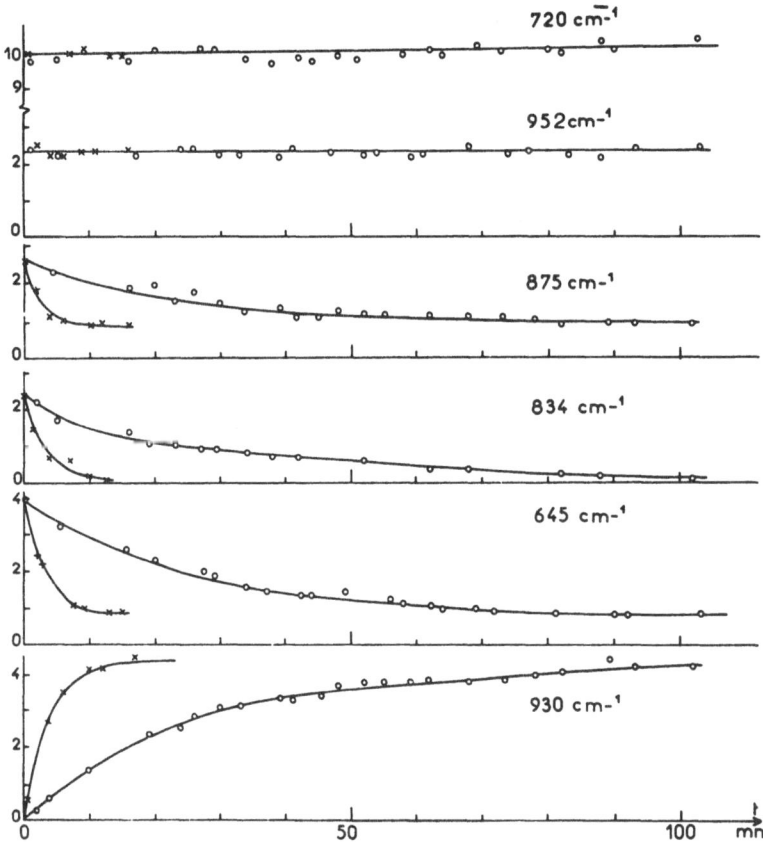

Fig. 6. Relative intensities of some Raman vibrations of Ach (0.25 M) in aqueous basic medium versus time: (○) — pH 10; (×) — pH 11 ('*Non-enzymatic hydrolysis of Ach*', Lautié *et al.*, *J. Raman Spectroscopy* [145], Copyright 1978. Reprinted by permission of John Wiley & Sons).

to 0.9. Thus, the reaction is also of the first order with respect to the (OH$^-$), with $k = 330 \pm 30$ L mole^{-1} min^{-1} or 5.5 ± 0.5 L mole^{-1} s^{-1}. This value is very close to the one obtained by the authors with the pH-stat method (5.4 L mole^{-1} s^{-1}).

7.1.3. Compounds Related to Acetylcholine

7.1.3.1. Nicotinics

Nicotine (Nic) = [CH:CHCH:NCH:CCH(CH$_2$)$_3$NCH$_3$]

The subject of investigation has been the Nic, pure and diluted in water, Nic-monochlorhydrate (NicHCl) and dichlorhydrate (Nic2HCl), solid and in aqueous solution, as well as the deuterated Nic monochlorhydrate (NicDCl) [146]. The main interest was raised by the Nic monohydrate which is the most active biological form in aqueous solution.

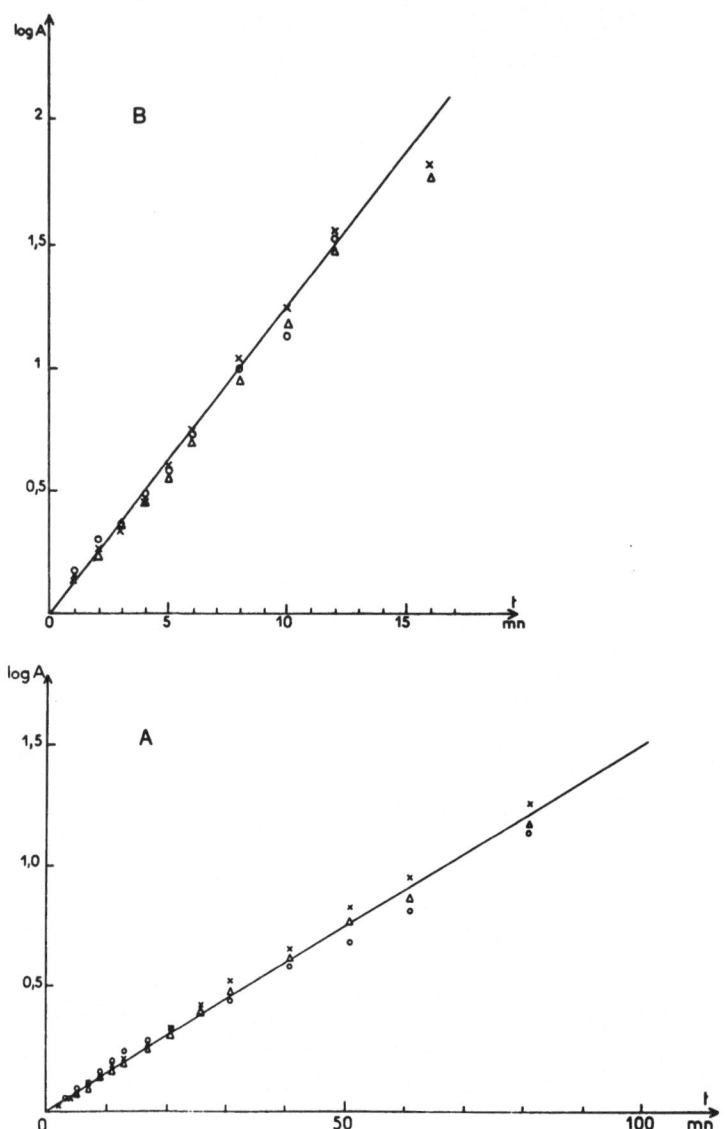

Fig. 7. Log A versus time at pH 10 (A) and pH 11 (B) for the Raman vibrations: 645 cm^{-1} (×), 834 cm^{-1} (O) and 930 cm^{-1} (△) ('*Non-enzymatic hydrolysis of Ach*', Lautié *et al.*, *J. Raman Spectroscopy* [145], Copyright 1978. Reprinted by permission of John Wiley & Sons).

The specific vibrations of each ring have been attributed using the Raman and IR spectra of NicDCl. The same spectra have also been useful to understand which of the two possible effects (mechanical coupling or electronic effect [147]) take place during protonation.

Figure 8 displays Raman spectra of pure Nic and NicHCl in aqueous solutions. In Table II we give the frequency variations between both spectra. These variations

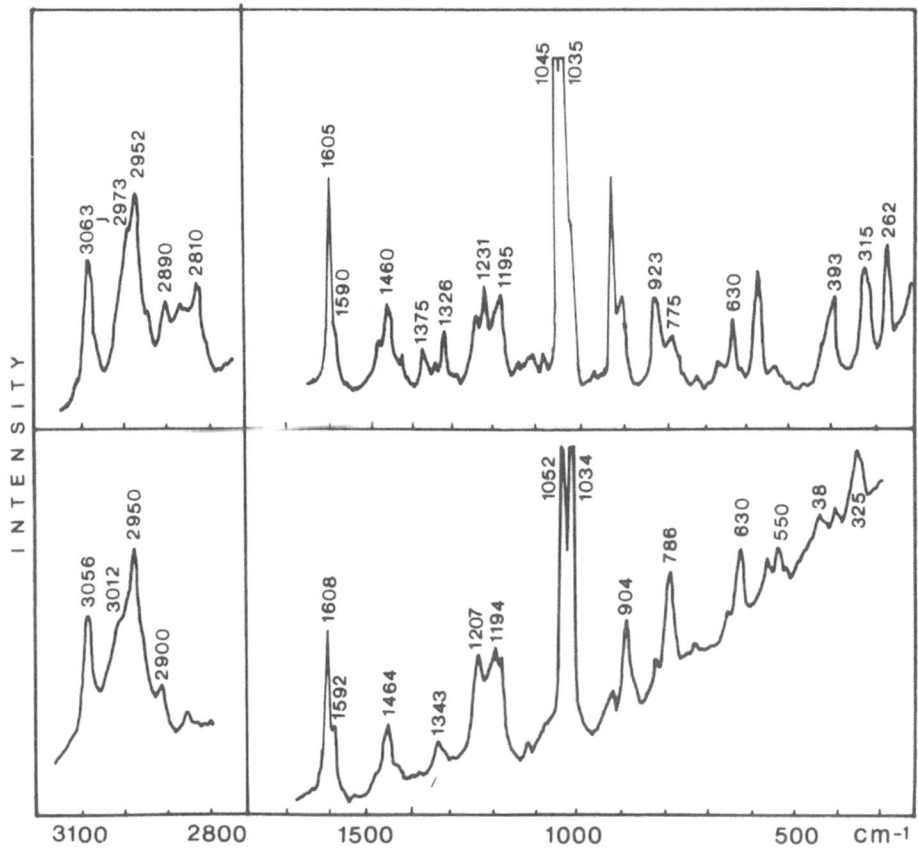

Fig. 8. Raman spectra of pure Nic and NicHCl in aqueous solution (Aslanian *et al.* [146]).

are more important for the pyrrolidine ring. Excepting two vibrations (530 and 2952 cm^{-1}) which do not change, all other vibrations vary considerably and two of them (2810 and 2890 cm^{-1}) disappear completely. Examination of the corresponding vibrations in the solid state spectra of NicHCl and NicDCl (Table III) shows that except for one frequency (902 cm^{-1}) the vibrations of the pyrrolidine ring do not change in the deuterated Nic monochlorhydrate. This indicates clearly that the variations in the pyrrolidine ring vibrations are due to the electronic effect of the protonation [147]. Consequently, one can admit that the electronic effect might be responsible for the high biological activity of monoprotonated Nic. Confirmation of this supposition has been found by the comparative study between Ach, Nic and NicHCl spectra [146]. The comparison concerns especially the stretching asymmetric vibrations of the methyl groups bonded to the positive nitrogen, $v_a(CH_3)N^+$. In pure Nic the frequency of this vibration at 2965 cm^{-1} is very far from that of Ach, situated at 3011 cm^{-1} [75]. In protonated Nic (NicHCl)

Table II. Frequency variation in the aqueous solution R spectra of NicHCl compared to the frequency of diluted in water Nic.*

Aqueous solution		Assignments
Nic	NicHCl	
Pyridine ring		
392	−13	ring (16a)
405	+15	ring (16b)
630	0	ring (6a)
667	−3	ring (6b)
720	+1	CH (11)
818	+4	CH (10b)
1035	−1	ring (1)
1045	+7	ring (12)
1195	−1	CH (15)
1375	disappears	ring (14)
1425	"	ring (19b)
1483	−3	ring (19a)
1590	+2	ring (8b)
1605	+3	ring (8a)
3063	−7	νCH (2, 13)
Pyrrolidine ring		
315	+10	ring
530	0	ring
571	−21	ring
775	+11	ring
923	−19	ring
1231	−24	CH_2
1326	+17	CH_2
2810	disappears	$\nu_a CH_2$
2890	"	$\nu_s CH_2$
2810	−18	$\nu_s CH_2$
2952	−2	$\nu_s CH_2$
2973	+39	$\nu_a CH_3$

* Reprinted with permission from *Life Sciences* **32**, D. Aslanian: 'Vibrational Spectroscopic Approach to Study of Ach and Related Compounds'. © 1983, Pergamon Journals Ltd.

the same vibration increases considerably in frequency (3002 cm^{-1}) and comes very close to that of Ach (3011 cm^{-1}). This vibration is not influenced by the deuteration as seen in Table III. Thus, the elevation of its frequency can be explained by the electronic effect of protonation which makes the electronic distribution of the protonated nitrogen in NicHCl to be of the same order of magnitude as in Ach. Therefore, one could admit that a common property of Ach

Table III. Some solid state R and IR frequencies of the pyrrolidine ring of NicHCl and NicDCl.*

NicHCl		NicDCl		Assignments
R	IR	R	IR	
312vs		313vs		ring
512s	513vw	514s	510vw	ring
548s	550w	548s	548w	ring
788s	787w	792s	785w	ring
902vs		884s	881m	ring
1204sh		1204sh		CH_2
	1300sh		1301vw	CH_2
2920sh		2925sh		$v_s CH_2$
2950s	2942w	2956s	2950m	$v_a CH_2$
3002s	3005m	3008s	3006m	$v_a CH_3$

* Reprinted with permission from *Life Sciences* 32, D. Aslanian: 'Vibrational Spectroscopic Approach to Study of Ach and Related Compounds'. © 1983, Pergamon Journals Ltd.

and Nic as chemical transmitters is the identical electronic density of their positive nitrogens.

7.1.3.2. *Muscarinics*

Two compounds of methylacetylcholine (β-MeAchCl and β-MeAchBr) and one compound of muscarine (MuCl) have been investigated by Raman spectroscopy. Their solid state Raman spectra have been studied in comparison to the solid state Raman spectra of AchCl, AchBr and AchI [148]. Three representative spectra are given in Figure 9. As shown in the figure, the vibrations which characterize choline (O_1—C_5—C_4—N^+) and acetyl (C_7—C_6—O_1) skeletons in Ach and b-MeAch molecules are very similar suggesting identical conformations of these skeletons in both molecules. On the contrary, the conformation of the methyl groups bonded to the nitrogen [$N(CH_3)_3$], exhibits differences. Figure 9 also shows that the solid state conformation of the choline skeleton and the methyl groups of Mu molecule is quite different from the respective conformations of the Ach and β-MeAch molecules. Figure 10 displays the aqueous solution Raman spectra of Ach, β-MeAch and Mu. The most striking phenomenon here is the presence of a very strong vibration in the β-MeAch and Mu spectra situated at 544 cm^{-1} and 520 cm^{-1}, respectively. This vibration has been attributed to the bending mode of the choline skeleton (O_1—C_5—C_4—N^+) in aqueous solution [148]. The same vibration disapeared completely in the Ach aqueous solution spectrum. The unsubstituted (O_1—C_5—C_4—N^+) chain of Ach is capable of relatively unhindered rotations around the (O—C) and (C—C) bonds while the b-methyl substitution of β-MeAch and the presence of the ring in Mu limit this possibility. It was concluded that in aqueous solution the conformation of the

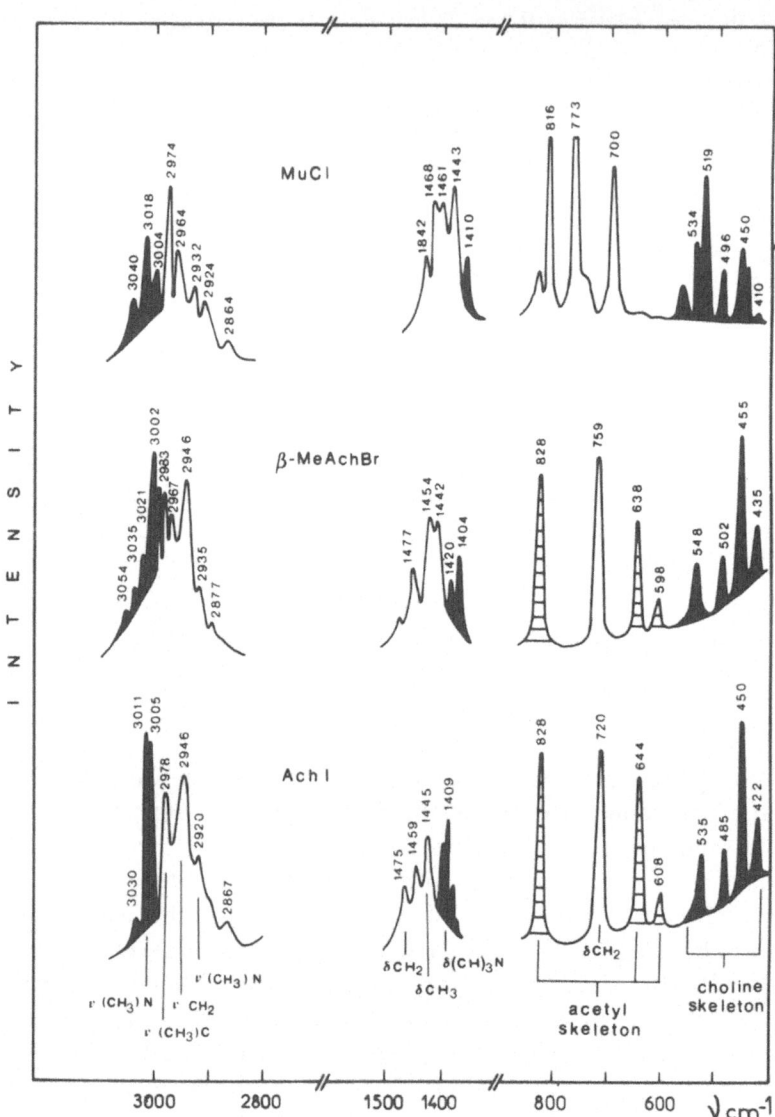

Fig. 9. Raman spectra of AchI, β-MeAchBr and MuCl in solid state (Reprinted with permission from *Life Sciences* **32**, D. Aslanian: '*Vibrational spectroscopic approach to study of Ach and related compounds*' [156]. Copyright 1983, Pergamon Journals Ltd.).

(O_1—C_5—C_4—N^+) skeleton, similar for β-MeAch and Mu, is different for Ach. Due to the higher rigidity of both analogues it can be considered that their conformation in solution is already closer to the structure which fits with the muscarinic receptor. On the contrary, Ach should take its 'muscarinic' form only in the neighborhood of muscarinic receptors.

The conformation of the methyl groups bound to the nitrogen, different in solid

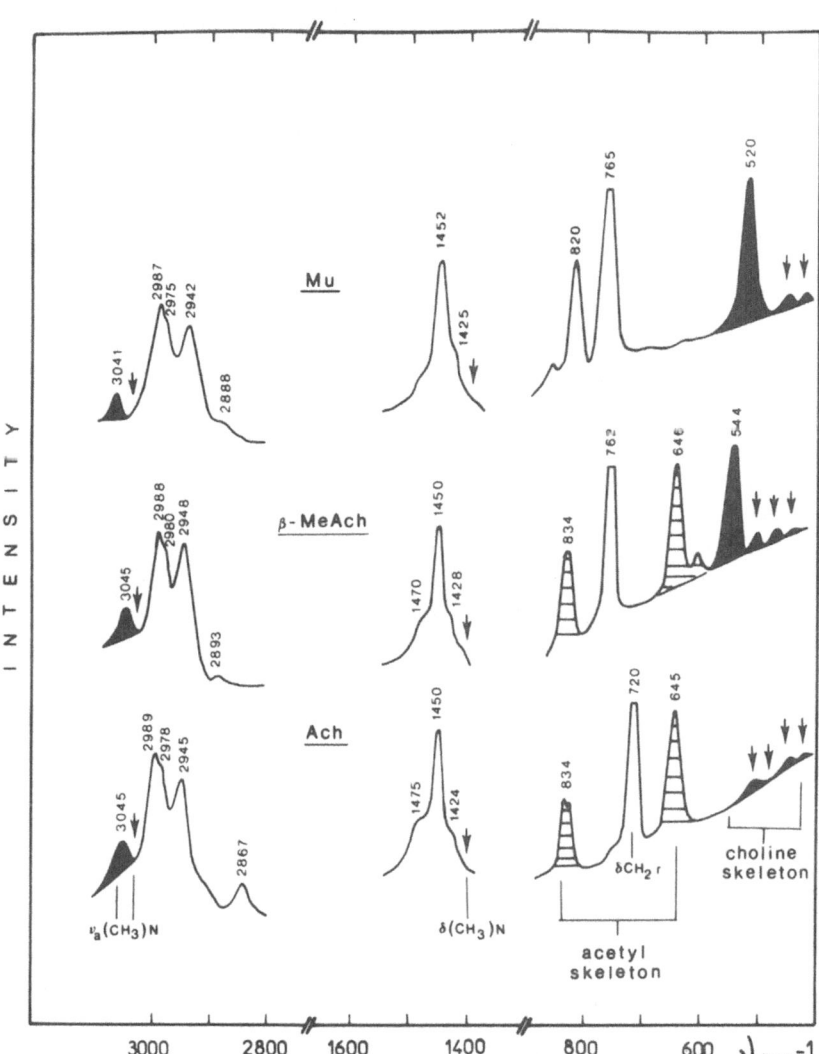

Fig. 10. Raman spectra of AchI, β-MeAchBr and MuCl in aqueous solution (Reprinted with permission from *Life Sciences* **32**, D. Aslanian: '*Vibrational spectroscopic approach to study of Ach and related compounds*' [156]. Copyright 1983, Pergamon Journals Ltd.).

state for the three molecules, appears to be identical in aqueous solution. Moreover, the similar values for the stretching vibrations of the methyl groups bound to the nitrogen in the spectra of Ach, β-MeAch, Mu, Nic indicate that in the neighborhood of the nitrogen atom the electronic density is almost the same for all these molecules.

7.1.4. *Interaction of Ach with a Model Receptor*

Models for the events which take place on the postsynaptic membrane have been suggested on the basis of X-ray studies and theoretical calculations of solid Ach and related compounds. According to Chothia [144] the quaternary nitrogen group and the carbonyl group seem to be involved in the interaction of Ach with nicotinic receptors. The same key groups have also been proposed by Kier [149]. In the early concept of the distribution of electron density in the Ach molecule [150] a dipole-dipole interaction between the Ach molecule and the receptor has been proposed: the dipole in Ach molecule is formed by the carbonyl oxygen which carries a partial negative charge and the ether oxygen which has a partial positive charge. Later calculations [151] indicated considerable negative charge of the ether oxygen atom. This suggested the idea [152] that dipole-dipole interaction of Ach with the esterophilic portion of the Ach receptor involves only the carbonyl group dipole of Ach without the participation of the ether oxygen. Experimental work to confirm this suggestion has been lacking.

In order to shed some light on the conformation of Ach when it interacts with a receptor, we applied a third vibrational spectroscopic method, the inelastic electron tunneling spectroscopy (IETS) [153—156]. Spectra recorded by this method possess both Raman and IR vibrations of the interacting molecule. Thus, using our previous Raman and IR spectra of Ach in solid state and aqueous solution as well as the spectra of the non-enzymatic hydrolysis of Ach, we were able to interpret the Ach inelastic electron tunneling spectra and to propose the first experimental model on the conformation of Ach interacting with an active surface.

In IETS the molecules to be studied are adsorbed on the oxide layer at a metal-oxide-metal junction. The system most frequently used is composed of an aluminium electrode (1000 Å thick) vapour deposited on a glass substrate, an alumina layer (15 Å) grown by glow discharge in an oxygen plasma and a lead counterelectrode (also vapour deposited, roughly 2000 Å thick). The dopant is adsorbed on the alumina prior to the deposition of the lead counterelectrode. When a bias V is applied to the junction, a current appears, formed by electrons crossing the insulating oxide layer by the tunnel effect. If their energy is sufficient, these electrons may excite a vibrational transition of energy in the molecule. The appearance of such a new current channel at a bias $V = \hbar w/e$ produces a step in the junction conductance $\sigma = dI/dV$ or a peak in the derivative of the conductance versus voltage — in the present case plots of $1/\sigma \, d\sigma/dV$ against energy (meV) or wave number (cm^{-1}) [154].

The spectra were taken at liquid He temperature (4.2 K) to assure that the energy of the electrons is well defined, and therefore the resolution is sufficient ($\sim 5 \, cm^{-1}$).

Both Raman and IR active vibrational modes are observed in tunneling spectroscopy. Peak positions coincide if the slight shift due to the dipole image in the counterelectrode is taken into account. This shift is usually of the order of a few tenths of 1%. When correct band mode assignments can be made, information about the perturbation due to adsorption may be derived. Tunneling spectra have been obtained for relatively small molecules as well as for large species such as amino acids, proteins and other biological materials. For a review see Hansma [157].

Important information about the interaction of Ach and its analogue β-MeAch with the surface of alumina (Al_2O_3) has been obtained. Although the alumina surface is definitely not a postsynaptic membrane it does possess complementary adsorption sites (Al^+—O^- acid-base pairs) which may provide a model for the complementary sites in Ach-receptor or in AchE enzyme that were suggested by Nachmansohn [33].

Ach and β-MeAch were deposited on the alumina under two different conditions. In the first case (Conditions A) a 10^{-4} molar aqueous solution of AchI or β-MeAchBr was left on an alumina surface for two minutes and the excess liquid was blown off with a jet of argon. In the second case (Conditions B) the molecules were deposited as indicated above but the sample was rinsed with distilled water and then dried with argon.

The analysis of the experimental results obtained under both conditions gives a picture of the behavior of these molecules in the vicinity of the alumina surface in terms of molecular configuration in the adsorbed state and illustrates the analogies with adsorption on a receptor site (non-hydrolytic interaction, Conditions A) or on AchE enzyme (hydrolytic interaction, Conditions B).

Representative spectra of both interactions are shown in Figure 11. Their interpretation is made using the solid state and aqueous solution Raman and IR spectra of Ach [75] and β-MeAch [148] as well as the spectra of non-enzymatic hydrolysis of Ach [145].

7.1.4.1. *Non-hydrolytic interaction (Figure 11-A)*

The most important peak positions and apparent peak intensities of AchI and β-MeAchBr tunneling spectra are given in Table IV. The table is completed with the Raman and IR spectra of AchI and β-MeAchBr in solid state and aqueous solution.

The analysis of spectra is made successively for both parts of the Ach and β-MeAch molecules, the choline part and the acetyl part, as well as for the (C—O—C) bond which binds them together.

Choline part. In the course of the nonhydrolytic interaction of Ach and β-MeAch with alumina surface the positive trimethylammonium group is probably attracted by the negatively charged oxygen atom of the surface. This interaction does not seem to influence the conformation of the choline skeleton of β-MeAch which remains close to that of its solid state and aqueous solution as indicated by the strong vibration at 443 cm^{-1} in the tunneling spectrum. The same vibration is also very strong in the solid state and aqueous solution spectra of β-MeAch at 450 cm^{-1} and 544 cm^{-1}, respectively. This is probably due to the great distance between the two opposed charges forming an ionic bond as suggested by Nachmansohn [33].

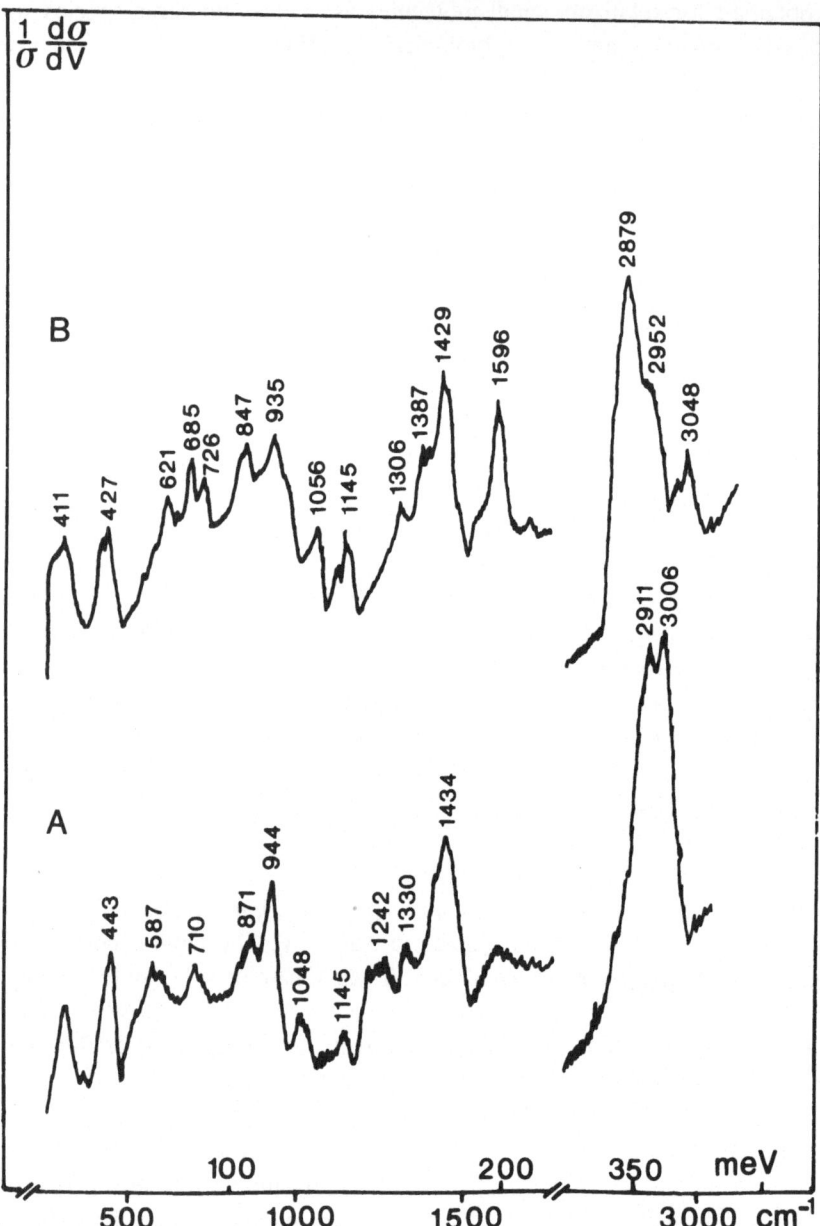

Fig. 11. Inelastic electron tunneling spectra of AchI: A — non-hydrolytic interaction; B — hydrolytic interaction (Reprinted with permission from *Life Sciences* **32**, D. Aslanian: '*Vibrational spectroscopic approach to study of Ach and related compounds*' [156]. Copyright 1983, Pergamon Journals Ltd.).

In the tunneling spectrum of Ach the choline skeleton is characterized by the same very strong vibration at 443 cm^{-1}. The solid state Raman spectrum of Ach also showed a strong vibration at 450 cm^{-1}, but it is barely visible in the aqueous

Table IV. Raman and IR tunneling frequencies of AchI and β-MeAchBr (Non-hydrolytic interaction).*

Tunneling frequencies

	β-MeAchBr aq. solution		β-MeAchBr solid		β-MeAchBr tunneling		AchI tunneling		AchI solid		AchI aq. solution		Assignments
	R	IR	R	IR	cm^{-1}	meV	cm^{-1}	meV	R	IR	R	IR	
Choline part													
	420vw, 450vw		435m, 455vs, 502m	454vw, 468vw	443s, 484sh	55s, 60sh	443s	55s	422m, 450vs, 485m	423w, 453m, 582	418vw	450vw	δ(O—C—C—N) choline skeleton
	544vs, 762vs		548m, 759vs	546w	548sh, 750m	68sh, 94m	532sh, 710m	66sh, 88m	535m, 720vs	532vw	525vw, 720vs		δ(CH$_2$) r
	895s	891m	892s	893s	895s	111s	871m	108m	863m	864m	875s	874m	δ(CH$_3$)N r
	909sh	907m	913m	912m					913m	912m	950sh	960a	ν$_a$(C—N)
	1329w	1321vw, 1403vw	1327s, 1404s, 1442sh	1321w, 1398w, 1440s	1315w, 1442s	163w, 178s	1330s, 1434s	165s, 177s	1304w, 1409s, 1445s	1303vs, 1410m, 1448vw	1450vs	1445sh	δ(CH$_3$)N, t or w
			2935s, 2967s	2968m	2911sh	361sh	2911sh	361sh	2920m				ν$_s$(CH$_3$)N
			3002vs, 3008s	3000s	3006sh	372sh	3006vs	373vs	3005vs, 3011vs	3002m			ν$_a$CH$_3$ (β-MeAch)
	3045m		3012sh, 3035vw, 3054vw	3019sh, 3050vw					3030vw	3029w	3045w	3040vw	ν$_a$(CH$_3$)N

Table IV (Continued)

Tunneling frequencies

Section / Assignment	β-MeAchBr aq. sol. R	β-MeAchBr aq. sol. IR	β-MeAchBr solid R	β-MeAchBr solid IR	β-MeAchBr cm^{-1}	β-MeAchBr meV	AchI meV	AchI cm^{-1}	AchI solid R	AchI solid IR	AchI aq. sol. R	AchI aq. sol. IR	Assignments
C—O—C bond	964s	962sh	958s	957m	960s	119s	117vs	944vs	953m	955vs	952s	951s	ν_s(C—O) + ν(Al—O)
	1245w	1249s	1245sh	1234vs	1234m	153m	154m	1242m	1221w	1223vs	1240vw	1248vs	ν_a(C—O)
Acetyl part	646s, 834s		598w, 638vs, 828vs	599m, 636w, 827w	605s, 635sh, 839w	75s, 81sh, 104w	74s, 80sh	587s, 645sh	608w, 644vs, 828vs	607m, 644m, 827s	645vs	645s	δ(C—C—O) acetyl skeleton
	1725m	1728vs	1730s	1732vs					1740m	1737m	1735m	1734vs	νC=O
	2980sh, 2988s		2983s, 2966sh	2982w, 2977s					2987s	2975m	2978sh, 2989vs	2973m	ν_a(CH₃)C

* Reprinted with permission from *Life Sciences* **32**, D. Aslanian: 'Vibrational Spectroscopic Approach to Study of Ach and Related Compounds'. © 1983, Pergamon Journals Ltd.

solution Raman spectrum of AchI. This indicates that when the flexible Ach molecule approaches the surface, the aqueous solution conformation of the choline skeleton changes by a rotation around the $(O_1—C_5)$ bond and adopts the solid state conformation which is very similar to that of the β-MeAch skeleton. It seems that Ach interacts with the alumina surface in the same way as does β-MeAch because both tunneling spectra are very similar [155].

Acetyl part. The tunneling spectra show that the acetyl part of both molecules interacts with positive Al^+ sites on the alumina surface via the strongly polarized oxygen of the (C=O) group. This is indicated by the absence of the (C=O) group characteristic vibration expected at about 1740 cm^{-1}. It was admitted that in the course of interaction the (C=O) double bond is broken and the oxygen atom is covalently bound to an Al^+ cation. The acetyl skeleton of both molecules does not present radical modifications during the interaction. The $\nu_a(CH_3)C$ vibration which has been identified in the Raman and IR spectra of both molecules at about 2980 cm^{-1} [75] cannot be distinguished in the tunneling spectra probably because it is masked by the neighboring (C—H) vibrations; this being an indication of its low intensity on the alumina surface.

$(C_6—O_1—C_5)$ *bond.* This bond binds together the choline and the acetyl part of both Ach and β-MeAch. It is characterized by two vibrations: $\nu_s(C—O)$ and $\nu_a(C—O)$. In the tunneling spectra of both molecules the $\nu_s(C—O)$ vibration is mixed with the $\nu(Al—O)$ mode. The asymmetric vibration $\nu_a(C—O)$ was found at 1242 cm^{-1} for Ach as a wide peak. In the β-MeAch spectrum it is situated at 1234 cm^{-1}. The solid state IR spectrum showed it at the same frequency and the aqueous solution IR spectrum at 1249 cm^{-1}. Thus, on the basis of the β-MeAch tunneling spectrum one could conclude that the $(C_6—O_1—C_5)$ bond does not change its solid state conformation on the alumina surface.

A model of the Ach non-hydrolytic interaction with an alumina surface is shown in Figure 12.

7.1.4.2. *Hydrolytic interaction (Figure 11-B)*

Table V lists the most important peak positions and peak intensities of AchI and β-MeAchBr tunneling spectra for the 400—1800 cm^{-1} region. The table is completed with:

1. Raman and IR vibrations of AchI in aqueous solution, before and after non-enzymatic hydrolysis [145];
2. Raman and IR vibrations of ChI in aqueous solution [75];
3. tunneling vibrational modes of acetic acid and ChI [155].

The significant modifications in the tunneling spectra of choline and acetyl skeleton vibrations together with the appearance of the predominant peak at 1596 cm^{-1} and the disappearance of the vibration $\nu_a(C—O)$ at 1251 cm^{-1} provide strong evidence that in this case a reaction takes place on the alumina surface leading to structural changes in the adsorbed molecules with formation of new

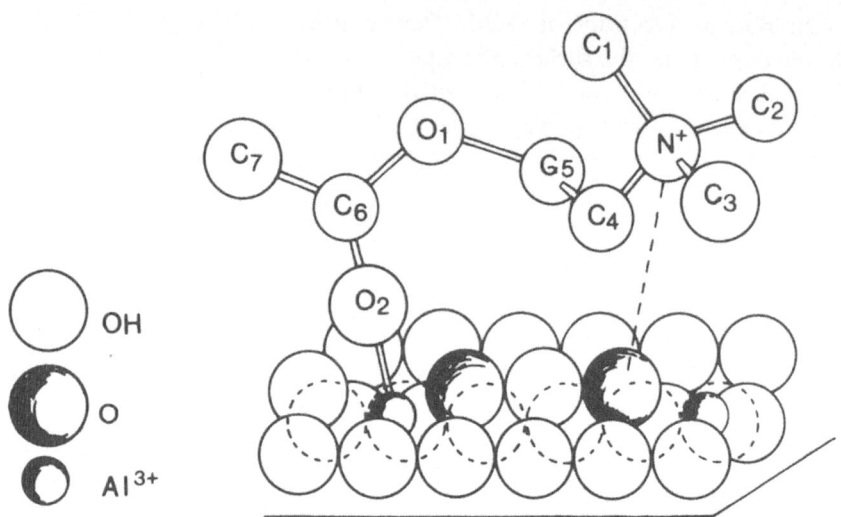

Fig. 12. A model of Ach non-hydrolytic interaction with an alumina surface (Reprinted with permission from *Life Sciences* **32**, D. Aslanian: '*Vibrational spectroscopic approach to study of Ach and related compounds*' [156]. Copyright 1983, Pergamon Journals Ltd.).

species. The products of the hydrolysis of Ach and β-MeAch on the surface (choline and acetate) have been identified using, on one hand, Raman and IR spectra of Ach before and after hydrolysis in the basic medium [145] and, on the other hand, tunneling spectra of acetic acid and choline chemisorbed on alumina [155].

The presence of both products of hydrolysis on the surface complicated the interpretation of the lateral group vibrations. Nevertheless some comments concerning the methylene groups are possible. Thus, the conformation of the (CH_3) groups bound to the N^+ becomes identical for Ach and β-MeAch, similar to the aqueous solution conformation. The conformation of the (CH_2) groups does not show significant modifications.

7.2. ACH RECEPTOR-RICH POSTSYNAPTIC MEMBRANES

7.2.1. *Native Postsynaptic Membranes*

In our first work on the Raman spectrum of the Ach receptor-rich postsynaptic membranes, the classical single scan technique has been used [158]. Although a significant fluorescence background was observed in the low frequency range $(400-800 \text{ cm}^{-1})$ very intense and well resolved peaks were recorded between $800-1700 \text{ cm}^{-1}$ and $2800-3100 \text{ cm}^{-1}$ (Figure 1a and 2a in [158]).

Following this work, we succeeded in obtaining more pure samples removing completely the sucrose which, used in the purification process of membranes,

Table V. Raman and IR tunneling frequencies of AchI, β-MeAchBr, CH$_3$COOH and ChI (Hydrolytic interaction).*

R and IR frequencies							Tunneling frequencies								Assignments
AchI: aqueous solution				ChI			AchI		β-MeAchBr		CH$_3$COOH		ChI		
before hydrolysis		after hydrolysis		aq. sol.	solid										
R	IR	R	IR	R	R	IR	cm⁻¹	meV	cm⁻¹	meV	cm⁻¹	meV	cm⁻¹	meV	
418vw				419vw	420m	410vw	411vs	51vs	411vs	51vs	403vs	50vs	440s	55s	δ(O—C—C—N) choline skeleton
450vw				477vw	449s	450w	427vs	53vs	427vs	53vs					
525vw				526vw	468sh	468m									
					531m	531v									
645s		645vw					621m	77m	613m	76m	613w	76w			δ(C—C—O) acetyl skeleton
834s							685s	85s	677w	84s	677w	84w			
		930vs					935m	116m	935m	116m	935s	116s	935vs	116vs	ν(C—C) + ν_a(Al—O)
	1251vs	1251vs	1251vs		1251vs	1321vw									ν_a(C—O)
		1420s	1413vs				1429vs	177vs	1435vs	178vs	1468s	182s	1444s	179s	ν_sCOO⁻
			1566vs				1596vs	198vs	1596vs	198vs	1597s	198s	1605w	199w	ν_aCOO⁻
1735m	1734vs														νC=O

* Reprinted with permission from *Life Sciences* **32**, D. Aslanian: 'Vibrational Spectroscopic Approach to Study of Ach and Related Compounds'. © 1983, Pergamon Journals Ltd.

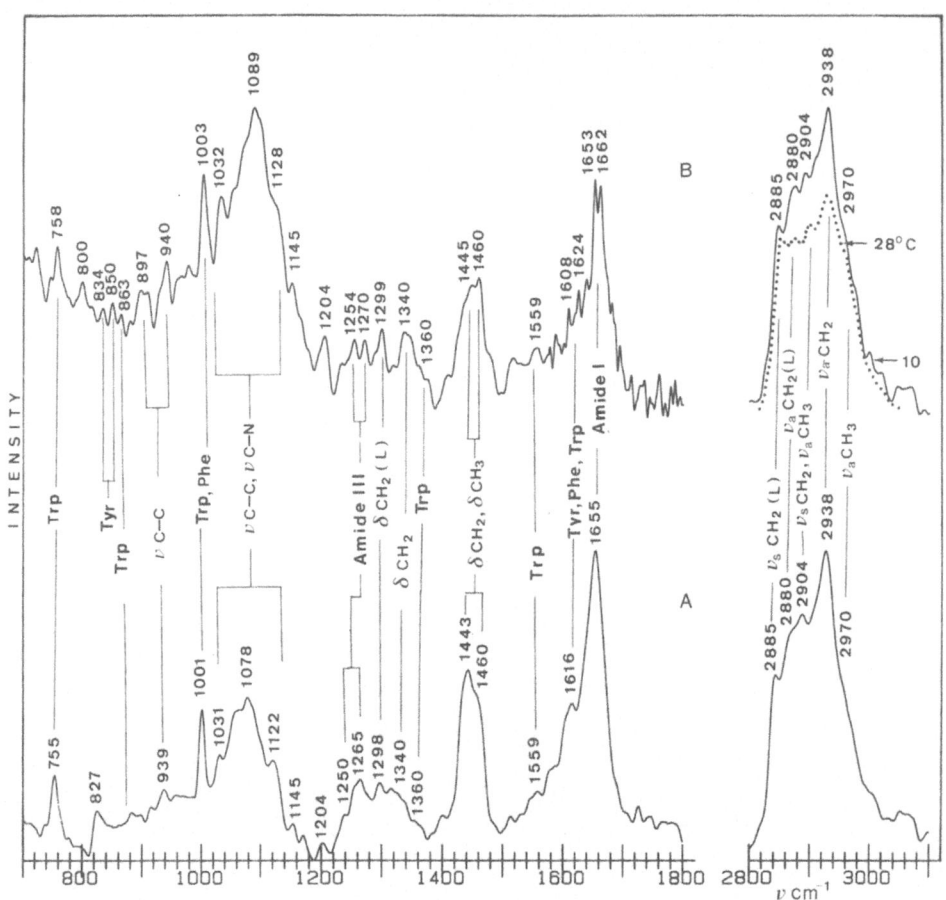

Fig. 13. Raman spectra of native postsynaptic membrane (A) and membrane with alkaline extracted 43 kDa protein (B), after subtraction of buffer spectra and fluorescence background and smoothing by Fourier transformation. Excitation wavelength 488 nm, power of sample level 150 mW, slits 8 cm⁻¹, temperature 10°C.

remains attached to them. Applying multiple scanning, subtracting the water and buffer spectra, the fluorescence background as well as smoothing of the spectra by Fourier transformations, we obtained a very well resolved spectrum of the native postsynaptic membranes as shown in Fig. 13-A. The difference between this spectrum and the spectrum in [158] have been explained with the absence of sucrose in the present spectrum.

Raman spectra of native postsynaptic membranes have been analysed in relation with the protein and lipid components of the samples.

In the (C—H) stretching region (2500–3120 cm⁻¹) spectra disclose two very strong bands at 2904 and 2938 cm⁻¹ and a shoulder near 2970 cm⁻¹. These frequencies correspond to the protein vibrations in this region as it was confirmed by the spectra of the purified Ach-receptor protein (see below Fig. 14). Two other shoulders are observed near 2855 and 2880 cm^{-1} which have been assigned to the symmetric $(\nu_s CH_2)$ and asymmetric stretching $(\nu_a CH_2)$ vibrations of the lipid methylene chains [133]. The great intensity of the protein vibrations compared to the intensity of the lipid vibrations can be related to the high protein/lipid ratio of the Ach-receptor-rich membranes [67]. The spectrum in Fig. 13-A, taken at 10°C shows also that lipids are in an ordered crystalline state as indicated by the relatively high intensity of the $(\nu_a CH_2)$ vibration at 2885 cm^{-1} compared to the intensity of the $(\nu_s CH_2)$ vibration at 2855 cm^{-1}. Increasing temperature does not perturb the conformation of the membrane components since there are no changes in intensity and frequency of the protein and lipid vibrations [159]. This is not surprising as far as the membrane proteins are concerned. A differential scanning calorimetric study of Ach receptor-rich membranes has indicated that the major protein transition takes place at 59°C and no protein modifications have been observed up to 50°C [160]. The absence of the modification in the intensity of the lipid asymmetric stretching vibrations at 2880 cm^{-1} [159] indicates that phase transition does not take place in the postsynaptic membrane lipids in this temperature range. Thus, it seems that lipids in these membranes preserve their ordered crystalline state in the temperature interval of 0°C to 35°C. This result agrees with the ESR study of March and Barantes [161] which pointed out an immobilization of postsynaptic membrane lipids. Using the same method Rousselet et al. [162] also noted a strong immobilization of the fatty acid lipid chains in these membranes. Both results, by Raman and ESR methods, show an interesting difference in the temperature behaviour between the subsynaptic membranes and other cell membranes. Indeed, Raman studies of erythrocyte [115], thymocyte [163], lymphocyte [164] and human blood platelet [165] membranes indicate a thermotropic state transition in this temperature range.

In the course of our studies on the molecular characteristics of the native postsynaptic membranes two kind of samples have been used: samples which after preparation have been stored in liquid nitrogen and thawed at room temperature before use, and samples which were used immediately after preparation. In the spectra of both samples, the lipid asymmetric vibration $(\nu_a CH_2)$ appeared at the normal vibration frequency (near 2885 cm^{-1}), but symmetric stretching band appeared at higher frequency in the samples used immediately after the preparation (near 2862 cm^{-1} instead of 2855 cm^{-1}) [159]. Since it is known that the (C—H) stretching region is sensitive to the crystalline packing of lipids [124, 132] the modified frequency of the $(\nu_s CH_2)$ vibration could be related to the different ordering in the lipid chain due to the different thermal treatments of samples. As it was pointed out first by Forrest [166], the thermal history of the sample has an important effect on the lipid packing. Indeed, the presence of a vibration near 2860 cm^{-1} in the Raman spectra of lipids has been shown as evidence to the existence of a latteral chain interaction typical for a triclinic lattice whereas the vibration at 2855 cm^{-1}, characterizes a hexagonal lattice [131, 133]. Thus, the vibration near 2860 cm^{-1} in the Raman spectra of native Ach receptor-rich postsynaptic membranes suggests that lipids in these membranes are predominantly in a triclinic crystalline lattice.

In the region 600—1800 cm^{-1} Raman spectra in Fig. 13-A discloses an Amide I vibration centered at 1655 cm^{-1} strongly suggesting a predominant α-helical conformation [91]. This is further strengthened by the Amide III region in which the band at 1265 cm^{-1} is also characteristic for an α-helical conformation [92]. However, the shoulder near 1250 cm^{-1} that accompanied this band indicates the presence of some fraction of a random-coil conformation [92].

The characteristic tyrosine doublet at 838—848 cm^{-1} appears weak and not well resolved in the spectrum of the postsynaptic membranes. The similar appearance of the conformational sensitive tryptophane vibrations near 880 and 1360 cm^{-1}, suggests a great fraction exposed residues [103].

The Raman frequencies associated to the coupled skeletal stretching vibrations ν(C—C) and ν(C—N) are sensitive to the amino acid incorporation into polypeptide chains [91] as well as to the conformational state of the lipid hydrocarbon chains [119, 120]. A vibration at 1130 cm^{-1} is indicative for all-trans chain segments (ordered state) and a vibration at 1060 cm^{-1} (for lipids) and near 1070 cm^{-1} (for proteins) for chains containing gauche conformations [100, 117]. According to the spectrum in Fig. 13 the gauche conformation of the side chains seems to be a preponderant one.

The vibration near 1443 cm^{-1}, accompanied by a shoulder near 1460 cm^{-1}, which belongs to the (C—H) deformation modes (δCH_2, δCH_3) is a mixture of protein and lipid vibrations. It is sensitive to the chain conformation [133] and becomes very intense in the spectrum of isolated Ach-receptor protein (see below Fig. 14) whereas it is broad and weaker in the spectrum of the membranes without 43 kDa protein (Fig. 13-B).

7.2.2. Postsynaptic Membranes Without 43 kDa Protein

In order to extract the 43 kDa protein, the native membranes are treated in alkaline medium as described above. Raman spectra of these samples were always accompanied by a very high fluorescence background which obscured almost all vibrations in the 400—1800 cm^{-1} spectral range. Fortunately, the 2800—3100 cm^{-1} region was free of the fluorescence and thus, in our earlier work, only this region has been analysed [158, 159]. Lately, successful work in the 400—1800 cm^{-1} region has been achieved (Fig. 13-B) using computerized data acquisition and evaluation which facilitate the assignments in this region. This spectrum compared to the spectrum of native membranes (Fig. 13-A) reveals some differences in both spectral regions.

In the region 2500—3120 cm^{-1} the spectrum of membranes without 43 kDa protein, at 10°C, reveals that a decrease in intensity of the protein vibrations at 2900 and 2942 cm^{-1} takes place as a consequence of the extraction. The lipid asymmetric stretching vibrations ($\nu_a CH_2$) at 2882 cm^{-1} preserves its higher intensity at 10°C suggesting that after extraction, lipids remain in their ordered crystalline state. However, increasing temperature produces a progressive decrease in intensity of the ($\nu_a CH_2$) vibration (Fig. 13-B, spectrum at 28°C) which indicates that a transition from ordered crystalline state to liquid-crystalline or to disordered state takes place. It was found that this transition is not very cooperative one [159]. Using the $I_{\nu_a CH_2}/I_{\nu_s CH_2}$ ratio, a transition temperature near 21°C has been indicated [159]. Thus, removal of 43 kDa and other non-receptor proteins, induced important changes in the lipid-protein interactions. It was found that after removal of all membrane proteins, i.e., leaving the lipid extract alone, a much more cooperative transition of postsynaptic membrane lipids takes place [159]. It appears, therefore, that the proteins of these membranes play an important role for the preservation of the crystalline state of membrane lipids. This is in accordance with many other recent spectroscopical measurements on lipid-protein systems [5].

In the region 400—1700 cm^{-1}, the following characteristic features have been observed:

— a strong Amide I band which appears as a doublet at 1665 cm^{-1} (α-helix) and 1661 cm^{-1} (random coil). Both vibrations are with comparable intensities. In the Amide III region the characteristic band at 1270 cm^{-1} (α-helix) and 1253 cm^{-1} (random coil) have also comparable intensities. Thus it appears that after extraction of 43 kDa protein the random coil conformations increase considerably in the postsynaptic membranes;

— an increase in intensity of the lipid vibration at 1300 cm^{-1} (Fig. 13-B);

— considerable modifications in the 1000—1200 cm^{-1} region. The maximum of this region is situated at 1082 cm^{-1} in the spectrum of alkaline treated membranes, indicating most likely a larger contribution from the PO$_2$ symmetric stretching vibrations of the phospholipid head groups [131];

— finally, the 1400—1450 cm^{-1} region also changes considerably.

Thus, important changes in the lipid-lipid and/or lipid-protein interaction take place at the level of the postsynaptic membranes as a consequence of the release of 43 kDa protein.

7.2.3. *Isolated Ach-Receptor Protein*

Well resolved spectra of the Ach-receptor protein, extracted from native membrane using a dimethyldodecylamine oxide (DDAO) detergent, have been obtained by single scan technique for the spectral regions 400—1700 cm^{-1} and 2800—3100 cm^{-1}, as shown in Fig. 14.

In the 2800—3100 cm^{-1} (C—H) stretching region, beside the characteristic protein vibrations at 2900, 2938, 2970 cm^{-1}, clearly appears a very intense vibrations at 2855 cm^{-1}. The presence of this vibration strongly suggests that lipids remain associated with the Ach-receptor purified with this particular detergent. This interpretation is also supported by the presence of two other lipid vibrations at 1304 cm^{-1} and 1082 cm^{-1}. It was also confirmed by the direct assay of

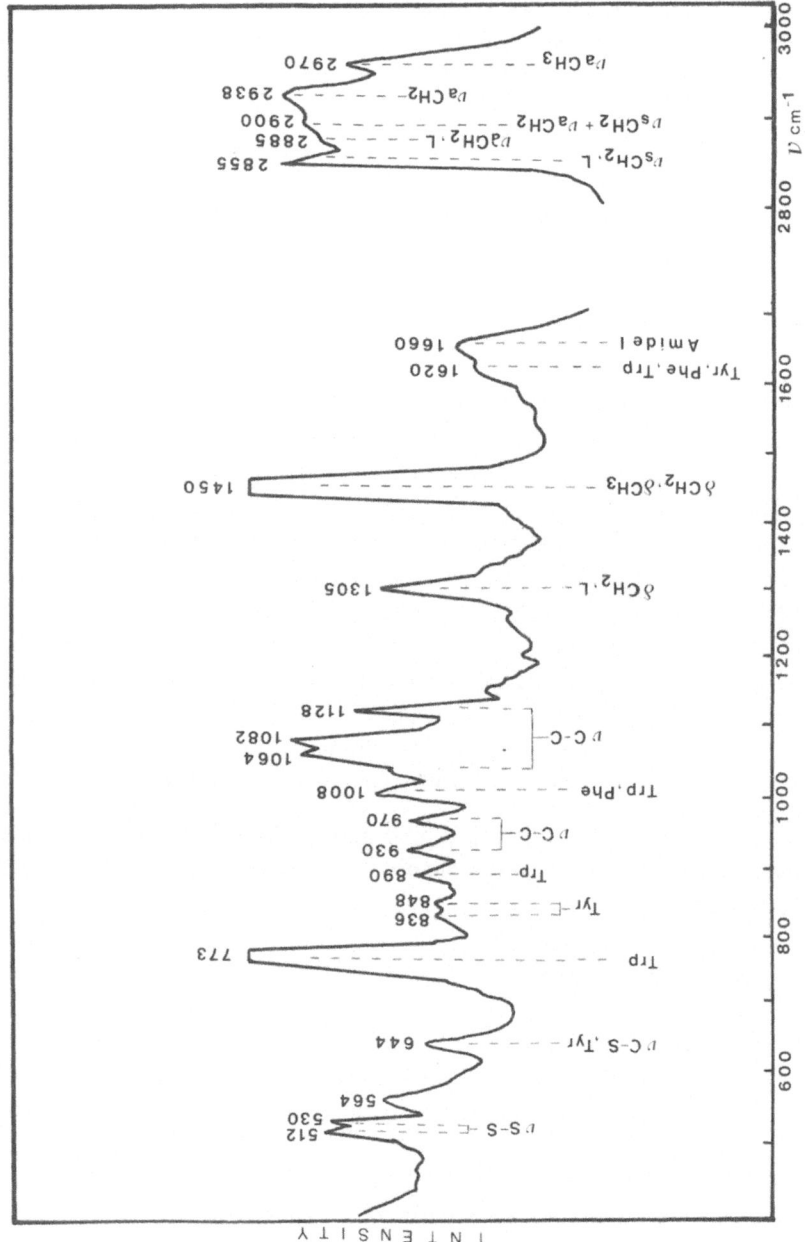

Fig. 14. Raman spectrum of purified Ach receptor protein. Recording conditions as in Fig. 13.

phospholipid content in the purified preparation of Ach-receptor. About 50 mol phospholipid/mol protein has been found [158].

The hydrocarbon chains of lipids associated with the Ach-receptor protein seem to be in a disordered state as indicated by the much lower intensity of the asymmetric stretching vibrations ($\nu_a CH_2$) near to 2885 cm^{-1}. Different factors, such as presence of detergent, disruption of the membrane bilayer, change in lateral pressure et cetera, would prevent lipids from crystallizing into the gel phase, even at the temperature of the experiment which was as low as 8°C. As expected, no change could be detected when temperature was raised up to 40°C.

The observation that lipids remain associated with Ach-receptor in presence of DDAO detergent, can be related to the finding that in reconstitution experiments lipids are essential to maintain the Ach-receptor in the functional state [167]. Also, low affinity binding of non-competitive blockers of the permeability response have been postulated to occur at the interface between the Ach-receptor protein and membrane lipids [167]. Thus, the lipids which remain associated with the Ach-receptor may play a critical role in the regulation of its functional properties.

The Raman spectra of Ach-receptor protein disclose the Amide I band at about the same frequency (1660 cm^{-1}) as in the native membranes (1655 cm^{-1}) but the peak is broader and less intense. Broadening of Amide I band without shift in frequency has already been observed on lysozyme in the presence of interacting phospholipids [168]. This result might thus be accounted for by changes in the interactions between the Ach-receptor and still associated phospholipids. The broadening of the Amide I band — in terms of the secondary structure — also means a more wide distribution of different structures without remarkable transition from one definite structure to another [72, 73].

The (S—S) stretching vibrations associated with disulfide bridges can be observed close to the 510 and 530 cm^{-1}. The intense band at 773 cm^{-1} has been attributed to Trp, but its very strong intensity remains poorly understood. The 644 cm^{-1} band might be a mixture of tyrosine and (C—S) vibrations.

8. Conclusion

The experimental results presented in this paper clearly show that Raman and IR spectroscopy are particularly well suited to the study of structural features characterising the acetylcholine and the postsynaptic membranes.

The most important fundamental vibration of Ach were assigned separately for the choline and acetyl part of the molecule. The differences between solid state and aqueous solution spectra were explained by an internal rotation of the acetyl-choline molecule around the O_1—C_5 bond.

The comparative study between Ach and its analogues, β-MeAch, Mu and Nic, indicated that the electronic density in the neighborhood of their nitrogen atoms is similar. This could be an important factor for the similarity of their biological activity. The interpretation of the vibrational spectra of these molecules has led to the first proposed experimental model of the interaction of Ach with an Al_2O_3 surface which possesses complementary adsorption sites analogous to the sites of the Ach-receptor protein or the AchE enzyme.

Further, Raman spectra of native membranes have been analyzed in relation with the protein and lipid components of the samples. In particular, the method reveals the predominancy of α-helix in the case of native postsynaptic membrane proteins and the isolated Ach-receptor protein. After extraction of the 43 kDa protein, a considerable increase of the random coil conformation has been found. The analysis of the characteristic lipid vibrations confirmed that in the native membranes the lipid/protein ratio is particularly low. It was shown that lipids remain associated with DDAO extracted Ach-receptor protein. It was also found that phase transition does not take place for native membrane lipids in the temperature range from 0°C to 35°C. However, for the membrane without 43 kDa protein, a transition has been observed near to 21°C. Thus, removal of the 43 kDa protein induced important changes in the lipid-protein interactions. Further studies on the precise molecular mechanism at the synaptic level could be developed by application of the resonance Raman and time resolved Raman spectroscopies.

Acknowledgements

The author thanks Dr P. Grof and M. Négrerie for reading the manuscript and making useful suggestions.

References

1. F. R. Dollish, W. F. Fateley, and F. F. Bentley: *Characteristic Raman Frequencies of Organic Molecules*, John Wiley & Sons (1974).
2. M. Avram and Gh. D. Mateescu: *Spectroscopie Infrarouge, Application en Chimie Organique*, Dunod, Paris (1970).
3. T. M. Theophanides, Ed.: *Infrared and Raman Spectroscopy of Biological Molecules*, Proceedings of the NATO Advanced Study Institute held at Athens, Greece, August 22—31 (1978).
4. D. F. H. Wallach, S. P. Verma, and J. Fookson: *Biochim. Biophys. Acta* **559**, 153 (1979).
5. P. R. Carey: *Biochemical Application of Raman and Resonance Raman Spectroscopies*, Acad. Press (1982).
6. F. S. Parker: *Application of Infrared, Raman and Resonance Raman Spectroscopy in Biochemistry*, Plenum Press, N.Y. (1983).
7. J. T. Edsall: *J. Chem. Phys.* **4**, 1 (1936).
8. J. T. Edsall: *J. Chem. Phys.* **5**, 508 (1937).
9. J. T. Edsall, J. W. Otvos, and A. Rich: *J. Amer. Chem. Soc.* **72**, 474 (1950).
10. D. Garfinkel and J. T. Edsall: *J. Amer. Chem. Soc.* **80**, 3807 (1958).
11. D. Garfinkel and J. T. Edsall: *J. Amer. Chem. Soc.* **80**, 3818 (1958).
12. D. Garfinkel and J. T. Edsall: *J. Amer. Chem. Soc.* **80**, 3823 (1985).
13. D. Garfinkel: *J. Amer. Chem. Soc.* **80**, 3827 (1958).
14. H. A. Szymanski: *Raman Spectroscopy, Theory and Practice*, Plenum Press, N.Y. (1967).
15. O. Loewi: *Pfluegen Arch. Ges. Physiol.* **189**, 239 (1921).
16. S. Korkes, A. del Campillo, S. R. Korey, J. R. Stern, D. Nachmansohn, and S. Ochoa: *J. Biol. Chem.* **198**, 215 (1952).
17. D. Nachmansohn and A. L. Machado: *J. Neurophysiol.* **6**, 397 (1943).
18. D. Nachmansohn, H. M. John, and H. Waelsch: *J. Biol. Chem.* **150**, 485 (1943).

19. L. T. Potter: *Handbook of Neurochemistry*, Vol. 4, Ed. A. Lajtha, Plenum Press, N.Y. (1970).
20. J. Bremer and D. M. Greenberg: *Biochim. Biophys. Acta* **37**, 173 (1960).
21. G. B. Ansell and S. Spanner: *Biochem. J.* **110**, 201 (1968).
22. T. Hada and H. Noda: *Biochim. Biophys. Acta* **291**, 564 (1973).
23. H. I. Yamamura and S. H. Snyder: *J. Neurochem.* **21**, 1355 (1973).
24. J. del Castillo and B. Katz: *Coll. Int. Cent. Natl. Rech. Sci.*, Gif-sur-Yvette, France, **67**, 245 (1957).
25. B. Katz: *Nerve, Muscle and Synapse*, McGraw-Hill (1966).
26. L. Tauc: *Physiol. Rev.* **62**, 857 (1982).
27. J. B. Cooper and E. M. Meyer: *Neurochem. Int.* **6**, 419 (1984).
28. Y. Dunant and M. Israel: *Scientific American* **252**, 40 (1985).
29. P. Waser: *Experientia* **17**, 1 (1961).
30. D. J. Trigle and B. Belleau: *Can. J. Chem.* **40**, 1201 (1962).
31. A. Bebblington and R. W. Brimbelcomble: *Advances in Drug Research*, vol. 2 (Eds. H. J. Harper and A. B. Simmonds), pp. 143—172. Acad. Press (1965).
32. D. Aslanian: The Acetylcholine Molecule — review, submitted for publication.
33. D. Nachmansohn: *Chemical and Molecular Basis of Nerve Activity*, Acad. Press, N.Y. (1959).
34. A. Karlin: *Cell Surf. Rev.* **6**, 191 (1980).
35. J. P. Changeux: *Harvey Lecture*, Series 75, pp. 85—254. Acad. Press (1981).
36. R. D. Keynes: *Discovery* **13**, 172 (1952).
37. M. Kasai and J. P. Changeux: *FEBS Lett.* **7**, 13 (1970).
38. M. Kasai and J. P. Changeux: *J. Mol. Biol.* **6**, 1 (1971).
39. J. P. Changeux, M. Kasai, M. Huchet, and J. C. Meunier: *C.R. Acad. Sci. (Paris)* **270D**, 2864 (1970).
40. J. P. Changeux, M. Kasai, and C. Y. Lee: *Proc. Natl. Acad. Sci. U.S.* **67**, 1241 (1970).
41. R. Miledi, P. Molinoff, and L. T. Plotter: *Nature (London)* **229**, 554 (1971).
42. J. Hazelbauer and J. P. Changeux: *Proc. Natl. Acad. Sci. U.S.* **71**, 1479 (1974).
43. H. Schindler and U. Quast: *Proc. Natl. Acad. Sci. U.S.* **77**, 3052 (1980).
44. A. T. Tu: *Ann. Rev. Biochem.* **42**, 235 (1974).
45. J. Cartaud, J. Benedetti, A. Sobel, and J. P. Changeux: *J. Cell Sci.* **29**, 313 (1978).
46. J. Cartaud: *Ontogenesis and Functional Mechanisms of Peripheral Synapses*, Ed. J. Taxi, pp. 199—210. North-Holland (1980).
47. J. Kistler and R. M. Stroud: *Proc. Natl. Acad. Sci. U.S.* **78**, 3678 (1981); *Biophys. J.* **37**, 371 (1982).
48. M. W. Klymkowsky and R. M. Stroud: *J. Mol. Biol.* **128**, 319 (1979).
49. M. J. Ross, M. W. Klymkowsky, D. A. Agard, and R. M. Stroud: *J. Mol. Biol.* **116**, 635 (1977).
50. R. M. Stroud and D. A. Agard: *Arch. Biochem. Biophys.* **181**, 484 (1977).
51. J. Reynolds and A. Karlin: *Biochemistry* **17**, 2035 (1978).
52. J. Lindstrom, J. Merlic, and G. Yogeeswaren: *Biochemistry* **18**, 4465 (1979).
53. M. A. M. Raftley, M. W. Hunkapiller, C. D. Strader, and L. E. Hood: *Science* **208**, 1454 (1980).
54. L. Wennogle and J. P. Changeux: *Eur. J. Biochem.* **106**, 381 (1980).
55. A. Karlin: *Cell Surface and Neuronal Function* (Eds. C. Cotman, G. Post, and G. L. Nicolson) pp. 191—260. Elsevier/North-Holland Biomedical Press, N.Y. (1980).
56. L. Vennogle, R. Ostwald, T. Saitoh, and J. P. Changeux: *Biochemistry* **20**, 2492 (1981).
57. M. Noda, H. Takahashi, T. Tanabe, M. Toyosato, S. Kikyotani, Y. Fututani, T. Hirose, H. Takashima, S. Inayama, T. Miyata, and S. Numa: *Nature (London)* **302**, 528 (1983).
58. K. Sumikawa, M. Houghton, J. C. Smith, L. Bell, B. M. Richards, and E. A. Bernard: *Nucleic Acid Res.* **10**, 5809 (1982).
59. A. Devillers-Thiery, J. Giraudat, M. Bentaboulet, and J. P. Changeux: *Proc. Natl. Acad. Sci. U.S.* **80**, 2067 (1983).
60. T. Claudio, M. Ballivet, J. Patrick, and S. Heinman: *Proc. Nat. Acad. Sci. U.S.* **80**, 1111 (1983).
61. R. R. Neubig, E. K. Krodel, N. D. Boyd, and J. B. Cohen: *Proc. Natl. Acad. Sci. U.S.* **76**, 690 (1979).

62. J. P. Changeux, T. Heidmann, J. L. Popot, and A. Sobel: *FEBS Lett.* **105**, 181 (1979).
63. A. Rousselet, J. Cartaud, P. F. Devaux, and J. P. Changeux: *EMBO J.* **1**, 439 (1982).
64. E. J. Barantes, W. C. Neugebauer, and H. P. Zingsheim: *FEBS Lett.* **111**, 73 (1980).
65. J. Cartaud, A. Sobel, A. Rousselet, P. F. Devaux, and J. P. Changeux: *J. Cell Biol.* **90**, 418 (1981).
66. M. Klimkovsky, J. E. Heuser, and R. M. Stroud: *J. Cell Biol.* **85**, 823 (1980).
67. J. L. Popot, R. A. Demel, A. Sobel, L. L. M. Van Deenen, and J. P. Changeux: *Eur. J. Biochem.* **85**, 24 (1978).
68. D. M. Michaelson and M. A. Raftery: *Proc. Natl. Acad. Sci. U.S.* **71**, 4768 (1974).
69. T. Saitoh, R. E. Oswald, L. P. Wennogle, and J. P. Changeux: *FEBS Lett.* **116**, 30 (1980).
70. J. B. Cohen, M. Weber, M. Huchet, and J. P. Changeux: *FEBS Lett.* **26**, 43 (1972).
71. O. Lowry, N. Rosebrough, A. Farr, and R. Randell: *J. Biol. Chem.* **193**, 265 (1951).
72. R. W. Williams: *J. Mol. Biol.* **166**, 581 (1983).
73. R. W. Williams: *Methods in Enzymology* **130**, 483 (1986).
74. F. R. Dollish, W. G. Fateley, and F. F. Bentley: *Characteristic Raman Frequencies of Organic Compounds*, John Wiley & Sons (1974).
75. D. Aslanian, A. Lautié, and M. Balkanski: *J. Chim. Phys.* **7−8**, 1028 (1974).
76. B. Nolin and R. N. Jones: *Can. J. Chem.* **34**, 1392 (1956).
77. G. Durgaprased, D. N. Sathynaryana, C. C. Patel, H. S. Ranohave, A. Goel, and C. N. Rao: *Spectrochim. Acta* **28A**, 2311 (1972).
78. J. Anhouse and M. C. Tobin: *Spectrochim. Acta* **28A**, 2141 (1972).
79. J. Kress and J. Guilleremet: *J. Chim. Phys.* **70**, 374 (1973).
80. H. Wieser, J. P. Laidlow, P. J. Kruger, and H. Fuhrer: *Spectrochim. Acta* **24A**, 1055 (1968).
81. J. M. Schachschneider and R. G. Snyder: *Spectrochim. Acta* **19**, 117 (1963).
82. T. Miyazawa: *J. Chem. Phys.* **32**, 1647 (1960).
83. S. Krimm and Y. Abbe: *Proc. Natl. Acad. Sci. U.S.* **69**, 2788 (1972).
84. B. G. Frushour and J. L. Koenig: *Adv. Infrared and Raman Spectroscopy* **1**, 35 (1975).
85. T. G. Spiro and B. P. Gaber: *Ann. Rev. Biochem.* **46**, 553 (1977).
86. R. C. Lord: *Proc. 23rd. Int. Congr. Pure Appl. Chem. Suppl.* **7**, 179 (1971).
87. N. T. Yu, C. S. Liu, and D. C. O'Shea: *J. Mol. Biol.* **70**, 117 (1972).
88. M. L. Chen, R. C. Lord, and R. Mendelsohn: *Biochim. Biophys. Acta* **328**, 252 (1973).
89. M. C. Chen, R. C. Lord, and R. Mendelsohn: *J. Amer. Chem. Soc.* **96**, 3038 (1974).
90. R. C. Lord and N. T. Yu: *J. Mol. Biol.* **50**, 509 (1970).
91. R. C. Lord and N. T. Yu: *J. Mol. Biol.* **51**, 203 (1970).
92. N. T. Yu and C. S. Liu: *J. Amer. Chem. Soc.* **94**, 3250 (1972).
93. N. T. Yu, B. H. Jo, R. C. Chang, and J. D. Huber: *Arch. Biochem. Biophys.* **160**, 614 (1974).
94. S. L. Hsu, H. W. Moore, and S. Krimm: *Biopolymers* **15**, 1513 (1976).
95. B. G. Frushour and J. L. Koenig: *Biopolymers* **13**, 1809 (1974).
96. M. Pezolet, M. Pigeon-Gosselin, and L. Coulombe: *Biochim. Biophys. Acta* **453**, 502 (1976).
97. L. J. Lippert, D. Tyminski, and P. J. Desmeules: *J. Amer. Chem. Soc.* **98**, 7075 (1976).
98. N. T. Yu, B. H. Jo, and C. S. Liu: *J. Amer. Chem. Soc.* **94**, 7572 (1972).
99. N. T. Yu and B. H. Jo: *J. Amer. Chem. Soc.* **95**, 5033 (1973).
100. B. G. Frushour and J. L. Koenig: *Biopolymers* **14**, 649 (1975).
101. M. C. Chen and R. C. Lord: *J. Raman Spectrosc.* **9**, 304 (1980).
102. N. T. Yu, B. H. Jo, and D. C. O'Shea: *Arch. Biochem. Biophys.* **156**, 71 (1973).
103. Siamwiza *et al.*: *Biochemistry* **14**, 4870 (1975).
104. R. C. Lord and R. Mendelsohn: *J. Amer. Chem. Soc.* **94**, 2133 (1972).
105. N. T. Yu: *J. Amer. Chem. Soc.* **96**, 4664 (1974).
106. I. Harada, T. Miura, and H. Takeuchi: *Spectrochim. Acta* **42A**, 307 (1986).
107. T. Kitagawa, T. Azuma, and K. Hamaguchi: *Biopolymers* **18**, 451 (1979).
108. H. Sugeta, A. Go, and T. Miyazawa: *Bull. Chem. Soc. Jap.* **46**, 3407 (1973).
109. H. E. Van Wart and H. A. Scheraga: *J. Phys. Chem.* **80**, 1823 (1976).
110. L. T. Maxfield and H. A. Scheraga: *Biochemistry* **16**, 4443 (1977).
111. V. J. Hruby, K. K. Deband, J. Fox, J. Bjarnason, and A. T. Tu: *J. Biol. Chem.* **253**, 6060 (1978).
112. A. T. Tu, J. Lee, K. K. Deband, and V. J. Hruby: *J. Biol. Chem.* **254**, 3272 (1979).

113. G. S. Bailey, J. Lee, and A. T. Tu: *J. Biol. Chem.* **254**, 8922 (1979).
114. H. Ishizaki, R. H. McKay, T. R. Norton, K. T. Yasunobu, J. Lee, and A. T. Tu: *J. Biol. Chem.* **254**, 9651 (1979).
115. S. P. Verma, D. F. H. Wallach, and R. Schmidt-Ullrich: *Biochim. Biophys. Acta* **394**, 633 (1975).
116. C. H. Huang: *Biochemistry* **8**, 344 (1969).
117. J. P. Lippert and W. L. Peticolas: *Proc. Natl. Acad. Sci. U.S.* **68**, 1572 (1971).
118. J. P. Lippert and W. L. Peticolas: *Biochim. Biophys. Acta* **282**, 8 (1972).
119. K. Larsson and R. P. Rand: *Biochim. Biophys. Acta* **326**, 245 (1973).
120. K. Larsson: *Chem. Phys. Lipids* **10**, 165 (1973).
121. R. Mendelsohn: *Biochim. Biophys. Acta* **290**, 15 (1972).
122. B. J. Bulkin: *Biochim. Biophys. Acta* **274**, 649 (1972).
123. B. J. Bulkin and N. Krishnamachari: *J. Amer. Chem. Soc.* **94**, 1109 (1972).
124. K. G. Brown, W. L. Peticolas, and E. Brown: *Biochim. Biophys. Res. Commun.* **54**, 358 (1973).
125. M. Tasumi, T. Shimanouchi, and T. Miyazawa: *J. Mol. Spectr.* **9**, 261 (1962).
126. R. G. Snyder and J. H. Schachtschneider: *Spectrochim. Acta* **19**, 85 (1963).
127. R. G. Snyder: *J. Chem. Phys.* **47**, 1316 (1967).
128. B. C. Spiker and I. W. Levin: *Biochim. Biophys. Acta* **388**, 361 (1975).
129. R. C. Spiker and I. W. Levin: *Biochim. Biophys. Acta* **433**, 457 (1976).
130. N. Yellin and I. W. Levin: *Biochemistry* **16**, 642 (1977).
131. B. P. Gaber, P. Yager, and W. L. Peticolas: *Biophys. J.* **21**, 161 (1978).
132. B. P. Gaber and W. L. Peticolas: *Biochim. Biophys. Acta* **465**, 260 (1977).
133. R. G. Snyder, L. S. Hsu, and S. Krimm: *Spectrochim. Acta* **34A**, 395 (1978).
134. J. H. Fellman and T. S. Fujita: *Biochim. Biophys. Acta* **56**, 227 (1962).
135. J. H. Fellman and T. S. Fujita: *Biochim. Biophys. Acta* **71**, 701 (1963).
136. F. G. Canepa and E. F. Mooney: *Nature (London)* **207**, 78 (1965).
137. J. H. Fellman and T. S. Fujita: *Biochim. Biophys. Acta* **97**, 590 (1965).
138. J. H. Fellman and T. S. Fujita: *Nature(London)* **211**, 848 (1966).
139. F. G. Canepa, P. Pauling, and H. Sorum: *Nature (London)* **210**, 907 (1966).
140. I. K. Herdklotz and R. L. Sass: *Biochim. Biophys. Res. Commun.* **40**, 583 (1970).
141. T. Svinning and H. Sorum: *Acta Cryst.* **B31**, 1581 (1975).
142. S. Yagner and H. Sorum: *Acta Cryst.* **B33**, 2757 (1977).
143. C. Chothia and P. Pauling: *Chem. Commun.*, 746 (1969).
144. C. Chothia: *Nature (London)* **225**, 36 (1970).
145. A. Lautié, D. Aslanian, J. C. Merlin, A. Dupaix, and M. Balkanski: *J. Raman Spectr.* **7**, 337 (1978).
146. D. Aslanian, A. Lautié, Ch. Mankai, and M. Balkanski: *J. Chim. Phys.* **72**, 1052 (1975).
147. R. Figlizzo and A. Novak: *J. Chim. Phys.* **66**, 1539 (1969).
148. D. Aslanian, A. Lautié, and M. Balkanski: *J. Amer. Chem. Soc.* **96**, 1974 (1977).
149. L. Kier: *Mol. Pharmacol.* **3**, 487 (1967).
150. B. Pullman and A. Pullman: *Quantum Biochem.*, p. 679, Interscience, London (1963).
151. B. Pullman, Ph. Courrière, and J. L. Coubeilis: *Mol. Pharmacol.* **7**, 397 (1971).
152. M. J. Michelson and E. V. Zeimal: *Acetylcholine — an Approach to the Molecular Mechanism of Action*, p. 93, Pergamon Press (1973).
153. S. de Cheveigné *et al.*: *Biochim. Biophys. Res. Commun.* **94**, 29 (1980).
154. S. de Cheveigné: Ph.D. Thesis, Univ. Paris VII, Chap. 5 (1981).
155. D. Aslanian and S. de Cheveigné: *Mol. Pharmacol.* **22**, 678 (1982).
156. D. Aslanian: *Life Sciences* **32**, 2809 (1983).
157. P. K. Hansma: *Physics Rep.* **30**, 145 (1977).
158. D. Aslanian, T. Heidmann, M. Négrerie, and J. P. Changeux: *FEBS Lett.* **164**, 393 (1983).
159. D. Aslanian and M. Négrerie: *EMBO J.* **4**, 965 (1985).
160. M. C. Farach and M. Martinez-Carrion: *J. Biol. Chem.* **258**, 4166 (1983).
161. D. March and E. J. Barantes: *Proc. Natl. Acad. Sci. U.S.* **75**, 4329 (1978).
162. A. Rousselet, P. F. Devaux, and K. W. Wirtz: *Biochim. Biophys. Res. Commun.* **90**, 871 (1979).

163. S. P. Verma and D. F. H. Wallach: *Biochim. Biophys. Acta* **436**, 307 (1976).
164. S. P. Verma, R. Schmidt-Ullrich, W. S. Thomson, and D. F. H. Wallach: *Cancer Res.* **37**, 3490 (1977).
165. D. Aslanian, H. Vainer, and J. P. Guesdon: *Eur. J. Biochem.* **131**, 555 (1983).
166. G. Forrest: *Chem. Phys. Lipids* **21**, 237 (1978).
167. T. Heidmann, A. Sobel, J. L. Popot, and J. P. Changeux: *Eur. J. Biochem.* **110**, 35 (1980).
168. J. L. Lippert, R. M. Lindsay, and R. Schultz: *Biochim. Biophys. Acta* **32**, 599 (1980).

Molecules as Signal Transmitters in Biological Systems

P. MacLEOD
Laboratoire de Neurobiologie Sensorielle de l'E.P.H.E. et GRECO 'Sens Chimiques' du C.N.R.S.,
1, Avenue des Olympiades, 91305 Massy, France.

1. Introduction

Some molecules are of biological importance not as components of biological structures, but rather as signal transmitters. This signalling function can tentatively be distributed into four main categories by respect to its range: information can be transferred between different points in a cell, between different cells, between different individuals or between individuals and their environment.

Intracellular messengers are often referred to as 'second messengers'. Their role is to bridge a gap between two macromolecular sites in the same cell, a receptor tuned to some physical or chemical input and the generator of some electrical, mechanical or chemical output. Being medium sized, they can diffuse across the cell in a matter of milliseconds, speeding up by two or three orders of magnitude the interaction between distant macromolecules. Moreover, this process usually is an amplifying one, as a single messenger molecule can undergo several cycles of reversible association with its target molecules.

Intercellular chemical communication is achieved by neurotransmitters in the nervous system and by hormones in the whole body. Release of the transmitter results from a highly evolved and well controlled process. Each of the few dozens of presently well known hormones and neurotransmitters has its unique natural history. The highly specialized biological meaning of these molecules contrasts with the more ubiquitous second messengers which most often are the last link in their action. Actually, the specificity of their action completely depends on the specificity of the membrane receptors on their target cells.

Interindividual chemical messages are conveyed by airborne molecules for air-breathing animals and by water soluble molecules for aquatic ones. The name of 'pherormone' or 'pheromone' has been coined for stressing the fact that, very much as hormones do within the body, they induce highly specific behavioural responses in their recipients. Whereas airborne pheromones usually are rather small molecules, as otherwise too low a vapor pressure would prevent them from being efficiently transported, water soluble pheromone bulkiness may be anything up to large polypeptides. The first identified pheromones were pure compounds; more recently, many mixtures with strictly defined proportions have been discovered [1].

Conveying information on the chemical environment is the privilege of a host of odorants and tastants, most of them having a molecular weight in the range of 30

Jean Maruani (ed.), Molecules in Physics, Chemistry, and Biology, Vol. IV, 281—298.
© 1989 *by Kluwer Academic Publishers.*

to 300. Olfactory and gustatory chemosensory systems, their live counterparts, are endowed with the utmost detecting and identifying power: the detecting power has reached the quantal limit of a single molecule per receptor cell [2] while the identifying power makes it possible to recognize any pure chemical or mixture thereof [3, 4]. Such an amazing performance results from the phylogenic optimization of a series of hierarchically organized amplifying and processing devices, up to the highest brain level.

This review will focus primarily on chemosensory systems since they display every feature of the other categories of chemical signalling systems, acting either in series (second messengers, neurotransmitters) or in parallel (hormones, pheromones) with them. Otherwise explicitly stated, all the reviewed material is relevant to human chemical senses.

2. Transport

2.1. AIRBORNE ODORANT STIMULI

In order to effectively fulfil their function, odorant molecules must be chemically stable and volatile enough to withstand atmospheric conditions while travelling from their source to their destination at concentrations in the ppm range. The required lifetime may be hours, days or more: examples are well documented of nocturnal Primates making no mistakes in identifying individual marks deposited more than two months before [5].

Smelling being associated with breathing, odorant molecules penetrate the nose during inspiration and are eventually lost for olfaction if they go directly to the lungs: no more than a few percent of incoming molecules reach the olfactory mucosa during normal breathing. Only sniffing ensures that all of them are dynamically forced to flow over sensory cells [6, 7, 8]. Even so, they are still separated from membrane receptors by a watery mucus layer of some 20 μm thickness. The last step before reaching the appropriate receptors is therefore to dissolve in water, finally abolishing a seemingly obvious difference between aquatic and terrestrial animals: olfaction is without exception a water phase process, and therefore requires that odorant molecules are at least water-soluble in the submicromolar range, which is not very stringent indeed.

2.2. UNDERWATER OLFACTION

Certain molecules have such a low vapour pressure that they cannot be odorant in the gaseous phase [9]. Being soluble and stable in water solution, they may be used as odorants by aquatic animals: this is actually the case of aminoacids [10] and bile salts [11] which are the main features of odorous landscapes in seas and fresh waters.

In fish, olfactory sacs communicate with the outer environment through external nares, and with mouth cavity through internal nares. Water flowing is achieved either by an appropriate coupling with the coordinated mouth and gill movements or by the pumping action of special muscles; in the latter case, nares provide an active, phase-locked valve function [12, 13].

An interesting example of underwater olfaction is fetal olfaction, where odorant molecules present in amniotic fluid are drawn through nasal cavities by frequent swallowing movements. As many food related odorants are known to diffuse in blood and therefrom to amniotic fluid, it is quite likely that some early odour learning might occur before birth [14]. A somewhat related phenomenon is haematogenic olfaction where intravenous injection of odorous material results in odour perception after about 20 seconds latency, which is the time needed for circulating the blood up to the nose and exchanging the odorants with nasal mucus in which olfactory processes always originate [15].

3. Reception

A general description of membrane bound receptors emerged from a few fortunate cases where it was possible to extract and characterize them; even fewer were fully sequenced [16]. They are all proteins, with three parts: an external portion is specifically architectured for binding the appropriate ligands; a transmembranal, highly hydrophobic segment anchors the receptor into the lipid bilayer; an internal portion is transconformed into an activated form as soon as the external portion binds a ligand and initiates a series of intracellular events [17, 18]. The most variable part of receptors is their external portion which is responsible for their specificity and diversity. Only two designs seem possible for their transmembranal portion: either a single hydrophobic helix, or a bulky seven segment folded hydrophobic helix crossing the membrane seven times. The internal, effector portion is, without exceptions, involved in more or less sophisticated phosphorylation processes [19].

Most of the receptor relevant data are indirect evidence. Surprisingly enough, much more valuable information came from psychophysics, behavioural or electrophysiological experiments than from biochemical approaches. A plausible reason for this is that ligand-receptor binding is usually so weak that the lifetime of the complex seldom exceeds a millisecond, which makes the usual labelling strategies quite ineffective.

3.1. A PROTOTYPAL CHEMORECEPTOR, THE NICOTINIC ACETYLCHOLINE RECEPTOR

Although it is a highly specialized neurotransmitter receptor, it deserves special attention as it is presently so well known that it is a model for understanding other receptors. The purified material can be dissociated into four different glyco-peptides of 50 to 65 kilodaltons that are arranged in a pentameric stoechiometry, one of the subunits being repeated. Some variations do occur from animal to animal, but aminoacid sequencing shows that the four coding genes arose by duplication from a single ancestral gene [20, 21].

The intact receptor simultaneously displays the functions of a receptor and of a ionic channel non-selectively permeable to cations. Its receptor function depends on two binding sites located on the two repeated subunits. The channel function results from the symmetrical organization of the five subunits around a central pore which is wide open (2.5 to 3 nm in diameter) at its outer extremity and much narrower (about 0.5 nm in diameter) at its inner extremity (Figure 1). This channel is closed at rest. It opens for a few milliseconds when two ligands simultaneously bind both receptor sites [22]. This is an extreme example of reception and transduction coupling in one and the same molecular edifice.

3.2. BACTERIAL CHEMORECEPTORS

Bacterial chemotaxis offers a unique opportunity for studying receptors: one can take advantage simultaneously of frequent mutations for genetic studies and of unlimited growing for biochemical studies and combine them with recording easily quantifiable behavioural responses [23, 24]. A thorough study on *E. coli* resulted in the characterization of six different receptor proteins; three are 30 kilodalton soluble periplasmic proteins, three others are 60 kilodalton integral membrane proteins. The soluble proteins specifically bind galactose [25], maltose [26] and ribose [27] respectively. Only when associated to their ligand, they can bind to membrane receptors. In addition membrane receptors can bind directly to indole, serine, aspartate or protons [28].

When bound to a ligand, membrane receptors undergo a conformational change which facilitates methylation of 3 or 4 glutamate carboxyles on the inner side of the membrane. Methylation and demethylation are in dynamic equilibrium and the set point is shifted in favour of methylation by attractants and in the opposite direction by repellents. This covalent modification of the membrane receptor has a much longer time constant than ligand binding, resulting in a rudimentary memory [29].

3.3. OLFACTORY CHEMORECEPTORS

In order to be capable of efficiently monitoring the largest possible variety of ligands, these receptors must combine a very low selectivity and a very high sensitivity, two obviously opposite qualities. As far as sensitivity is concerned, many well documented experiments show that binding a single odorant molecule may result in a useful response at the level of a single olfactory nerve axon: this is the maximum expectable sensitivity since it is the absolute quantal threshold [30]. Selectivity can be derived from electrophysiological experiments where a wide variety of stimuli have been applied to each of a large number of olfactory units.

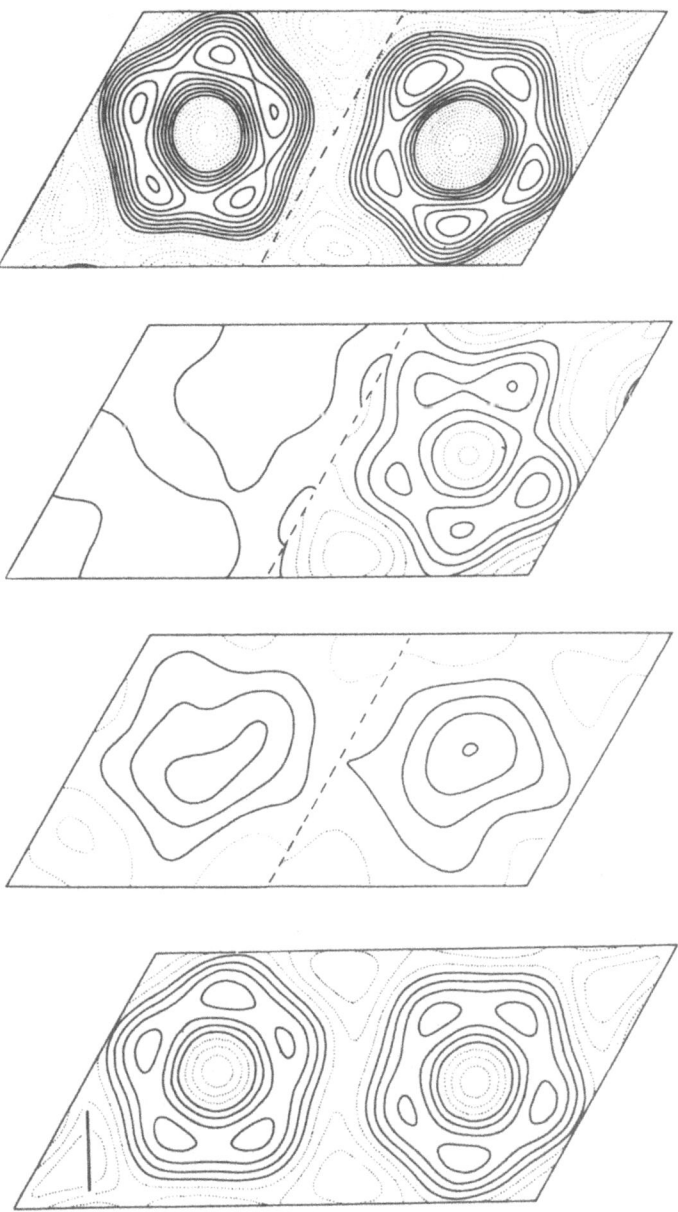

Fig. 1. Pentagonal symmetry of the acetylcholine receptor. Isoelectronic density contours with a 5 Å resolution were computed from electron diffraction on frozen tubes prepared from Torpedo marmorata electric organs (see [21]). Four sections at different levels of the three-dimension map are presented. From top to bottom, the successive planes are at +25, −15, −40 and −60 Angströms by respect to the outer surface of the lipid bilayer. The resolution of the analysis is insufficient to disclose structural details of the subunits and of the cytoplasmic end of the channel. The vertical scale bar (bottom left) corresponds to 25 Å.

Factor analysing the recorded response profiles showed that interfiber variance requires 10 to 20 different receptor types randomly distributed on neuroreceptor cells to be accounted for [31, 32].

A survey of the olfactory sensitivity of large population samples was undertaken by J. Amoore and resulted in the finding that the olfactory thresholds of some compounds are bimodally distributed (Figure 2), contrasting with the usual gaussian pattern of anthropometric data [33]. This pattern was obtained with reasonable credibility for at least six different odorants and the corresponding 'specific anosmias' were statistically independent [34]. The obvious hypothesis that each of these deficits could, similarly to daltonism in colour vision, result from the lack of a given receptor, was never seriously challenged.

Despite many attempts, it has not yet been possible to isolate olfactory receptors. The most promising results came from Lancet and his group who succeeded in extracting from olfactory cilia a unique group of glycoproteins of about 95 kilodaltons (GP 95) which could very well comprise the receptor material [35, 36, 37].

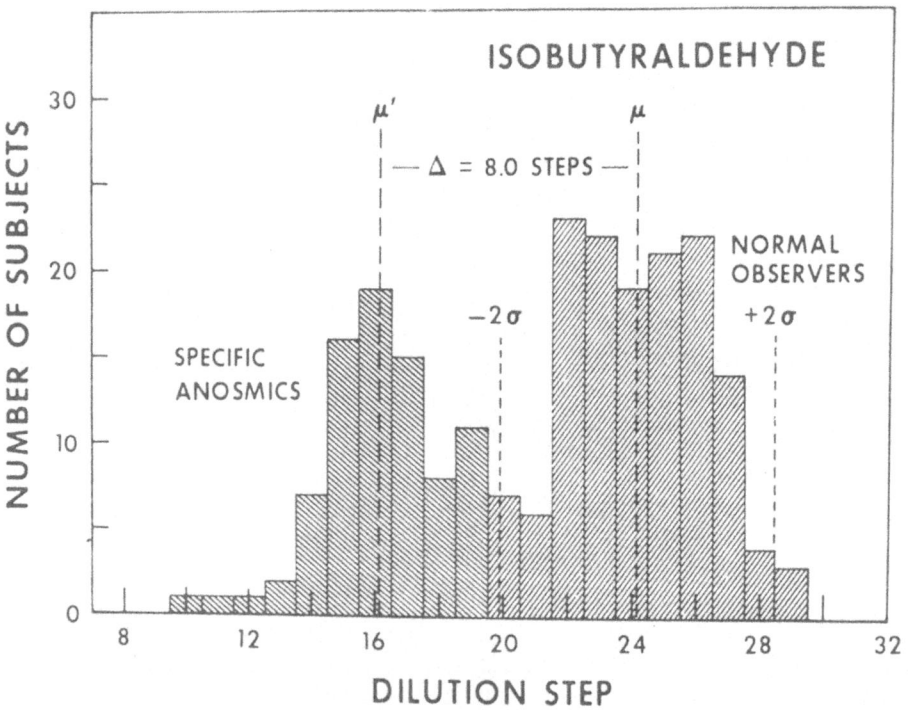

Fig. 2. Bimodal distribution of human sensitivity to an odorant. Olfactory detection thresholds were measured on a sample of 222 subjects (see [33]). Abscissae: binary steps of dilution of the test solution. Ordinates: number of subjects found having their threshold at each dilution. The histogram fits a bimodal distribution with a concentration ratio of 200—300 between both modal values.

3.4. TASTE CHEMORECEPTORS

Taste perception is semantically tagged by the concept of the four primaries: acid, salty, sweet and bitter. This was misleading for a whole generation of pioneer electrophysiologists who satisfied themselves by using only four stimuli, each of them representing a primary [38]. This approach resulted in an underestimate of the taste receptor variety. It was only in the '70s, under the influence of R. Erickson, A. Faurion and S. Schiffmann, that the real complexity of the gustatory system was envisioned [39, 40, 41].

Gustatory stimuli are better separated in two classes: ions (i.e. H^+, 'acid' and Na^+, 'salty') on the one hand, and organic compounds (i.e. 'sweet' and 'bitter') on the other hand. Very little is known concerning the receptor system for ions. S. Schiffman showed that sensitivity to Na^+ and not to H^+ is reduced in man by local application of amiloride, a potent inhibitor of sodium transport, on the tongue [42]. The classical and puzzling observation that acid taste depends not only on pH but also on anions was recently elucidated by taking in account mouth pH and buffering potential which had been consistently overlooked by earlier investigators [43].

When recording responses to organic gustatory stimuli, one soon realizes that their variety is far greater than expected from the combinations of only two primaries, that no clear similarity grouping is apparent and, finally, that a multidimensional continuum is the only acceptable model [44]. Factor analysing relevant data resulted in estimating to about 6 or 7 the number of poorly selective receptors necessary to account for the observed variance [45]. This points toward a greater than expected similarity between taste and smell.

Partial agueusia has been convincingly established only in the case of phenyl-thiourea by Fox, as early as 1932 [46]. It is inherited on a recessive mode, as a single genetic defect [47]. Many attempts at isolating 'taste receptor proteins' have resulted in only limited success and did never go far beyond the stage of impure fractions [48].

4. Transduction

The binding energy of a ligand on a receptor is very small, usually about 10^{-20} J. This is hardly enough to appreciably disturb the structure of a macromolecule and at least two orders of magnitude below the noise level of a cell. An efficient amplifying process is therefore requested. The previously mentioned nicotinic acetylcholine receptor is an amazing example of a complete amplifier in a single molecule: during a single opening, it can transfer 10^4 elementary charges under a 10^{-1} V gradient, featuring a 10^4 power gain [49].

4.1. BIOCHEMICAL TRANSDUCTION

Direct or indirect kinase activity is constantly associated with receptor activation and mediates responses to chemical signals. Proteins which are essential to cellular life are reversibly altered by covalent bonding of a phosphate residue on hydroxylated aminoacids: serine, threonine or tyrosine [17]. In the simplest schedule, the receptor itself is a kinase: this is the case of insuline receptor and of epidermal growth factor. More commonly, there is an intermediate G-protein, modulated by a GTP nucleotide, which activates or inhibits an adenylate cyclase. The end product, cAMP, is a second messenger which in turn activates a cytoplasmic protein kinase [50]. This is the case of glucagon receptor for instance. Alternatively, a G-protein can activate a phospholipase, hydrolysing membrane phospholipids into phosphatydylinositol and diacylglycerol which finally activate, with the help of Ca^{++}, a membrane bound protein kinase [51]. This is the case of adrenalin receptor. These indirect kinase processes have a tremendous amplifying potential as a single receptor event can trigger thousands of phosphorylations through multiple intermediary steps. Moreover, fine grain modulation easily takes place at each step.

4.2. ELECTRICAL TRANSDUCTION

When chemical transmission takes place in the nervous system, or between neurons and peripheral effectors (e.g. muscular endplates), the response typically consists in an electrical signal. The most elaborated electrical response is a two step process involving a graded local response triggering a self regenerating volley of axonal impulsions [52]. Such a system effectively conveys information over meters in tenths of seconds. It depends on the function of highly specialized transmembrane proteins called ionic channels.

4.2.1. *Ionic Channels*

Apart from the nicotinic acetylcholine receptor which cumulates in a single molecule both receptor and channel functions, an ever growing variety of ionic channels is being discovered. They can be classified either by their specific ionic conductance or by the nature of their activating signal [53].

   The central hydrophilic pore which is responsible for their ionic conductance can be coated by positively or negatively charged residues respectively corresponding with anionic or cationic conducting properties. The only transported anion is Cl^-; cationic channels are either more or less selective for Na^+, K^+ or Ca^{++} or non-selective at all.

   Modern patch-clamp electrophysiology have shown that channels oscillate between two conditions, open and closed, with sharp transitions of quantal nature (Figure 3). Opening probability is modulated either by transmembrane electrical field or by phosphorylation of the intra-cellular extremity of the channel. In both cases gated ion fluxes are controlled by membrane receptors through more or less direct pathways and display a considerable amplifying power [54].

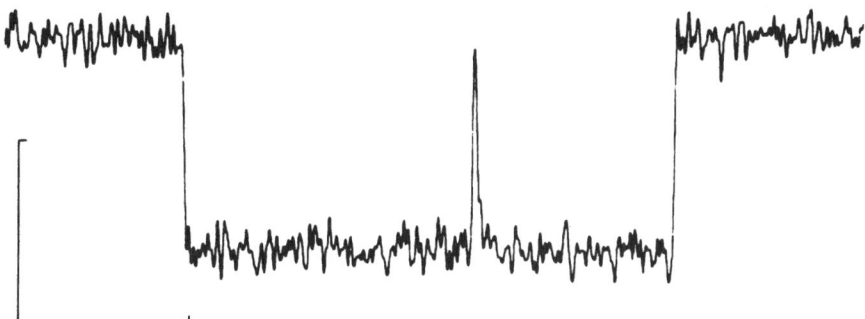

Fig. 3. Single channel current pulse activated by 0.1 mM Ach. Adult frog muscle end-plate single channel recorded in the cell-attached mode with patch-clamp technique (see [22]). This elementary current represents two resolved channel openings separated by a short closure. Membrane potential: −182 mV; horizontal bar: 5 ms; vertical bar: 5 pA.

4.2.2. Receptor Potentials

When recording a chemosensitive neuron by means of a fine intracellular micro-pipette, one finds a resting membrane potential of about 60 mV [55]. This results from two cooperating factors: the membrane permeability at rest is exclusively controlled by potential dependent K channels and a concentration gradient of potassium is maintained by metabolically driven pumps. If some spurious event alters the membrane polarization, it is immediately cancelled by an appropriate change of the outward current flowing through the K channels.

Upon adequate stimulation a new type of channels are opened under the influence of a second messenger released by receptor activation [56]. These channels are not potential dependent; they are either cationic non-selective or Na selective (Figure 4). In both cases, they carry an inward current, resulting in a graded depolarization, directly reflecting the intensity of the transduced stimulus [57].

4.2.3. Action Potentials

Axons are filamentous extensions of neurons, capable of transmitting information as far as one meter from cell body. They are highly specialized in generating electrical impulsions of constant amplitude and propagating them at high speed toward terminal synaptic buttons where neurotransmitter release initiates similar responses in postsynaptic neurones [58].

Better known under the name of action potentials, these impulsions result from a temporary activation of potential dependent Na channels carrying a strong inward current [59]. Such a current is made possible by metabolically driven pumps maintaining a very low intracellular Na^+ concentration [60]. An action potential is triggered as soon as the membrane is depolarized below a threshold where inward Na current cancels K outward current, starting a positive feedback

ISO 7µM + BUT 10µM

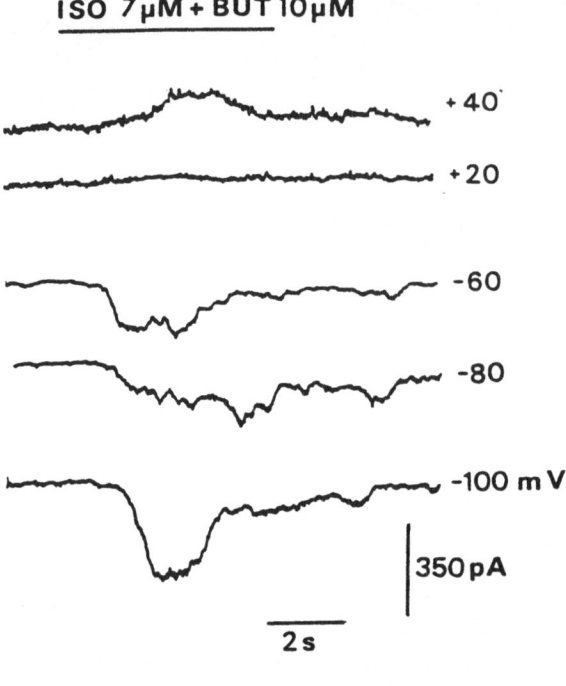

membrane potential (mV)

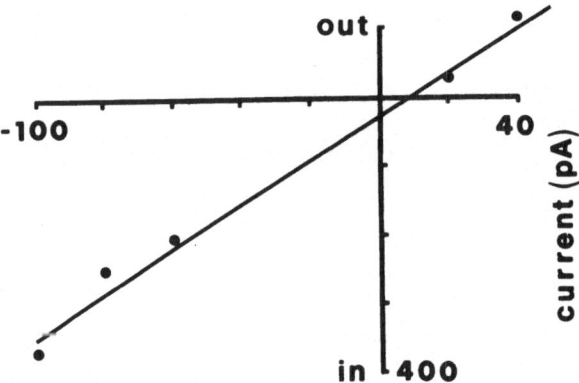

Fig. 4. Odour-induced current in an olfactory neuroreceptor. Patch-clamp recording in the whole-cell mode (see [56]). Upper traces correspond to various indicated membrane polarizations: the reversal potential is slightly below +20 mV. The lower current-voltage diagram of this response suggests a non-specific cationic conductance.

process and driving the membrane near to Na potential equilibrium. Na channels then quickly inactivate, allowing K channels to resume membrane control and to restore resting potential [61].

Action potential is a self-regenerating process, as it depolarizes vicinal points of axonal membrane under firing threshold, and so on down to axonal endings. The velocity of action potential varies from more than 50 m.s^{-1} for the largest myelinated fibres to less than 0.1 m.s^{-1} for the finest unmyelinated fibres. It is a constant parameter for a given fibre. The only variable left in the process is the firing frequency which faithfully reflects actual somatic membrane depolarization.

5. Coding

A fascinating issue in chemical information processing by higher animals is how physicochemical features associated with stimulus molecules are translated into a neural message that finally results in an adapted behaviour. A key for understanding such a process is to realize that the central nervous system is hierarchically organized in superposed layers and that each level "reads" the message it receives from its partner just below and sends a modified copy to the one just above. This information is transferred in a parallel mode by thousands of fibres and each stage combines two basic treatments; both resulting from the function of the same inhibitory interneurons (Figure 5). Firstly, autoinhibition, or negative feedback, reduces the overall gain proportionally to stimulus intensity; secondly, lateral inhibition enhances the contrasts between adjacent channels. The end product is a well stabilized, highly contrasted sensory image, optimally reduced in order to be properly handled by memory and cognitive functions [4].

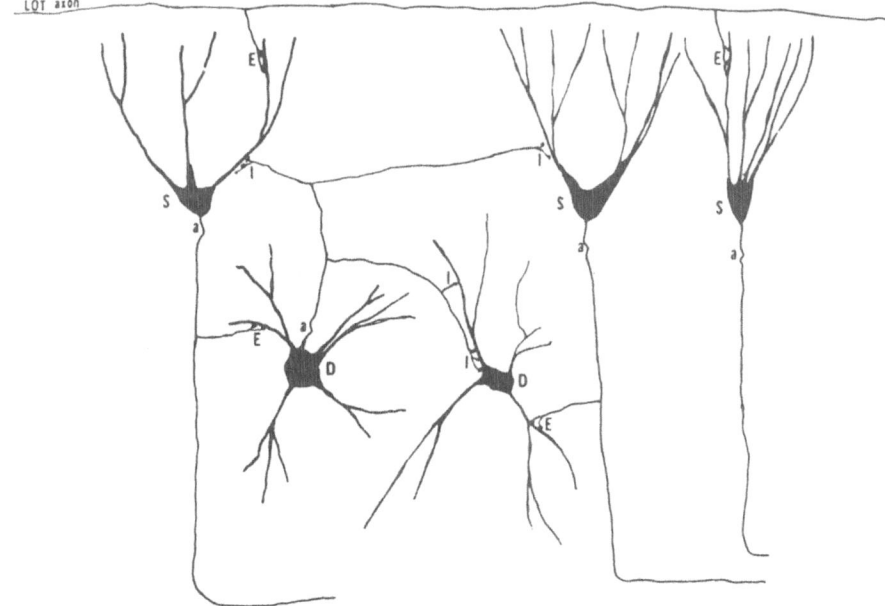

Fig. 5. Semi-schematic diagram of piriform cortex neurones. This basic pattern of synaptic connections in the piriform cortex simultaneously accounts for negative feedback and for contrast enhancement. S: main neurones; D: inhibitory interneurones; E: excitatory synapses; I: inhibitory synapses.

When it reaches the stage of conscious perception, the chemosensory image is normally associated with three main characteristics: intensity, quality and hedonic tone [62]. Intensity and quality are both relevant to coding processes whereas hedonic tone rather depends on integration processes.

5.1. INTENSITY CODING

Many psychophysical experiments have established that perceived intensity varies up to a maximum as a function of stimulus concentration, just as does the frequency of firing of first order axons involved in transduction in sensory organs. Recording of the electrical activity of a somatosensory nerve [63] or of the gustatory nerve chorda tympani during surgical procedures on conscious subjects [64, 65] have confirmed the perfect parallelism between electrophysiological and psychophysical intensity functions. Intensity perception is therefore directly related to the mass of neural activity elicited by a stimulus.

5.2. QUALITY CODING

When recording single unit responses of first order chemosensory neurons one is striked by two conspicuous features: these neurons are poorly selective and each of them has its own peculiar sensitivity spectrum; it is always possible to find two neurons discriminating a given molecule or to find two molecules discriminated by a given neuron [66]. When looking at the whole set of units in a sensory organ (Figure 6), it means that a quite different subset of responding units corresponds

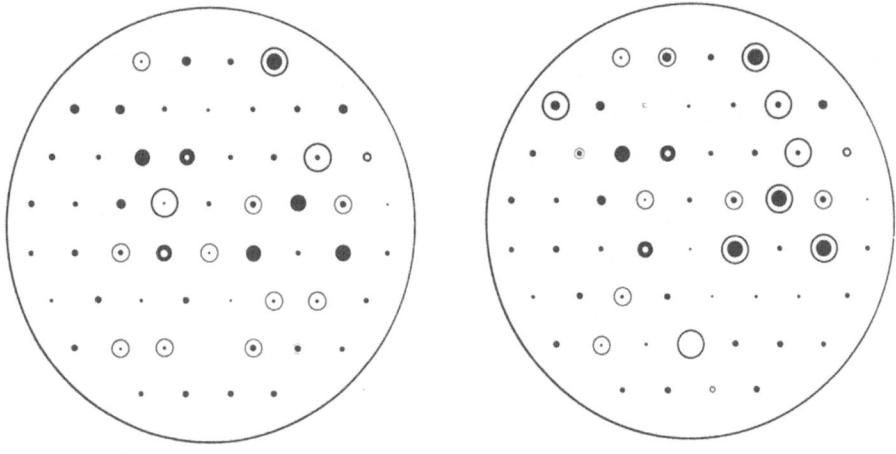

MENTHOL   GERANIOL

Fig. 6. Activation patterns of a sample of 56 olfactory nerve units. When subjected to two qualitatively different odour stimulations, the nerve units generate two distinct spatial activation patterns. Filled circles areas are proportional to resting spike frequencies. Open circles indicate responding units (see [31]). Stimulus name under each diagram.

to each possible pure chemical or mixture thereof [67, 68]. The expression of 'across fibre pattern' [69] was coined to refer to this coding mechanism.

It is therefore possible to think of the subset of units responding to a given stimulus as of a true sensory image, with the full topographical attributes that would be relevant to a visual image. That a quality invariant is topographically encoded was elegantly demonstrated by metabolic mapping of olfactory bulb activity with radioactively labelled 2-deoxyglucose. This glucose analogue is readily absorbed, but not consumed by excited neurons. After exposing an animal for 30—40 minutes to a given odour, brain slices generated clearly patterned and well delineated radioautograms, developing a typical, easily identified image for each tested odorant [70—74].

6. Integration

Behaviour of living organisms can be considered as an uninterrupted sequence of responses to ceaseless variations of their environment. Brain function consists in storing a repertoire of adapted responses and to constantly select the best one, taking in account the actual position of affairs. In lower organisms, such an optimization is merely synonymous with survival. In higher organisms, pleasure is required over and above survival.

6.1. PLEASURE CENTRE

Recent electrophysiological investigations disclosed, in the brain of conscious monkeys, in the far lateral hypothalamus, a specialized area the excitation of which exclusively precedes attractive responses, regardless of the actual sensory input [75]. Self stimulation of this area induces a deadly feedback, revealing its true function as a pleasure centre and opening to physiology a field of investigation previously reserved to psychology.

It is now easier to understand, without leaving the solid grounds of experimental science, how the pleasantness of a perception depends on integrating processes while its nature and strength depend on coding processes. Hedonic tone is actually not a stimulus attribute: it rather reflects the subject's previous learning about the pleasure potential associated with this stimulus [76]. Hence the inanity of structure activity relationship studies concerning pleasantness.

6.2. ODOUR SOURCE LOCALIZATION

It is not enough to identify a smell, to know how strong it is and whether it is a pleasant one: unless we also know where it comes from, we remain unable to adopt the proper behaviour. Unlike light and sound, odorant molecules have no rectilinear trajectories. Most higher animals, including humans, fortunately have a dual olfactory system with a system of reciprocal inhibition between both sides: it enables them to sense very accurately the slightest intensity difference between both nostrils. This device greatly facilitates head orientation in a concentration gradient and helps walking toward an attractant as well as away from a repellent [77].

6.3. FOOD INTAKE REGULATION

Feeding behaviour alternates with other activities, its occurrence is controlled by a regular alternance of hunger and satiety. Both are triggered by chemical signals: hunger depends on systemic signals such as blood glucose while satiety depends on olfactory and gustatory signals associated with ingested foods [78].

Satiety is fully an integrative process as it results from identifying foods, monitoring ingested quantities and looking at a memorized table to recall their satiating power [79]. These olfactory, gustatory and somesthesic data are processed in specialized hypothalamic centres, in a totally unconscious way; only the end product, i.e. satiety, emerges to consciousness. Unconsciousness does not mean fixedness; on the contrary, recent information updates old data so as to provide the system with a maximum of plasticity [80].

6.4. COGNITIVE PROCESSES

Chemical signals are undoubtedly perceived, identified, memorized and recognized, but they lack a specific reference system: one can only refer them to their putative source. It is quite impossible to conceive abstract words exclusively related to tastes or odours: only the level of straightforward similarities is available. It seems to work better in taste where many different molecules are said to be sweet or bitter than in olfaction where no tentative classification was ever successful [81].

Odours are like human faces: one can duplicate them using appropriate techniques, but not describe them using appropriate words. The fact that no standard observer can be defined for human chemoreception, associated with the high dimensionality of the chemosensory space, provides an explanation for the absence of abstract descriptors in relation to odours and tastes. Even the four 'primary' tastes, despite an apparent consensus, must now be considered as being convenient types rather than references or categories.

7. Conclusion

When they use molecules as signal transmitters, biological systems never rely on 'killing the messenger' by irreversibly binding it in a covalent mode: the energy balance of such an information transfer would be favourable only in appearance, as it implies a complete synthesis of a new receptor for each unique receptor event. It is much more efficient to rely on weak energy binding which is spontaneously and rapidly reversible at normal temperature and is therefore compatible with a rapid turnover of receptor molecules. As a consequence, physicochemical rather than chemical properties of signal molecules contribute to encode relevant information.

Trying to figure what kind of information transfer can be achieved through weak energy binding, one soon realizes that the topography of electron density at the surface of molecule envelope is the only available medium for 'writing' the message and that electrostatic forces are the only tool for 'reading' it. One is therefore facing a problem of pattern recognition at molecular level: Van der Waals forces can anchor a ligand on a receptor site and trigger an effective transconformation of a receptor molecule provided a reasonable complementarity of charge distributions do exist on both molecular envelopes. Intermolecular distances involved in weak energy bonding being large enough to minimize quantal effects that would prevail in covalent bonding, two important consequences ensue: the first one is that different molecules may eventually bind a given receptor site; the second one is that tedious *ab initio* quantal computation is not required for modelling approaches (Figure 7) and that simple empirical electrostatic formulae are quite satisfactory substitutes [82].

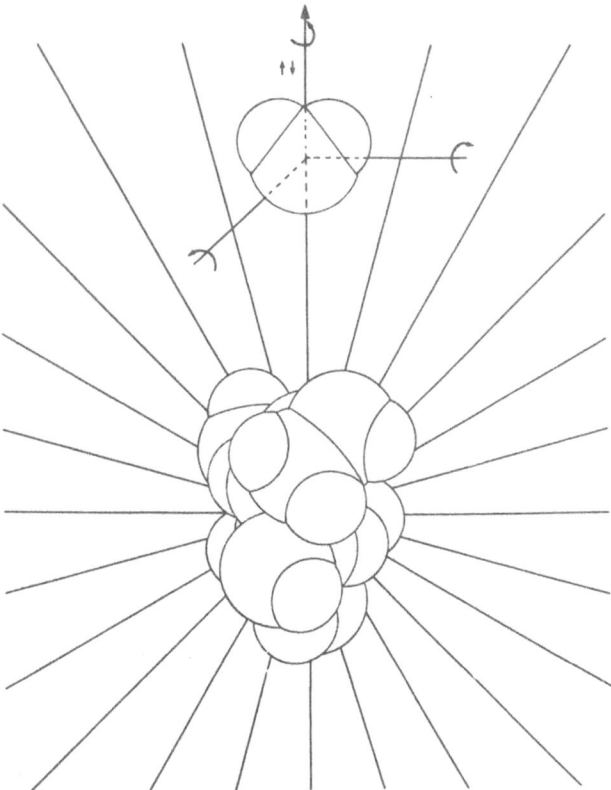

Fig. 7. Mapping out the molecular envelope. The lower molecular surface was drawn from crystallographic data of alpha-*d*-glucose, using Van der Waals radii. The upper envelope was similarly drawn for water. Water was used as a probe, sliding on one of the converging axes, to compute its local potential binding energy by means of simplified electrostatic formulae (see [44] and [82]). The same operation was repeated 188 times all around the molecule, in order to obtain a complete mapping of its potential binding energy with about 1 Å resolution.

When chemical information is used by higher organisms, spatial pattern recognition comes again into play: at a much larger scale, brain information processing also is pattern oriented. A well defined and highly specific sensory image corresponds to each particular stimulus, each pixel of this image being a neuron with its actual excitation level. Although a biunivocal relation is found between a given molecular pattern and its neuronal counterpart, neuronal similarities do not reflect molecular similarities. They reflect instead similarities of biological significance (Figure 8).

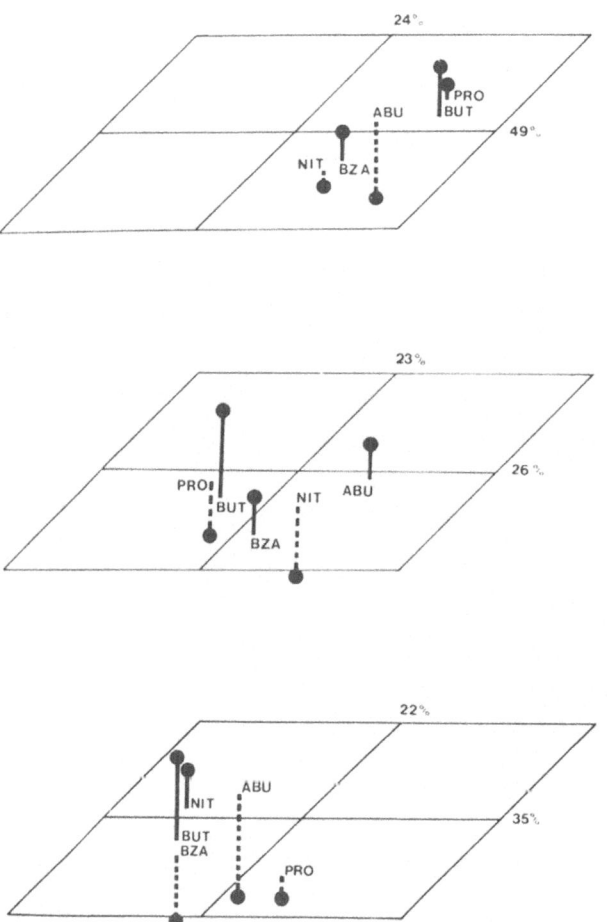

Fig. 8. Odour similarities at different cerebral levels. Each diagram represents the qualitative proximities of 5 odorants obtained by principal component analysis of response profiles of about 60 individual neurones. Top: olfactory neuroreceptors; middle: mitral cells of the olfactory bulb; bottom: piriform cortex neurones. Central proximities cannot be deduced from peripheral ones; this uncoupling results from the full resoltuion of proximal pairs (PRO-BUT and NIT-BZA) achieved at deutoneurone level. ABU: butyric acid; BUT: *n*-butanol; BZA: benzaldehyde; NIT: nitrobenzene; PRO: propanol.

References

1. G. Preti, A. B. Smith, and G. K. Beauchamp: *Chemical and Behavioral Complexity in Mammalian Chemical Communication Systems* (Chemical Signals in Vertebrates, Ed. D. Müller-Schwarze and M. M. Mozell), pp. 95—114, Plenum Press, New-York (1977).
2. D. A. Marshall, L. Blumer, and D. G. Moulton: *Chem. Senses* **6**, 445—453 (1981).
3. D. G. Moulton: *Physiol. Rev.* **56**, 578—593 (1976).
4. A. Holley and P. MacLeod: *J. Physiol. (Paris)* **73**, 725—828 (1977).
5. A. Schilling: *Olfactory Communication in Prosimians* (The Study of Prosimian Behavior, Ed. G. A. Doyle and R. D. Martin), pp. 461—538, Academic Press, New-York (1979).
6. H. Zwaardemaker: *Die Physiologie des Geruchs*, Engelman, Leipzig (1895).
7. D. Tucker: *J. Gen. Physiol.* **46**, 453—489 (1963).
8. D. G. Laing: *Perception* **12**, 99—117 (1983).
9. P. Laffort, F. G. Patte, and M. Etcheto: *Ann. NY Acad. Sci.* **237**, 193—208 (1974).
10. J. Caprio: *High Sensitivity and Specificity of Olfactory and Gustatory Receptors of Catfish to Aminoacids* (Chemoreception in Fishes, Ed. T. J. Hara), pp. 109—134, Elsevier (1982).
11. K. B. Döving, R. Selset, and G. Thommesen: *Acta Physiol. Scand.* **108**, 123—131 (1980).
12. J. W. Atz: *Qu. Rev. Biol.* **27**, 366—376 (1952).
13. K. B. Döving, M. Dubois-Dauphin, A. Holley, and F. Jourdan: *Acta Zool. (Stockholm)* **58**, 245—255 (1977).
14. P. MacLeod: *La formation d'une image chimiosensorielle* (L'Aube des Sens, Ed. E. Herbinet and M. C. Busnel), pp. 345—356, Stock, Paris (1981).
15. J. Marco, H. Morera, and J. A. Gimenez: *Acta Otolaryngol.* **46**, 114—126 (1956).
16. T. Kubo, K. Fukuda, A. Mikami, A. Maeda, H. Takahashi, M. Haga, T. Haga, A. Ishiyama, K. Kangawa, M. Kojima, H. Matsuo, T. Hirose, and S. Numa: *Nature* **323**, 411—416 (1986).
17. A. Danchin: *La Recherche* **18**, 262—264 (1987).
18. Z. W. Hall: *Trends Neurosci.* **10**, 99—101 (1987).
19. A. C. Dolphin: *Trends Neurosci.* **10**, 53—57 (1987).
20. J. L. Popot and J. P. Changeux: *Physiol. Rev.* **64**, 1162—1239 (1984).
21. A. Brisson and P. N. T. Unwin: *Nature* **315**, 474—477 (1985).
22. D. Colquhoun and B. Sackman: *J. Physiol. (London)* **369**, 501—557 (1985).
23. D. E. Koshland: *Physiol. Rev.* **59**, 812—862 (1979).
24. D. E. Koshland: *Bacterial Chemotaxis as a Model Behavioral System*, Raven Press (1980).
25. G. L. Hazelbauer and J. Adler: *Nature New Biol.* **30**, 101—104 (1971).
26. G. L. Hazelbauer: *J. Bacteriol.* **122**, 206—214 (1975).
27. R. Aksamit and D. E. Koshland: *Biochemistry* **13**, 4473—4478 (1974).
28. R. W. Reader, W. Tso, M. Springer, S. Goy, and J. Adler: *J. Gen. Microbiol.* **111**, 363—374 (1978).
29. D. E. Koshland: *Ann. Rev. Biochem.* **50**, 765—782 (1981).
30. K. E. Kaissling: *Molecular Recognition* (Biophysics, Ed. W. Hoppe, W. Lohman, H. Markl, and H. Ziegler), pp. 697—709, Springer-Verlag, Heidelberg (1983).
31. A. Holley, A. Duchamp, M. F. Revial, A. Juge, and P. MacLeod: *Ann. NY Acad. Sci.* **237**, 102—114 (1974).
32. A. Duchamp and G. Sicard: *Chem. Senses* **8**, 355—366 (1984).
33. J. E. Amoore, L. Forrester, and P. Pelosi: *Chem. Senses Flav.* **2**, 17—25 (1976).
34. J. E. Amoore: *Chem. Senses Flav.* **2**, 267—281 (1977).
35. Z. Chen and D. Lancet: *Proc. Natl. Acad. Sci. USA* **81**, 1859 (1984).
36. U. Pace, E. Hanski, Y. Salomon, and D. Lancet: *Nature* **316**, 255—258 (1985).
37. D. Lancet: *Ann. Rev. Neurosci.* **9**, 329—355 (1986).
38. C. Pfaffmann: *Chem. Senses Flav.* **1**, 61—67 (1974).
39. A. Faurion, S. Saito, and P. MacLeod: *Chem. Senses* **5**, 107—121 (1980).
40. S. S. Schiffman and R. P. Erickson: *Neurosci. Behav.* **2**, 109—117 (1980).
41. R. P. Erickson: *Definitions: a Matter of Taste* (Taste, Olfaction, and the Central Nervous System, Ed. D. W. Pfaff), pp. 129—150, The Rockefeller University Press, New-York (1985).
42. S. S. Schiffman, E. Lockhead, and F. W. Maes: *Proc. Natl. Acad. Sci. USA* **80**, 6136—6140 (1983).

43. C. M. Christensen and D. Malamud: *Chem. Senses* **10**, 454 (1985).
44. A. Faurion and P. MacLeod: *Sweet Taste Receptor Mechanisms* (Nutritive Sweeteners, Ed. G. G. Birch and K. J. Parker), pp. 247—273, Applied Science Publishers, London (1982).
45. A. Faurion: *MSG as one of the Sensitivities Among a Continuous Taste Space* (Umami: a Basic Taste, Ed. Y. Kawamura and M. R. Kare), pp. 387—408, Marcel Dekker, New-York (1987).
46. A. L. Fox: *Proc. Natl. Acad. Sci. USA* **18**, 115—120 (1932).
47. H. Kalmus: *Ann. Hum. Genet.* **22**, 222—230 (1958).
48. R. H. Cagan and A. G. Boyle: *Biochim. Biophys. Acta* **799**, 230—237 (1984).
49. E. Neher and B. Sackman: *Nature* **260**, 779—802 (1976).
50. A. M. Spiegel, P. Gierschik, M. A. Levine, and R. W. Downs: *New Engl. J. Med.* **312**, 26—33 (1985).
51. M. J. Berridge: *Sci. Am.* **235**, 142—151 (1985).
52. D. Ottoson and G. M. Shepherd: *Cold Spring Harb. Symp. Quant. Biol.* **30**, 105—114 (1965).
53. B. Hille: *Ionic Channels of Excitable Membranes*, Sinauer, Sunderland, Mass., USA (1984).
54. B. Sackmann and E. Neher: Ed. *Single-Channel Recording*, Plenum Press, New-York (1983).
55. D. Trotier and P. MacLeod: *Brain Res.* **268**, 225—237 (1983).
56. D. Trotier: *Pflüg. Arch.* **407**, 589—595 (1986).
57. T. Nakamura and G. H. Gold: *Nature* **325**, 442—444 (1987).
58. J. Dudel: *Excitation, Conduction and Synaptic Transmission* (Biophysics, Ed. W. Hoppe, W. Lohmann, H. Markl, and H. Ziegler), pp. 641—657, Springer-Verlag, Heidelberg (1983).
59. R. Horn: *Gating of Channels in Nerve and Muscle* (Ion Channels: Molecular and Physiological Aspects, Ed. W. D. Stein), pp. 53—97, Academic Press, New-York (1984).
60. L. Cantley: *Trends Neurosci.* **9**, 1—3 (1986).
61. A. L. Hodgkin and A. F. Huxley: *J. Physiol. (London)* **117**, 500—544 (1952).
62. R. H. Harper: *Human Senses in Action*, Churchill-Livingstone, Edinburgh (1972).
63. R. S. Johansson and A. B. Vallbo: *J. Physiol. (London)* **286**, 283—300 (1979).
64. H. Diamant, B. Oakley, L. Ström, C. Wells, and Y. Zotterman: *Acta Physiol. Scand.* **64**, 67—74 (1965).
65. B. Oakley: *Chem. Senses* **10**, 469—481 (1985).
66. M. F. Revial, G. Sicard, A. Duchamp, and A. Holley: *Chem. Senses* **8**, 179—190 (1983).
67. E. H. Polak: *J. Theor. Biol.* **40**, 469—484 (1984).
68. J. Kauer: *Coding in the Olfactory System* (Neurobiology of Taste and Smell, Ed. T. E. Finger), pp. 205—231, Wiley, New-York (1987).
69. R. P. Erickson: *Sensory Neural Patterns and Gustation* (Olfaction and Taste I, Ed. Y. Zotterman), pp. 205—213, Pergamon Press, New-York (1963).
70. L. C. Skeen: *Brain Res.* **124**, 147—153 (1977).
71. M. Verrier, I. Giachetti, J. Levêteau, and P. MacLeod: *C.R. Acad. Sci. (Paris)* **286**, 1293—1296 (1978).
72. F. Jourdan, A. Duveau, L. Astic, and A. Holley: *Brain Res.* **188**, 139—154 (1980).
73. D. Lancet, C. A. Greer, J. S. Kauer, and G. Shepherd: *Proc. Natl. Acad. Sci. USA* **79**, 670—674 (1982).
74. L. Astic and D. Saucier: *Dev. Brain Res.* **2**, 243—256 (1982).
75. E. T. Rolls and B. J. Rolls: *Activity of Neurones in Sensory, and Motor Areas during Feeding in the Monkey* (Food Intake and Chemical Senses, Ed. Y. Katsuki, M. Sato, S. Takagi, and Y. Oomura), pp. 525—550, Japan Scientific Societies Press (1977).
76. M. Cabanac: *Qu. Rev. Biol.* **54**, 1—29 (1979).
77. J. Levêteau, P. MacLeod, and G. Daval: *Physiol. Behav.* **4**, 479—482 (1969).
78. J. Le Magnen: *Prog. Physiol. Psychol.* **4**, 203—261 (1971).
79. J. Pager and J. P. Royet: *J. Comp. Physiol. Psychol.* **90**, 67—77 (1976).
80. W. J. Freeman and W. Schneider: *Psychophysiol.* **19**, 44—56 (1982).
81. M. Yoshida: *Chem. Senses Flav.* **1**, 443—464 (1975).
82. P. Claverie: *Elaboration of Approximate Formulae for the Interaction between Large Molecules* (Intermolecular Interactions: from Diatomics to Biopolymers, Ed. B. Pullman), pp. 69—305, Wiley, New-York (1978).

Molecular Cues in Salmonid Migration

KJELL B. DØVING
Department of Biology, P.O. Box 1051, Blindern, 0316 Oslo 3, Norway.

1. Introduction

Man is greatly attracted to salmon fishing in rivers as a sport, and this fish's renown to the gourmet has made it a luxury dish in fashionable restaurants the world over. Salmon is thus praised or highly priced, depending on a person's point of view or his particular interest in the fish.

In Norway, with its many rivers, salmon was earlier a common dish, and the fish's peculiar habits made the salmon a mysterious and even sacred fish (Figure 1). Many people asked themselves how it could possibly find its way back from the open sea into the rivers. Farmers and fishermen throughout Norway have always known that each river has its own salmon population.

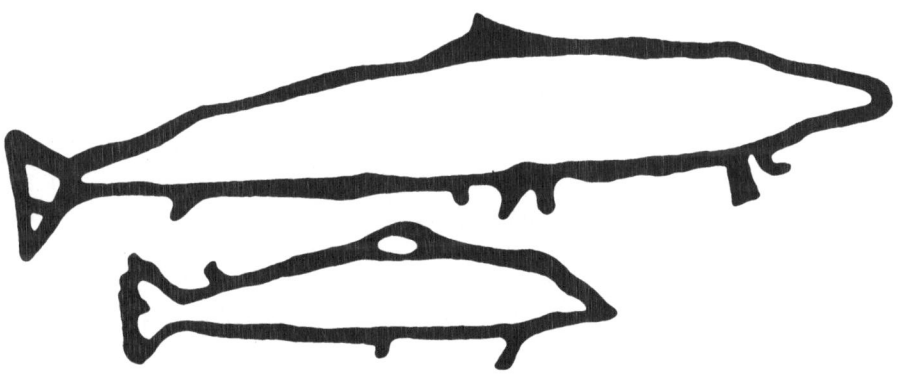

Fig. 1. Rock carving of two adult salmon at a stone age site near Alta, Northern Norway.

Peder Claussøn Friis (1545—1614) was a bishop in his home county of Agder in Southern Norway. He wrote a *Natural History of Norway*, described the topology of Iceland, the Faeroe Isles and Greenland. A man of outstanding intellect both as a cleric and humanist, he also translated folk tales, and liked to listen to the stories told him by the farmers and fishermen of his parish. He heard of the wonderful wanderings and peculiarities of the salmon both from his own parishioners and from his curate, who was from Nordland, in Northern Norway. This must have held special interest for him, since he recorded it as follows:

Jean Maruani (ed.), Molecules in Physics, Chemistry, and Biology, Vol. IV, 299—329.
© 1989 *by Kluwer Academic Publishers.*

What is most peculiar is that each salmon searches the stream to the place where it was born, here documented. Every river has a salmon with its own particular shape. Thus each of the rivers Lyngdal, Undal, Mandal, Torrisdal and Toppdal have salmon of a peculiar shape, even though the rivers run close to each other here in Agder County. Two rivers flow from a little narrow fjord at Egersund. There is not a bowshot between the river mouths, yet each river has its distinct salmon, so that people are able to tell the salmon in the one river from those of the other. The fishermen can see when the salmon come into the fjord together in shoals. They then separate and swim, each to its own river. If the salmon were to swim up the rivers at random, then there are hundreds of rivers between Nordland and here up which they could have steered . . . .

His curate had told him that the salmon gathered in great numbers in 'The Northern Ocean' (Norwegian Sea) outside the island of Senja. From there they were led back to the rivers of their birth by the even bigger fish opah (kingfish). In his opinion it was God that led the opah.

2. The Biologist's Confirmation

It took over 300 years before biologists started to question the observations recorded by Peder Claussøn Friis. It was painstakingly and tediously shown in a great number of experiments that P. Claussøn Friis had been right. Moreover, one river could have more than one population of salmon. The results from the many experiments on salmon and other salmonid fishes were obtained mainly by scientists in North America and Europe between 1920 and 1940. But these results seem to have been disregarded for a span of some 20 years. Losing faith in the opah, scientists started anew to search for the mechanisms that led salmon back to their rivers. The behavioral experiments demonstrated that the olfactory sense was essential for correct return, and it has been hypothesised that the odorants emanated from the fish itself. The present review includes a summary of the studies leading to the discovery of potent fish odorants and a recapitulation of the experiments leading to our understanding of the behaviour of the salmon in a fjord system. Two chapters describe the anatomy and physiology of the fish olfactory apparatus. Finally a brief account on the chemical properties of fish pheromones is given.

3. The Olfactory Sense and Selection of Tributaries

The experiments which convinced the scientific community that the olfactory sense is important for the salmon's ability to find its spawning grounds were started in the early 1950s. In 1954 Wisby and Hasler published a report [61] on their experiments on the effect of olfactory impairment on the river migration of coho salmon (*Oncorhynchus*) from two tributaries of Sqammamish Lake, Washington, USA. Of the 302 fishes that were trapped and recaptured, 153 had their olfactory cavities plugged with cotton, and in some, the olfactory nerves were sectioned. The results are summarized in Table I. Wisby and Hasler concluded that the control fishes were readily able to repeat their original choice at the juncture of the

TABLE I. The number of silver salmon (*Oncorhynchus kisutch*) released and recaptured in Issaquah Creek and East Fork. From Wisby and Hasler [61].

	Released	Homing	Non-homing
Controls	149	65	8
Olfactory impairment	153	42	28

streams. Those with olfactory occlusion were unable to select accurately. This study had an important impact on salmon migration research.

Hasler obtained exciting results from his experiments on salmon and other fish and they moved him to propose a hypothesis for salmon homing [19]. He suggested that the salmon recognized and identified the home river by its characteristic odour. He proposed that the characteristic scent originated from the unique plant community of the stream's drainage basin and the flora within the stream. On its way downstream the home river the smoltified salmon is 'imprinted' by these characteristic odorants. The salmon retained these unique memories of its home river. The fish smells its way home from the coastline of the sea, tracking a familiar scent like a fox hound.

This suggestion is known as the 'imprinting' hypothesis for salmonid migration. Tagging experiments where salmon smolts are released at another site than the home river tend to support this hypothesis. For a review, see Hasler and Scholz [20]. These experiments have been reconsidered by Stabell [48].

In the present article the studies which have been performed in the path of the pheromone hypothesis proposed by Nordeng [35] will be presented.

Groves *et al.* [15] reported the results of studies on the homing of adult Chinook salmon (*Oncorhynchus tschawytscha*) in Columbia River, Washington, USA. The fishes that entered the hatchery at Spring Creek were captured, marked, anesthetized for treatment and released on the same day. The treatments were olfactory occlusion by petroleum jelly and cotton tampons, visual occlusion by removal of the lenses, and both olfactory and visual occlusions. The controls and experimental fishes were treated alike except for the final procedures. A summary of the results is given in Table II. Of the 866 fishes that were released, 40% were recovered. The authors conclude that olfaction appeared to be the key sense that directed the return of these fish to Spring Creek. Only control and blinded fish returned from the upstream release site. Of the 482 fishes without olfactory occlusion released from all sites, 35% returned to Spring Creek. Of the 384 fishes released with olfactory occlusion only 8 returned. These eight fishes were released downstream and appeared in the hatchery so late that the conditions of their olfactory plugs could not be examined. In comparison with the controls, the blinded fish were handicapped during their return trip, as indicated by the fewer recoveries, delay and battered appearance on arrival at Spring Creek.

De Lacy *et al.* [5] reported the homing behaviour of Chinook salmon

TABLE II. The results of releasing 866 chinook salmon (*Oncorhynchus tshawytscha*) in Columbia River and subsequent recapture in and just outside home location. From Groves, Collins and Trefethen [15].

	Released	Homing	Non-homing
Controls	241	112	50
Olfactory impairment	193	6	46
Impaired vision	241	56	57
Olfactory and visual impairment	191	2	19

(*Oncorhynchus tschawytscha*). Fish whose olfactory sense had been destroyed, and control fish, were released 5 miles downstream from the home pond. No fish with the olfactory sense impaired returned, whereas 90% of the control fish reentered the pond. Of 25 fishes released 15 miles upstream 3 found their way back to the home pond. The authors suggest that the absence of home stream cues may have evoked a negative rheotaxis, and thus the fish returned to the location at which the appropriate olfactory stimulation again elicits positively rheotactic behaviour. Some chum salmon (*Oncorhynchus keta*) captured in a trap of the lower course of Otsuchi River, Japan, were displaced to a release station in the bay about 3 km from the estuary. Only 3 out of 10 blinded and none of the 10 anosmic salmon returned to the Otsuchi River [22].

Stuart [49] observed that the brown trout (*Salmo trutta*) showed definite and regular migrations in Dunalastair Reservoir and tributaries in Perthshire, Scotland. Evidence for a high degree of homing accuracy has been obtained in a reservoir receiving five distinct spawning streams and their tributaries. Experiments to investigate the role of the olfactory sense were made by plugging the olfactory cavity with cotton wool, cotton wool plus vaseline, or cotton wool plus chloroxylenol. Even though some of the plugged fishes returned to the home tributary, a significant decrease in return was noted. The pattern of movement of indigenous and displaced trout (*Salmo trutta*) was studied in Airthrey Loch, Scotland (Tytler and Machin [56]). The movements of displaced fish indicated that the return was principally a pattern of random turns. The nose plug influenced the distribution of the angles of turns, but did not significantly alter the time for return to the home area. Nose plugs made with Impressil (ICI) formed a cast of the olfactory rosette. Even though the plug formed a faithful cast of the rosette the authors had no other means of deciding the efficiency of the olfactory plug.

Cutthroat trout (*Salmo clarki*), after displacement, showed in-season homing to the Yellowstone Lake, Wyoming, USA [26]. When the fishes were blinded by injection of 3% aqueous benzethonium chloride they seemed to find their spawning grounds again just as well as the controls did. Fishes with impaired olfaction, accomplished by filling the nasal cavities with petroleum jelly, or impaired olfaction and vision, returned to the spawning tributaries in significantly lower numbers. Some of the fishes released in the lake were tracked in open water in directions usually coinciding with those of the sun azimuth. Fish taken from the east side of the lake went west-northwest when tracked in late afternoon, and fishes taken from the west side of the lake went east-southeast when tracked in the morning. These directions pointed away from the home streams.

Homing cutthroat trout (*Salmo clarki*) in two creeks entering Yellowstone Lake were further investigated by LeBar [33]. The fish that were to be displaced from their redds were taken by hook and line or by a fish shocker when a spawning pair was sighted. Fish were blinded under anesthesia by injecting 3% aqueous benzethonium chloride, or their olfactory cavity was filled with petroleum jelly. Significantly, more blind fish than other groups were categorized as non-homing fish. Both olfactory plugging and blinding appeared to affect homing. The ability of the blind fish to move upstream appeared to be impaired. Five of six 'anosmic' fishes that returned home had lost one or both of their nasal plugs.

4. The Olfactory Sense and Navigation at Sea

The first experiment intended to show that the olfactory sense is important as an aid for salmon to return to the spawning grounds was carried out by Craigie [4]. Sockeye salmon (*Oncorhynchus nerka*) were caught in Deep Water Bay, 125 miles north of the mouth of New Westminster (i.e. Fraser River estuary) Canada, and recaptured at the mouth of New Westminster, Deep Water Bay, in Fraser River, and locations in the Strait of Georgia. The results of the experiments are summarized in Table III. As shown, 14 fish with impaired olfaction remained in Deep Water Bay and were recaptured there, compared with only 3 of the controls. Only three of the salmon went up the Fraser River and were caught relatively far downstream. Craigie says that "the results are obviously inconclusive, but appeared to him to indicate that the cutting of the olfactory nerves did interfere considerably with the normal migration of the salmon subjected to this operation".

Later review articles have upheld Craigie's original opinion. However, calculation of the chi square of the recapture data as given in Table I, gives a value of

TABLE III. The numbers of sockeye salmon (*Oncorhynchus nerka*) released at Deep Water Bay and recaptured in the water system of the Strait of Georgia and Fraser River. Chi-square on the data of the recaptured fish gives a value of 22.3 ($p < 0.001$) 3 d.f. From Craigie [4].

	Released	Recaptured			
		Deep Water Bay	Mouth of New Westminster	Fraser River	Other sites
Normal fish	254	3	43	13	6
Operated fish	259	14	16	3	9

22.3 when the control and anosmic fishes are compared ($p < 0.001$). Thus, the operation certainly had a great influence on the return. It should be kept in mind that the destination of the fishes used for the experiment was unknown. Of particular interest is the large proportion of fishes remaining near the release site.

Toft [55] has published the results of extensive studies on the return of Atlantic salmon which normally spawn in 'Indalsälven' and 'Skellefteåälven', Sweden, and migrate in the Gulf of Bothnia. The fishes were anesthetised with MS222 (1/3000) and blinded by heat cautery of the corneas. Olfactory impairment was achieved by sectioning the olfactory nerves or heat cautery of the olfactory rosettes. His results are summarized in Table IV. An appreciable number of fishes were taken in the

TABLE IV. The numbers of salmon released and recaptured in the home rivers Indalsälven and Skellefteälven at the Gulf of Bothnia. From Toft [55].

	Released	Homing	Non-homing
Controls	546	70	104
Impaired olfaction	554	20	149
Impaired olfaction on one side	83	8	18
Impaired vision	75	13	10

home river estuaries due to intense fishing in that area, but that does not necessarily imply that the fishes were heading for the river. This criticism has little influence on the general finding, *viz.* that the olfactory sense, as shown by these studies, is important for the fishes' return to their home river. Impaired vision has little influence upon accurate homing; this was true even for the fishes displaced far out to sea. If the fishes were blinded or the olfactory organ was destroyed on one side only, their homing ability was similar to that of the control fishes.

Bertmar [2] reported experiments with 538 Baltic trout (*Salmo trutta* L.) from the stock of the River Indalsälven. The author concluded from these experiments that the subadults are guided mainly by olfactory stimuli over the entire migration route. Olfaction is also essential for mature and adult fish towards and within the estuary, but the fish are guided by visual stimuli in the river itself.

Tagging experiments give only indirect information on the fate of the salmon. The scientist knows the site and time of release, and with luck, the fish is recaptured and reported to the investigator. But what happens to the salmon in the sea? How does it use its olfactory sense and how does the loss of the sense of smell affect salmon behaviour? These questions were asked by Håkan Westerberg of Göteborg University, Sweden, when he equipped the salmon with depth-coded ultrasonic devices and recorded the swimming depth in relation, as far as possible, to the hydrographic structures [58–60].

(A) SALMON BEHAVIOUR IN A FJORD

Westerberg's experiments were extended to observe the swimming of salmon in a fjord system. Fish with intact olfactory systems showed preference for particular water layers by swimming at those depths for extended periods of time. These layers are interfaces between quasihomogeneous layers. The behaviour of Atlantic salmon observed in the fjord [12] is very similar to that demonstrated among Baltic salmon (*Salmo salar*) in the Gulf of Bothnia and land-locked salmon (*Salmo salar*) in Lake Vänern in Sweden [58, 59]. Typically, the salmon made continuous small-scale vertical movements in the fine-structure gradient layer with occasional larger excursions in the water column (Figure 2). However, these large amplitude movements never went below the sill depth, marking the separation of the stratified upper layers from the stagnant lower basin.

The behaviour of anosmic salmon differed in several aspects. They did not select specific depths and often made large amplitude movements up and down through the water column. These movements took them much deeper into the fjord, often well into the lower basin. Unlike the intact fish, which remained in mid-water, the anosmic fish were also observed to follow bottom contours. Behaviours of anosmic fish observed in this study are similar to those described by Westerberg [58, 59] for an anosmic salmon tracked in Lake Vänern.

Critics of sensory impairment studies [3, 18, 40, 43] argue that fish suffering such treatment may home with less precision because of generalized traumatic or inhibitory effects rather than loss of the olfactory sense. In spite of the trauma which may have been caused by the ablation procedure, the anosmic fish demonstrated large-scale vertical movements, although net horizontal progress was slow. Thus the surgery does not seem to affect the motivation for active swimming. The authors suggest that the large amplitude swimming behaviour of anosmic fish may be described as active, but unsuccessful, search for water layers with suitable odor characteristics. This argument is further supported by the behaviour of a salmon with unilateral ablation which ultimately entered the river of release [12].

(B) SALMON ORIENTATION

An explanatory model of salmon olfactory orientation in open water must take into account two major difficulties, *viz.*, the large-scale horizontal turbulence and the lack of visual points of reference making rheotactic orientation impossible [1].

The results of the tracking experiments by Westerberg [59] and Døving, Westerberg and Johnsen [12] suggest the following hypothesis for salmon orientation in the coastal zone. The basic concept is that information for orientation is found locally in the vertical distribution of odorants rather than in a complicated, large-scale horizontal distribution.

All naturally stratified bodies of water have a layered structure. There are almost homogeneous layers containing homestream odorants, and these are

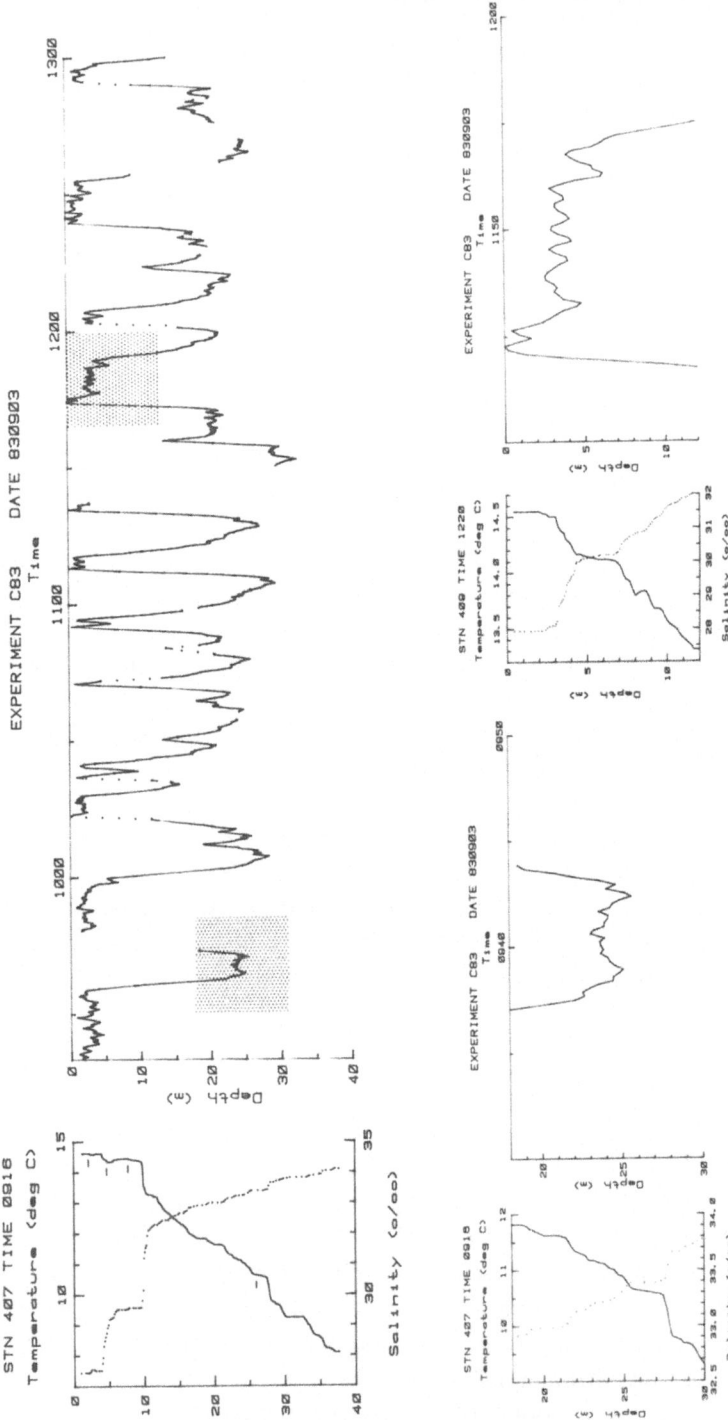

Fig. 2. Recordings of the swimming depth of a salmon in a Norwegian fjord, September 1983. Simultaneous temperature and salinity profiles are shown on the left. The shaded parts of the upper diagram are shown expanded in the lower diagrams (From Døving, Westerberg and Johnsen [12]).

assumed to be the layers of interest to the salmon. These layers containing specific odorants, and their adjacent layers, establish a reference system for oriented movement. In order to maintain its position with respect to the important information-giving layer, the salmon moves from the layer of interest to the adjacent one with low amplitude zig-zag movements, through the interface layer of strong vertical density gradient. Figure 3 shows an example.

An interesting similarity to this vertical zig-zag motion can be found in the horizontal zig-zag pattern of salmon in a stream. Upstream movements of migrating coho salmon (*Oncorhynchus kisutch*) have been shown to be rheotactic responses to the artifical homestream chemical signal to which the fish were exposed during smoltification [29]. Regions of confluence between two tributaries are characterized by discrete masses of water bearing different scents separated by a common interface boundary. Detailed observations of migratory behaviour indicate that the fish move along this interface boundary in a zig-zag pattern which permits the animal to monitor the edges of the two water masses with different chemical compositions [28]. This zig-zag behaviour in the horizontal plane allows the fish to follow the chemical trail into the correct tributary. In this case the directional cue is the visual contact with the bottom, and it is impossible to make a direct analogue to the vertical zig-zag notion.

If the interface of adjacent layers in the ocean becomes indistinct, the fish will begin large-amplitude, up and down movements to locate another layer of interest. Salmon without an olfactory sense will continue to swim up and down with large amplitude movements in search of a layer of interest. In the absence of odor information the vertical extent of the search is limited only by the physical boundaries; the surface, the bottom or a strong thermocline. This unsuccessful search behaviour has been observed for anosmic fish.

The directional cues that are used by the salmon to choose and maintain horizontal swimming direction are unknown. A likely clue is the local direction of the current shear, i.e., the direction of the relative motion of the layer of interest and the adjacent one. Statistically, this will be related to the direction of the source of the odorants. If and how this shear can be detected is unclear. One proposal [60] is that any stretching of microscale fluctuations in the gradient layer can be sensed by the fish. If the fluctuations are detected by the olfactory system this should imply a tropotactic orientation. But the successful homing of a salmon with a unilateral ablation of the olfactory nerve seems to contradict a mechanism of this kind [12].

At present, the tracking technique, including occasional hydrographic measurements in the vicinity of the fish, is too imprecise to provide more than some additional support for such an interpretation of the observed behaviour. The intention is rather to investigate the role of olfaction within the framework of the hypothesis. Physiologically it is seen that the salmon is able to discriminate the water in adjacent fine-structure layers by olfaction. Behaviourally the loss of the olfactory sense will cause large amplitude vertical movements that can be inter-

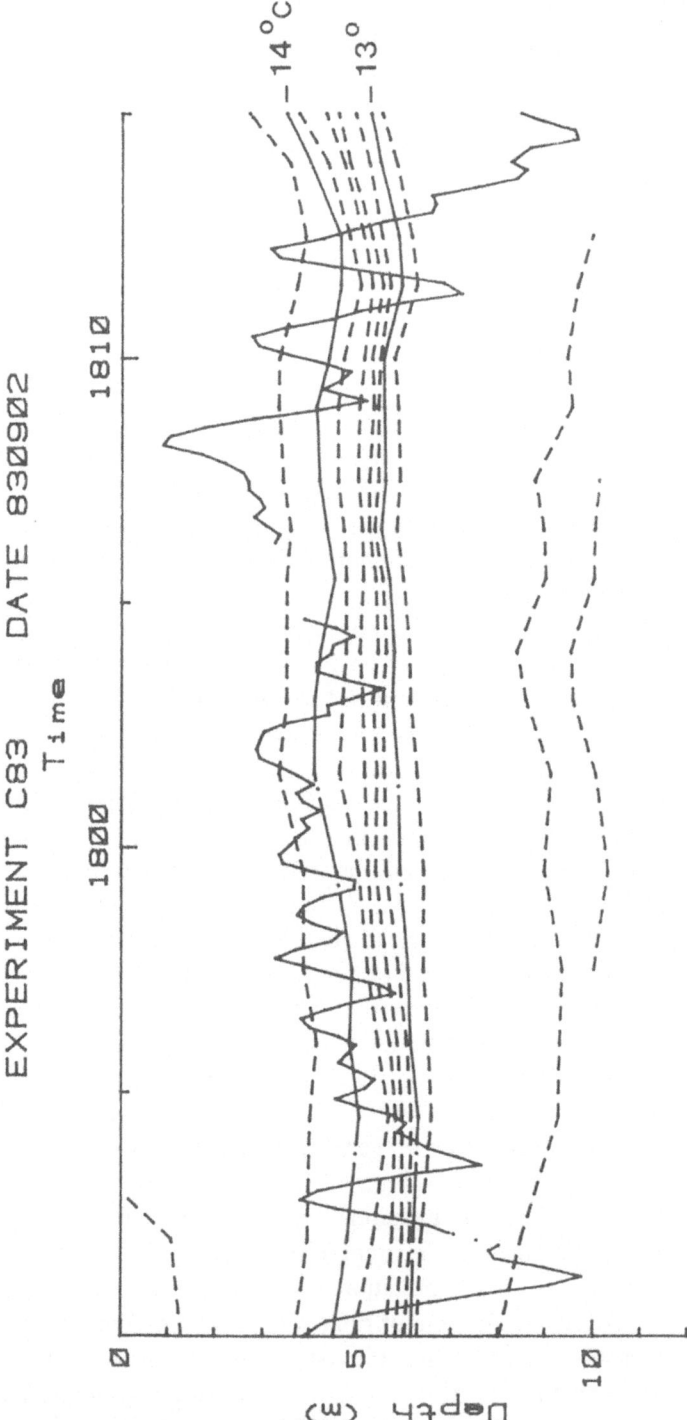

Fig. 3. Detail of the swimming depth of a salmon with intact olfactory organ in relation to the thermal structure along the track. The temperature soundings are made at approximately 2 min intervals, and the isotherms are drawn with a spacing of 0.2 °C (From Døving, Westerberg and Johnsen [12]).

preted as olfactory searches. In future studies it will be important to design experiments to determine which external directional cues are used, such as the current shear, and which sensory systems other than the olfactory system are required for successful salmon homing.

5. The Pheromone Hypothesis

On the basis of his observations and experiments Nordeng put forward the idea that the fishes sensed themselves. More precisely, that the salmon's homeward migration is initiated and directed by odour trails derived from the related descending smolt. The train of smolts establishes a trail of odours leading from their respective freshwater home localities out to the salmon at sea. The maturing salmon respond innately to the smolt odours and start homewards. Such odours, which originate from one animal and have an effect on other animals of the same species are well-known to biologists. They are called pheromones. The salmonid pheromones were thought to be released from the skin mucus. Never before had anybody suggested that salmon migration might be initiated by such population-specific odours. Nordeng formulated this hypothesis in 1960. When Nordeng published his preliminary theory in 1971 [35] he had already done field experiments to support his hypothesis. He had taken eggs from one population of sea-char, native to Salangen, Northern Norway, which makes a yearly seaward migration but limited to about one month of the summer. The eggs were hatched and the fry reared in a hatchery at Voss in Western Norway, 1100 km south of Salangen. The embryos hatched and the young were reared in the hatchery at Voss for four years. In 1968 he released 174 of these fishes in the Løksebotn estuary 10 km west of the Salangen river mouth. 10 of the fishes were found in Løksebotn river and 21 in the Salangen river. Of the 143 fishes released at the Salangen estuary 26 were recaptured in Salangen river but only one in the Løksebotn river [35].

Experiments were later carried out in an artificial situation at the Fish Experimental Breeding Station at Sunndalsøra. The fish investigated were Hammerfest sea-char. When placed in a maze, the fish were able to show their preference in a two-choice test. It was shown that the fish preferred the scent from smolt of their own population. Thus the results strongly support the pheromone hypothesis proposed by Nordeng [36]. These experiments have been supported by the Norwegian Fisheries Research Council from 1972 to 1980. Later studies have shown that migrating sea-char are specifically attracted to smolts of their own population. The attractant substances are found in the intestinal content of the smolts and are possibly of bile origin [44, 45].

The preference tests at Sunndalsøra also showed that the substances acting as attractants came from the faeces. The intestinal content contains much bile salts from the liver. Bile was also preferred in some experiments at Sunndalsøra. Physiological experiments have shown that bile is a potent source of odours for

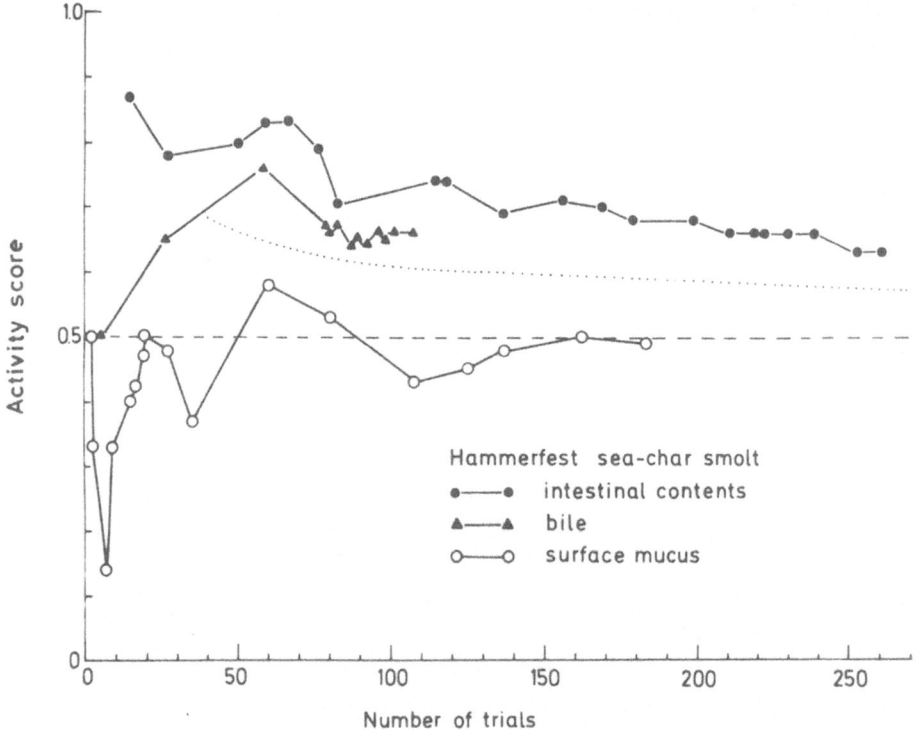

Fig. 4. The activity score from behavioural bioassays with mature sea-char demonstrating the attractiveness of the intestinal content and bile (From Selset and Døving [45]).

fishes like salmon and sea-char. The fish can sense one drop of bile (50 µl) diluted into 500 000 liters of water, i.e. one drop of bile in a swimming pool measuring 25 × 10 × 2 m.

The results of the experiments carried out under the guidance of Hans Nordeng have strengthened our belief that population specific odours make the trail that the salmon follow back home, that the odours come from the fish itself, and that if the fishes "memorize" to the odorants of the rivers, it must be these specific substances. We do not yet know the exact structure of these substances.

6. The Olfactory Epithelium

Figure 5 gives a schematic outline of the olfactory organs and their relation to the brain. The olfactory epithelium of a salmonid fish consists of a rosette of lamellae carrying the olfactory receptor cells. Figure 6 shows the general shape of a lamella. Each lamella is fastened to the floor of the olfactory pit and to a central raphe. Three compartments surround the lamellae. The central space over the raphe is denoted the vestibulum, the spaces between the lamellae are called corridors, and the peripheral compartment is called the gallery [13].

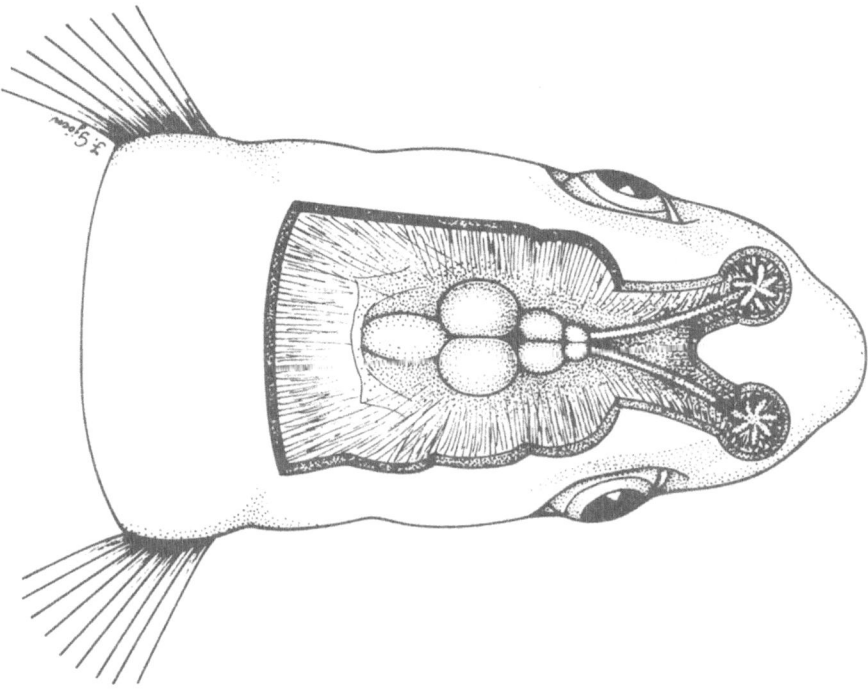

Fig. 5. Dorsal view of a salmonid head showing the olfactory rosettes and nerves running to the olfactory bulbs. The size of the brain is exaggerated.

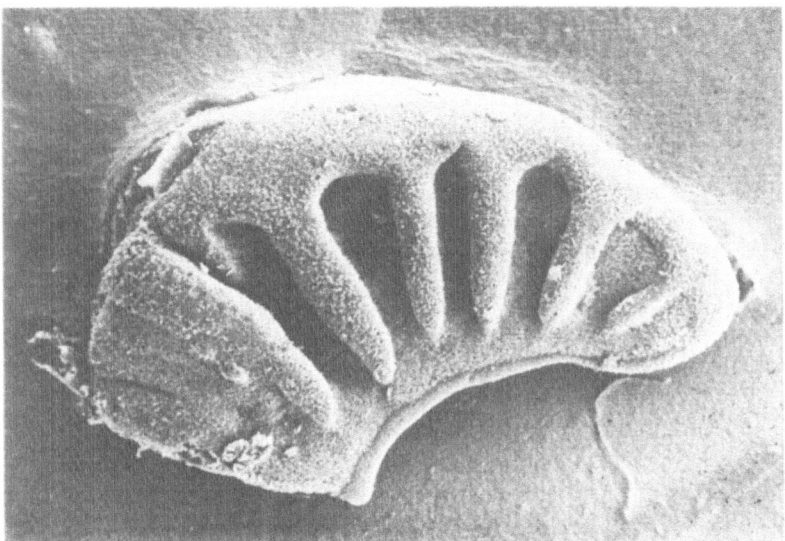

Fig. 6. A scanning electron-micrograph showing the surface of a lamella from an olfactory rosette of a rainbow trout. Its position in the rosette is demonstrated in Figure 7 (From Thommesen [52]).

The surface of each lamella is divided into a sensory area and a non-olfactory, or indifferent, area. The non-olfactory area covers both the peripheral and the central edge. It also extends from the peripheral edge into the sides of the lamella, thereby forming secondary lamellae. The olfactory area of each lamella is located in the secondary corridors between the secondary lamellae, and fuses into an uninterrupted field in the central part of it.

The indifferent area of the epithelium is covered by cells with microridges like fingerprints. Scattered among the microridge cells, and also comprising a signifi-cant part of the surface of the olfactory area, are ciliated cells of non-olfactory type. The receptor cells are of two types, *viz.*, cells with microvilli at the surface and cells with cilia.

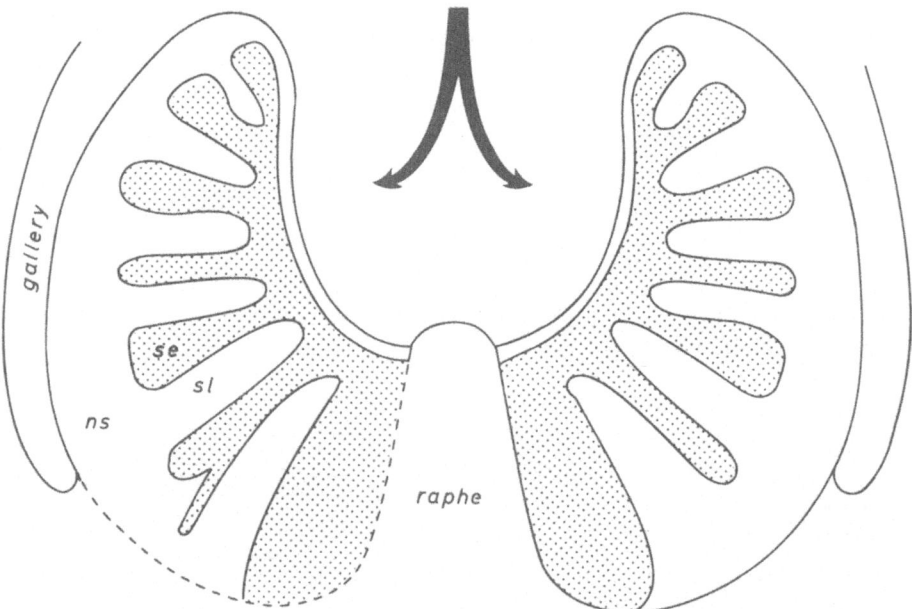

Fig. 7. Schematical drawing of a cross section of an olfactory rosette of a rainbow trout. The sensory epithelium (se, stippled) is found inbetween the secondary lamellae (sl) with non-sensory epithelium. The arrows indicate the direction of water flow. Scale as in Figure 6 (From Thommesen [52]).

(A) RECEPTOR CELLS

The functional significance of the distribution of olfactory receptor types has been studied by Thommesen [51, 52]. The olfactory receptor cells fall clearly into two categories, ciliated and microvillous. Figure 8 shows part of an olfactory lamella.

The ciliated receptor cells possess up to 10 comparatively short cilia extending from the periphery of a hemispherical apical knob about 1.5 μm in diameter. The apical parts of these receptor cells regularly resemble seed potatoes with long sprouts. In the microvillous receptor cells the apical surface of the cell dendrite is

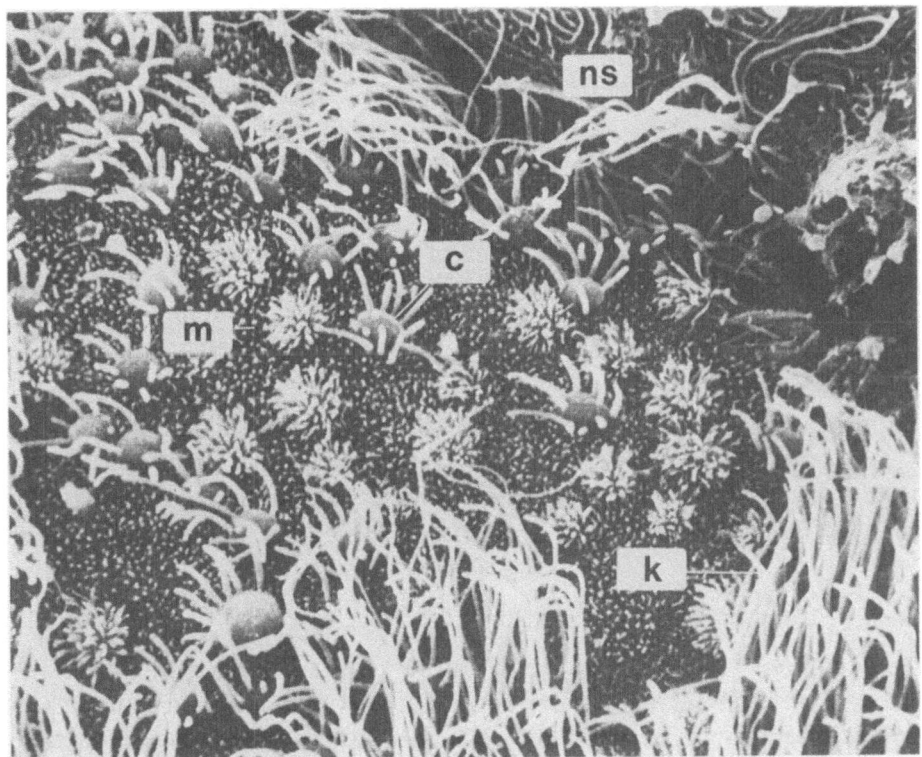

Fig. 8. Close-up scanning electron micrograph of the sensory epithelium of an olfactory rosette of a rainbow trout near the non-sensory epithelium (ns). The morphological distinction between the ciliated (c) and microvillous (m) receptor cells is evident. The receptor cells are found in clusters surrounded by supporting cells, some of which carry kinocilia (k) (From Thommesen [52]).

not visible. The cell apex is covered by a dense tuft made up of about fifty microvilli, longer than those on the non-olfactory cells, and resembling the end of a brush.

The cilia and microvilli can be distinguished by their diameters, cilia being about 0.25 μm across, microvilli about 0.1 μm, and the apex of a ciliated non-olfactory cell has an irregular outline. In the non-olfactory area they may be as much as 8 μm in diameter. In the olfactory area they are smaller, usually 3—4 μm across. The cilia are long, frequently 10 μm. The length of the microvilli is difficult to determine, but is probably 0.5 μm at most. They project from and stand scattered on the cell surface like the spines of a sea urchin. The density of these cells varies, being lowest in the indifferent epithelium of the narrow central edge of the lamella.

Several holes are found in the non-olfactory area. Some are the openings of goblet cells, some are obviously ruptured ciliated cells. There are few holes in the olfactory area. The boundaries between non-olfactory cells of the same type are often difficult to see. In general, the distribution of the olfactory receptor cells is restricted to what has been described above as the olfactory area. Only occasionally are a few receptor cells, predominantly of microvillous type, seen between the fingerprint cells of the central border zone.

(B) DENSITY OF RECEPTOR CELLS

Receptor cells of the same type tend to be located in clusters measuring about 10 μm across. The clusters of ciliated receptors are more concentrated than those of microvillous receptors. Clusters of ciliated receptors often contain more than 10 cells. Clusters of microvillous receptors are less conspicuous and always contain less than 10 cells. Even within a cluster, receptor cell apices do not seem to be in direct contact with each other.

The density of receptor cells, irrespective of type, is higher in the peripheral end of each secondary corridor than in the central field. Typical figures are about 40 000 receptors per mm^2 in the periphery vs. 20 000 receptors per mm^2 in the central part of the corridor.

(C) SPATIAL DISTRIBUTION OF THE EOG

The typical electro olfactogram (EOG) of the char, as obtained with electrodes close to the epithelium has negative polarity. It rises from the baseline to its peak amplitude between 1/2 s and 2 1/2 s after the turning of the stimulation valve. Then, gradually, it approaches a plateau level 20—60% below the peak value in about 10 s. After cessation of the stimulation the potential drops in a similar manner to baseline level with a half-time of about 5 s. The maximal peak amplitudes obtained with the specified stimuli were usually below 10 mV. Negative EOG potentials can be obtained from most parts of the olfactory pit. In the periphery the EOG is most pronounced near the exit of the corridors between the lamellae.

In order to reveal stimulus specific differences in the spatial distribution of the EOG, the principle used by Mustaparta [34] and by Thommesen and Døving [53] was applied: If the ratio between the amplitude at the 'mobile' electrode and that at the 'stationary' electrode varies significantly with the type of stimulus, this is the necessary and sufficient sign of different spatial distributions of the involved receptor cells.

The results of the experiments with dual stimulation using different classes of odorant stimuli indicate that the receptors sensitive to bile salts and those sensitive to amino acids are functionally independent receptor structures. The available evidence suggests that these structures are independent functional units, viz., receptor cells of the two types, those equipped with microvilli and those with cilia.

Simultaneous EOG recordings were made from one central and several peripheral positions in the olfactory rosettes [51]. The results show that in the three species examined, *viz.* rainbow trout, trout and char, the responses were to dilute bile located more towards the periphery than were the responses to the amino acid methionine.

A typical electrophysiological result is illustrated in Figure 9. To the left, the figure shows the outline of a couple of lamellae drawn in perspective with recording electrodes as the set-up would have appeared *in situ*. The left electrode (a) is placed in the central vestibule over the raphe, the right electrode (b) in the peripheral gallery at the exit of the corridor. The simultaneous recordings from these electrode positions show a generally higher response at the central electrode and, more importantly, a difference in the ratio of the peripheral to the central potential when stimulating with bile vs. methionine. The shift in ratio shows a relatively higher density of receptors sensitive to bile salt-like substances near the peripheral edge of the lamellae.

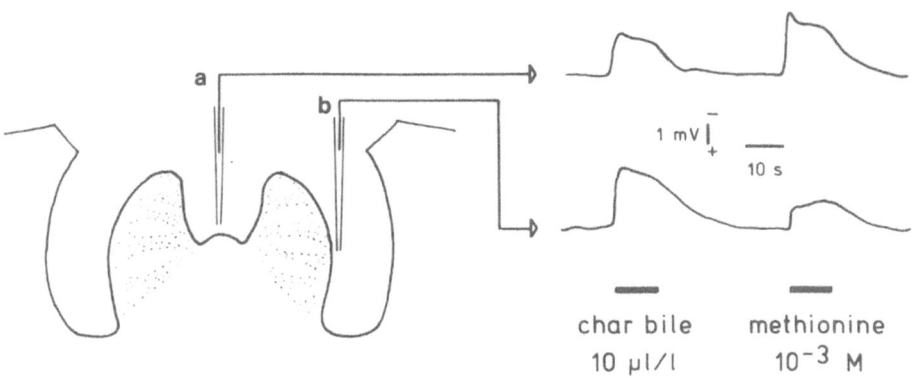

Fig. 9. Schematical drawing of a cross section of the olfactory rosette in char and the recording arrangement for observing differential EOG responses. Notice that when the sensory epithelium is stimulated with char bile the amplitude of the EOG response is largest at the electrode located in the peripheral corridor (b). Stimulation with the amino acid induced the largest response from the electrode situated above the raphe (a) (From Thommesen [51]).

To sum up, by 1983 Thommesen's experiments [52] demonstrated that the olfactory receptor cells of the ciliated type were more densely distributed in the peripheral parts of the sensory epithelium than in the central part. The density of the microvillous receptor cells was higher at the central part of the sensory epithelium than in the peripheral parts. On the basis of his experiments, where he related the density distributions of the receptor cells and the relative size of the EOG amplitudes at the central and peripheral parts of the olfactory rosette, Thommesen suggested that it is the receptor cells with microvilli that are sensitive to amino acids.

(D) DUAL STIMULATION

The degree of receptor specificity was investigated by dual stimulation recording the electro-olfactogram (EOG). Spatial distribution of the receptors was studied, partly using EOG recordings, partly by studying the change in bulb surface responses during progressive lesioning of the olfactory epithelium.

The EOG is a very slowly adapting signal once it has reached the plateau level. A second stimulus given in addition to one already present will produce an additional EOG on top of the first one. If the two EOGs are generated in entirely independent receptor cell populations, the second EOG will simply add to the first one, using its plateau as a baseline. If the two EOGs originate in the same cell population, as is the case, for example, when the second stimulus merely increases the concentration of the first one, the second EOG will only rise to a level above the true baseline corresponding to the combined stimulus concentration. Such 'cross-adaptation' studies were carried out by Thommesen [52]. One of them is displayed in Figure 10. His results clearly demonstrate that the receptor cell population giving rise to responses to bile salts is independent of the cell population that responds to amino acids.

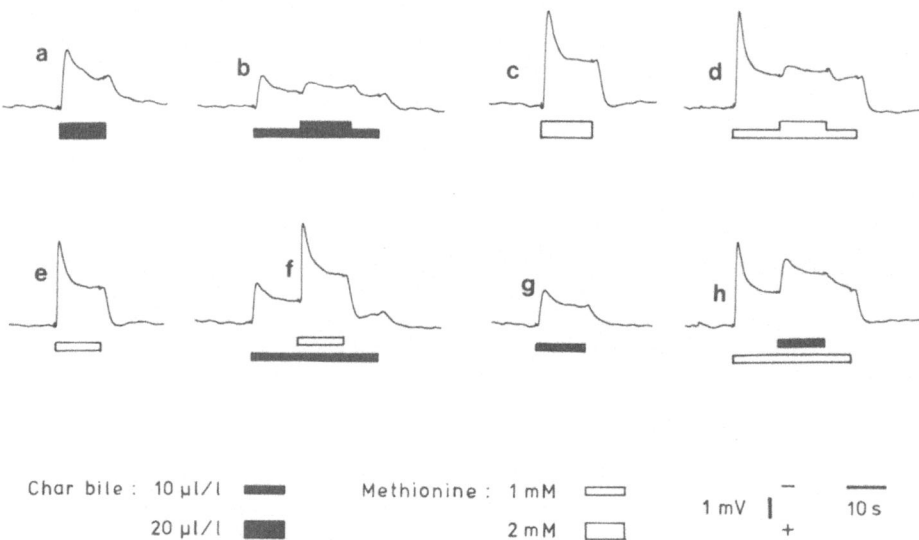

Fig. 10. Dual stimulation. EOG recordings from the olfactory epithelium of char. The upper traces (a—d) show the responses when stimulating with a single odorant, either char bile or the amino acid methionine. Notice the modest increase in response amplitude when the stimulus intensity is increased (b and d). The lower traces (e—h) show the responses when one stimulus is given on top of the other (dual stimulation). The response to stimulation of methionine on top of bile (f) is the same as methionine alone (e), and stimulation of bile on top of methionine (h) is the same as for bile alone (g) (From Thommesen [51]).

7. The Olfactory Bulb

The olfactory bulb is the first relay station in the olfactory system of vertebrates. Here the axons of the sensory cells terminate, and the transmission of olfactory information is conveyed by a new set of secondary neurones called relay cells. In some fishes the bulb can be found close to the olfactory epithelium, connected to the brain by a long central tract, as for example, in the cod-fish. Such long tracts are also found in catfish and carps. In many other fishes the olfactory nerve is long, for example in pike, garfish, salmon, and perch.

(A) ANATOMY

The early works on the fish brain are landmarks in the descriptive anatomy of the brain and the olfactory system of fishes. The most extensive works are those by Sheldon [46] on the carp *Caprinus carpio*, Holmgren [23] on the smelt *Osmerlanus erlanus*, and Jansen [27] on the hagfish *Myxine glutinosa*.

Starting from the surface, the layers in the olfactory bulb can be distinguished as: (a) the olfactory nerve layer, (b) the layer of olfactory glomeruli, (c) the layer of relay cells, and (d) the internal cell layer. The last has also been called the granule cell layer or anterior olfactory nucleus. Granule cells have not been definitely described, however, so the name internal cell layer seems appropriate.

Most of the cellular elements constituting the olfactory bulb in higher vertebrates are found in the bulb of fishes. Their organization is less distinct, but is still appreciable. The fibers coming from the olfactory epithelium are distributed in the superficial layer of the anterior part of the bulb. Here the fibers are sorted and the axons terminate in restricted regions of the bulb, so-called glomeruli, structures devoid of cell soma where axons make synaptic contact with dendrites of secondary neurones of the bulb. In contrast to the relay cells of the mammalian olfactory bulb, i.e., the mitral and tufted cells, the dendrites of these cells in the fish branch to reach several glomeruli. The secondary neurones constitute a third layer inside the layer of glomeruli.

Scattered among the mitral cells are ruffed cells. They were first discovered and described by Kosaka and Hama [30, 31] in the goldfish *Carassius auratus*.

(B) SYNAPTIC ORGANIZATION

Few studies have been conducted on the ultrastructure of the fish olfactory bulb. Apart from the studies on the ruffed cells, Ichikawa [24] has described the fine anatomy of the goldfish *Carassius auratus*. In the glomeruli, asymmetrical synapses are found, characterized by a dense thickening of the postsynaptic membrane and spherical synaptic vesicles. These synapses are thought to be excitatory. Dendro-dendritic synapses have also been observed in the glomeruli. Pictures taken by

Ichikawa [24] show that a sensory receptor axon terminates onto several dendrites of secondary neurones. If these dendrites belong to different cells, this implies that the receptor cells branch to make contact with several relay cells, as shown in the mammalian olfactory bulb [42]. In the relay cell layer, dendrodendritic synapses between relay cells and cells thought to be granule cells have been observed. According to Pinching and Powell, the polarity of the synapses is such that the mitral cells are excitatory to the granule cells and the granule cells inhibitory to the relay cells.

Both asymmetrical and symmetrical synapses are found in the internal cell layer. The neuronal elements, however, have not been identified. The centrifugal fibers originating in the telencephalon terminate in this layer [25]. In the degenerating terminals both types of synapses are described. These studies emphasize the similarity of the synaptic organization to that of the mammalian olfactory bulb.

In a recent study on the synaptic organization of the teleost olfactory bulb, Kosaka and Hama [32] stressed the discrepancies between the functional organization of the olfactory bulb in teleosts and in mammals. The mixed-synapse cell is postsynaptic both to the relay and to the granule cells, and receives asymmetrical synapses from dendrites of the relay cells and symmetrical synapses from granule cells. The perinest cell, not found in the mammalian bulb, has protrusions in the nest region and receives symmetrical synapses from the granule cell only. The output from these two types of cells is still unknown. Based on their recent findings, Kosaka and Hama made a diagrammatic representation of the synaptic connections in the olfactory bulb of teleosts: this diagram is reproduced in Figure 11.

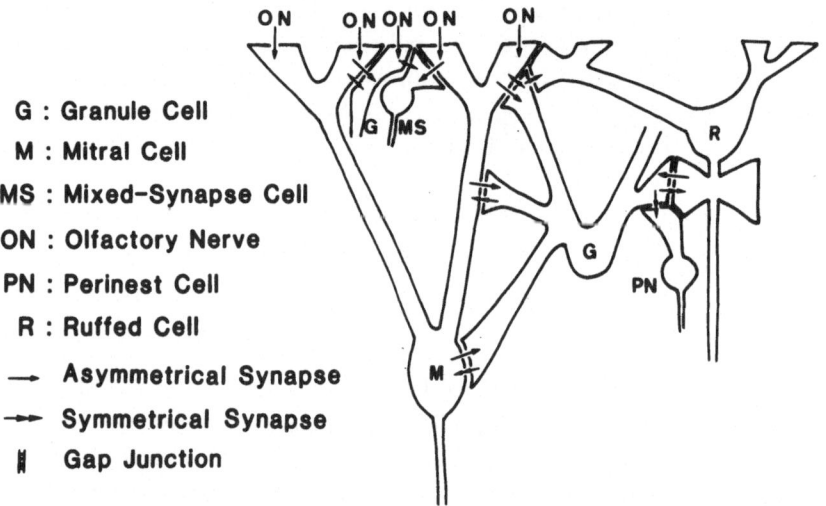

G : Granule Cell
M : Mitral Cell
MS : Mixed–Synapse Cell
ON : Olfactory Nerve
PN : Perinest Cell
R : Ruffed Cell
→ Asymmetrical Synapse
→ Symmetrical Synapse
⫲ Gap Junction

Fig. 11. Schematical drawing of the neurones in the olfactory bulb of fish. Notice the reciprocal synapses between the granule and mitral cells, and the ruffed and mtiral cells (From Kosaka and Hama [32]).

Convergence Ratio

The number of axons making synapses in the olfactory bulb is in the order of 10^7, while the number of fibers in the olfactory tract is in the order of 10^4 [9]. Thus, the nominal convergence ratio is about $1000:1$. As shown by van Drongelen *et al.* [6], this convergence ratio increases the probability of response in the relay cells. Since there is reason to believe that the axons branch and make synapses with several secondary neurones, the functional ratio will be greater than indicated by the counted number of receptor and relay cells, which further increases the response probability.

An increase in axonal branching and terminal sprouting results in increased sensitivity or 'gain' by the olfactory system. There will be greater sensitivity towards those odorants to which the sprouting receptors respond. Such mechanisms may explain a specific odour preference exhibited by animals that have previously been exposed to the odour.

(C) BULBAR SURFACE ACTIVITY

A convenient method for recording the electrical activity of the olfactory system is to place electrodes at the surface of the readily accessible olfactory bulb. The surface EEG activity in the fish olfactory bulb resembles that seen in the mammalian and amphibian bulbs. Stimulation of the sensory epithelium with odorants gives rise to two bulbar components, a slow shift in DC potential upon which are superimposed regular oscillations, frequently referred to as "induced waves". Both the amplitude of the slow potential and the induced waves increase with increasing concentrations. The magnitude of the slow component may reach a few millivolts. The induced waves are usually of smaller amplitude.

(D) SPATIAL DISTRIBUTION OF RESPONSES

Figure 12 gives an example of a variation in the amplitude of the bulbar responses with the quality of the stimulus. The electrode positions are in the anterior region, one lateral and the other medial. As seen, the medial part of the bulb responded only to the taurolithocholic acid. The lateral part of the bulb responded to methionine. Responses increased approximately linearly with the logarithm of the stimulus concentration.

It could be argued that the EOG responses obtained by dual stimulation of the olfactory epithelium with bile salts and amino acids are due to receptor site distinction, and not to different receptor cell populations (see above). In this connection it is important to realize that the responses to amino acids and the responses to bile salts are evoked in two regions of the olfactory bulb. Since it is not conceivable that the axons of receptor cells could give responses at different regions of the bulb depending upon what type of odorant is used to stimulate the

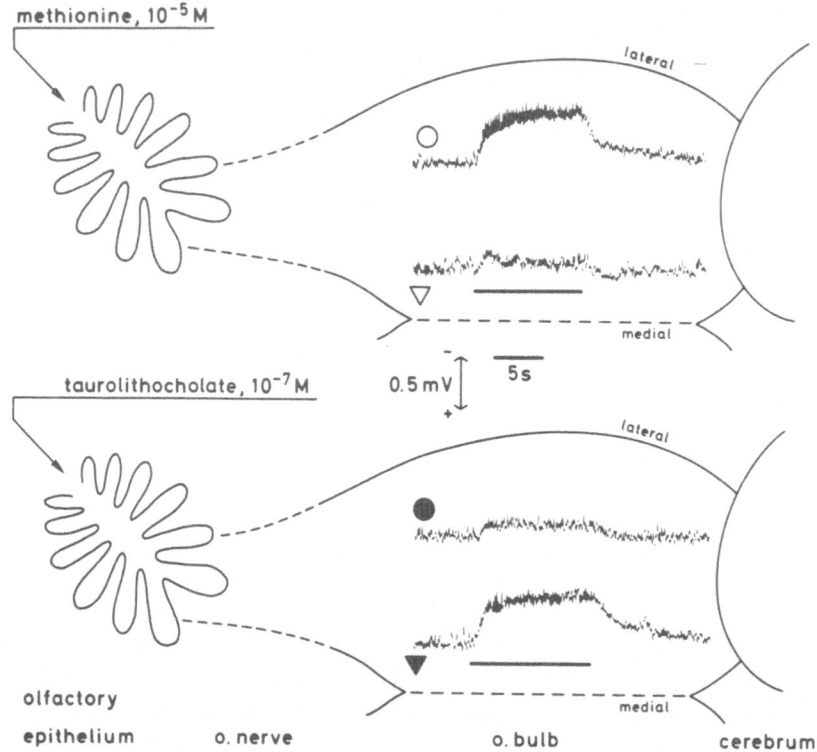

methionine, 10⁻⁵M

taurolithocholate, 10⁻⁷M

0.5 mV

5 s

olfactory

epithelium   o. nerve   o. bulb   cerebrum

Fig. 12. Dorsal views of the olfactory epithelium and bulb of the char, right side. The recordings of surface potentials of the olfactory bulb were made in the lateral (●, ○) and medial (▼, ▽) aspect of the bulb. Stimulation with methionine, upper drawing, gave responses mainly in the lateral part of the bulb. Stimulation with the bile salt, taurolithocholate, lower drawing, evoked response mainly from the electrode situated at the medial aspect of the bulb (From Døving, Selset and Thommesen [11]).

epithelium, it becomes even more convincing that there are at least two distinct receptor cell populations, one responding to amino acids and the other to bile salts. In light of the experiments performed by Thommesen [51, 52] it seems likely that these receptor populations are the two morphologically distinct receptor cells, viz., the microvillous and the ciliated cell types, the first responding to amino acids and the second to bile salts.

The spatial isolation of the bulbar responses to the two types of stimuli thus suggests that there are two groups of olfactory receptors which respond either to amino acids or to bile acids. Dual stimulation experiments were made using 10^{-4} M of methionine and 10^{-5} M of taurocholic acid. The bulbar responses were recorded when switching directly from one stimulating substance to the other. Even after 11 min of continuous exposure to one of these stimuli, the immediate response to the other remained unaltered. Thus, it seemed that there was no significant cross-adaptation between the bulbar responses to methionine and to taurocholic acid.

(E) CHEMICAL STRUCTURE AND RESPONSE THRESHOLD

The different unconjugated bile acids were not so potent as odorants as the conjugated ones were. Table V gives mean sensitivity for the different compounds. For purposes of comparison, the table includes the results for methionine and taurine as well. The threshold for eliciting responses in the olfactory system was regularly lower for the taurine conjugates, especially the sulfonated one, than for any of the unconjugated derivates. The absolute sensitivity varied for the different preparations, but the ranking followed nearly the same order in all experiments. Some of the bile acids derivatives have so low solubilities that their threshold could not be satisfactorily determined. These were therefore excluded from further investigations. These substances are marked with an *a* in Table V.

TABLE V. Threshold concentration for a number of compounds tested by electrophysiological methods. From Døving, Selset and Thommesen [11].

Common name	Position of groups			Mean threshold value (M)	Threshold value range (M)
	OH	SO$_4$H	Tau		
Cholic acid	3, 7, 12	—	—	8.0×10^{-8}	$10^{-6} - 10^{-8}$
Taurocholic acid	3, 7, 12	—	24	2.0×10^{-8}	$10^{-6} - 10^{-9}$
Deoxycholic acid[a]	3, 12	—	—	—	—
Taurodeoxycholic acid	3, 12	—	24	6.3×10^{-8}	$10^{-5} - 10^{-9}$
Chenodeoxycholic acid[a]	3, 7	—	—	—	—
Taurochenodeoxycholic acid	3, 7	—	24	4.0×10^{-8}	$10^{-5} - 10^{-9}$
Lithocholic acid[a]	3	—	—	—	—
Taurolithocholic acid	3	—	24	6.3×10^{-9}	$10^{-6} - 10^{-9}$
Sulfotaurolithocholic acid	—	3	24	1.0×10^{-8}	$10^{-6} - 10^{-11}$
Methionine	—	—	—	1.3×10^{-6}	$10^{-5} - 10^{-8}$
Taurine	—	—	—	1.0×10^{-5}	

[a] Threshold not determined due to low solubility.

Water samples collected during the tracking studies were used as stimuli for electrophysiological experiments designed to determine if the olfactory system of Atlantic salmon is capable of discriminating fine-structure features of hydrographic stratification. Typically, the vertical separation of adjacent samples was 2 m, with temperature and salinity differences of a few tenths of a degree and some parts per thousands, respectively. Sixty-six per cent of all olfactory bulb neurons did, in fact, display differential changes in discharge rates in response to test samples. However, only 72% of all neurons responded to any of the water samples. Of these responding neurons, fully 90% could discriminate among the water samples. Of the two different water sample series that were tested, Series I was collected near the mouth of the home river, and 85% of the neurons tested with that series responded to the samples with 91% of the responding cells able to discriminate among samples. Series II samples were collected 30 km down the fjord from the mouth of Imsa River, and 63% of the cells responded to those water samples. Yet of those responding, 90% were able to discriminate among the samples collected at different depths [12].

(F) DISCRIMINATION OF POPULATIONS

Authorities on salmon migration have hypothesized that the odorants which guide homing fish emanate from the plants and minerals characteristic of the home stream waters [19]. For a long time the possibility of fish pheromones was not adequately considered, even though various behavioural experiments suggest species and perhaps individual recognition based on odour. For example, Wrede [62] demonstrated that minnows (*Phoxinus phoxinus* L.) were attracted to and recognized other members of their species on basis of odorants in skin mucus. The results obtained by Wrede were confirmed and extended in minnows by Göz [16] and in yellow bullhead (*Ictalurus natalis*) by Todd *et al.* [54]. Alarm substances are also well-known in fishes [14, 41].

Nordeng [35] has shown that anadromous char from northern Norway (Salangen) which were reared for 3 years in southern Norway (Voss) went back to their parent river when released in the Salangen fjord. He suggests that young char resident in the spawning area of the parent river release a pheromone in their mucous which attracts the migrating fish. Electrophysiological experiments by Døving *et al.* [8] have shown that substances emanating from the fish are an effective olfactory stimulus and may be a source of such pheromones. Consequently, it would be useful to make an electrophysiological examination of the capacity of char to detect and distinguish among appropriate fish odours. Certain attempts to study the electrophysiological correlates of homing in salmonids have relied upon EEG responses of the olfactory bulb. These initial studies concluded that for spawning salmon among the several natural waters tested, home water was the only effective stimulant for EEG arousal [17] or at least the most effective stimulant [57]. This work has been criticized by Oshima *et al.* [38] because of the deteriorated condition of the fish and the lack of fresh samples of the waters used for stimulation. They found that with healthy fish and new water samples, all natural waters tested were effective stimulants of the EEG; sometimes as effective as home water.

In further tests it was determined that there was no systematic increase in EEG arousal to samples of waters obtained along the migratory route [39]. The chemical from one particular stream will of course increase up the migratory route. But EEG arousal to natural waters is a response to the totality of all effective odorants present. It is clear that a more discriminating electrophysiological measure would be helpful in experiments on salmonid homing. Oshima *et al.* ([39], p. 2132) were ultimately led to the conclusion that the EEG "yields data the significance of which is mostly cryptic". In their experiments they examined the possibility that various geographical populations of char release characteristic odours that elicit differential responses in the char olfactory system.

In our laboratory we used as odorant producers young chars originating from the two Salangen and the two Voss populations mentioned above. They were reared separately for two years before being used in our experiments. In addition we used adult chars from Løna and Vangsvatn caught with beach seine. This gave six different char groups.

The results demonstrate that bulbar neurones are similarly affected by the six stimuli [10]. The diverse combinations of response patterns suggest that the odours from the different char populations have different effects upon the olfactory system. A more rigorous Cochran Q test appropriate for repeated measures (6 stimuli per neuron) led to the rejection of the hypothesis that these neuronal responses are the result of repeated presentations of the same odour ($p < 0.01$, $Q = 20.2$, assigning $-$ to the zero responses $N = 45$). The statistical analysis gives a $p < 0.01$ whether one deletes the units with one or more '0' responses ($N = 30$) or $+$ or $-$ is assigned at random to the zero responses ($N = 45$). Thus, there is sufficient information for physiological discrimination by the olfactory cells of the odorants from the various fish populations even when the quantitative data are omitted from the analysis ($+$ and $-$ only).

It is possible to examine the effects of particular stimuli more closely. For each of the 15 pairs of stimuli (combinations of 6 stimuli like $A - B$, $A - C$, $A - D$, etc. It was possible to determine the percentage of single unit responses (units with one or more 0 entries were deleted) that were the same ($++$ or $--$) or different ($+-$ or $-+$). In a typical pair of stimuli nearly half the neurons showed opposite responses ($M = 45.6\%$; Mdn $= 47.2\%$). Young fish from Vangsvatn and adult fish from Løna tended to cause opposite (69%) effects, whereas young Løna and young migratory char tended to have the same (16% opposite; 84% the same) effects upon the neuron. It is important to stress that even a small percentage of units with differential responses may have behavioral significance, because these patterns of excitation and inhibition were highly reliable. Most of the entries in the data were replicated with a second presentation of the stimuli. Since only 5 of 151 entries changed with replication ($p < 0.0001$ of random fluctuation; normal approximation to the binomial distribution, $p = q = 0.5$) it can be concluded that the difference between pairs of stimuli is reliably reflecting differences in the properties of the odorants. It is a pertinent comparison that the odours from young migratory and young stationary char elicited differential responses in 47.4% of the single units.

The results of Døving et al. [10] suggest that these batches of char release different odours and that a substantial portion of the olfactory bulb cells are differentially tuned to mediate physiological and presumably behavioral discrimination among these naturally occurring fish odours.

To determine whether the discriminatory power was sufficiently strong to differentiate between individuals of the same population, two new plastic tanks (120 l) were arranged with identical water flow rates (6 l/min). There was no detectable difference in the olfactory cell response to the 2 tank waters without fish. Within 25 min of the introduction of one adult male of Vangsvatn char into each tank (same age and length) these tank waters, A and B, evoked reactions in the olfactory neurons and in some cases (4 of 17 cells) the effects were differential. Figure 13 gives an example of a unit being excited by the A, but inhibited by the B Vangsvatn char.

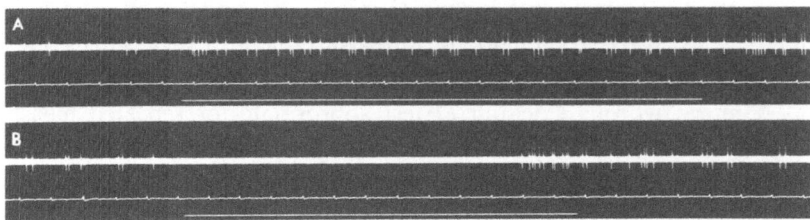

Fig. 13. The nervous discharges from a single neuron in the olfactory bulb in char in response to stimulation with samples of water containing males of two different populations. This demonstrates the ability of the fish olfactory system to discriminate the scent of gentes (From Døving, Nordeng and Oakley [10]).

The presence of excitation and inhibition in single unit activity provides evidence of qualitatively different odorants in situations where the concentration of the active chemical of the stimulating waters is unknown. This is an important advantage over the integrated EEG response, which simply increases in magnitude. In order to determine the functional significance of differentially responding cells it is first necessary to establish that the differential responses are reliable and not a result of chance fluctuation. In *our* experiments, the excitatory or inhibitory character of a response is highly reliable, e.g. if a stimulus inhibits at the first trial it will inhibit at subsequent trials. Thus, even a small percentage of cells responding differentially to a pair of odorants would be physiologically significant.

It is almost certain that the concentrations of all tank chemicals were initially low, inasmuch as the tank water was rapidly replaced. In almost all cases, dilution of the tank water reduced the vigor of the response, but did not seem to change its tendency to either excite or inhibit the activity. It is known from single unit work in the olfactory bulb of frog that a few units can be excited by a low concentration and inhibited by a higher concentration of odorant [7]. In the present experiment described above there were large numbers of cells with differential responses.

Comparisons of electrophysiological data from frog making use of such differential responses with psychophysical data obtained from human subjects, certainly indicate that the neurophysiological reactions excitation and inhibition are relevant to olfactory discrimination. The presence of excitatory and inhibitory responses should be interpreted as evidence that the odours released by the various fish populations are qualitatively different and do not merely reflect differences in concentration. Our studies suggest that the olfactory system of the char is capable of detecting and discriminating between the various char odours released.

Electrophysiological studies of fish olfaction related to the natural biology of the fish have used the summated response [17]. In the series of experiments on EEG activation of the olfactory bulb the home water always produced a large response. This is also understandable, since the home waters are associated with salmon ponds and hatcheries with their extensive aggregation of fish (e.g. Figure 21 of [57]). It has been assumed that the olfactory EEG arousal to home water is a direct electrophysiological manifestation of olfactory memory underlying homing [37, 39, pp. 2131—2132]. In order to prove that the EEG arousal or any other sensory response is learned, it would be necessary to show a change in responsiveness to waters as a result of experience. What has been shown in a series of tests is that if fish are left in various waters, this increases the effectiveness of each type of water as a stimulatory agent [38]. The most reasonable interpretation is that the stimulus properties of the water are changed as a result of chemicals secreted by the fish; not that the responsiveness of the fish olfactory system is changed, i.e., the response is not a process of learning.

The results suggest that the char olfactory receptors are sensitively tuned to biological meaningful odour and that the olfactory system in char is capable of detecting and discriminating between various char odours. This means that the migratory populations of char during upstream runs have the potential ability to orientate along odour tracks released from specific populations of standing young relatives in the nursery areas in their parent river system. This view may also hold good for other anadromous species orientating in local waters.

It is not a prerequisite that the fish must migrate up a gradient. The presence of the right char odour may induce a positive rheotaxis in the migratory fish, provided it is properly motivated. When the char odour is absent, the fish will drift down until it finds its right clue. This behaviour may explain the relatively large proportion of 'correct returns' in the experiments by Groves *et al.* [15] of normal and blinded salmons displaced upstream in Columbia River.

8. Odorants from the Fish

Through his presentation of the pheromone hypothesis, Nordeng [35, 36] moved the concept of salmonid migration from a level of conjecture on the nature of the substances which guide the fish back home, to a level at which these problems could be attacked by well established chemical methods. The experiments by Døving *et al.* [10] demonstrated that the fish produce highly potent odorants. These substances were later shown to have a bile salt character [44, 45]. The substances were so potent that introducing a fish in a new tank of 120 l with a flow of 6 l/min made the water an effective inducer of responses in the bulb by 25 min.

Bile has been shown to be of much higher potency than any other olfactory substance known for fish. Taurine conjugated bile acids were up to 1000 times more potent as olfactory stimuli than methionine. The physiological responses show a spatial distribution in the olfactory bulb, indicating that bile acids and amino acids have separate neurological pathways in the olfactory system of fish [11]. The odorant potency of the bile acids, their evolutionary history and variability, together with their renowned adherent properties, make them interesting candidates for specific signals in aquatic environments.

Bile acids are large molecules that can take a multitude of shapes. They have properties which make them adhere to almost anything. They are certainly difficult to get rid of. The evolutionary history of bile salts is an interesting chapter of biology that has been studied in detail by G. A. D. Haslewood [21] and his fellow-workers in England. Each species of fish has a number of particular bile salts that are in the majority, and are important for the digestive function. But there are also a large number of other bile substances which can be characterized as chemical 'noise'. The function of these substances is unknown.

Crude bile taken from the gall bladder of the char and trout in the bulb gave responses in dilutions down to about $1:10^9$. Basin water containing fish gave responses both in the medial part of the olfactory bulb, i.e. at the same area as the bile acids, and in the lateral part [50]. The presence of the bile acids in basin water was investigated by holding a char in 10 l of water for 4 h. The water was then analyzed for its content of 3-beta-hydroxy-steroids in relation to a control (Sterognost 3-beta, Nyegaard & Co. Oslo). The analysis showed an effluence of 3-OH steroids of 34 nmol kg^{-1} h^{-1}. This result indicates that the quantity of bile acids emanating from the fish is sufficient to be detected by the olfactory system of the fish.

Unfortunately little progress has been made identifying the substances in salmonids. The reasons for this need not be discussed here.

According to Stacey and his collaborators [47] the reproductive physiology and behaviour of the male goldfish appear to be influenced by at least three pheromones:

(1) An unknown substance possibly related to estrogen that serves to identify the gender.
(2) 17-alpha, 20-beta-dihydroxy-pregnen-3-one, a steroid hormone that induces the final maturation of the oocyte. This substance seems to be synthesized by the female 4 to 8 hours before ovulation and acts as a 'primer' pheromone, which induces a rapid increase in gonadotropic hormone in the male. It also acts as a 'releaser' pheromone by inducing increased social and locomotor activity.
(3) An unknown compound released by the ovulated females which induces typical male spawning behaviour and possibly serves to synchronize male-female spawning readiness.

9. Conclusions

During the last decade new avenues have been opened for our understanding of salmonid homing and navigation. The chemical substances important for recognizing the home rivers have been connected to the evolutionary biochemistry of the bile salts in vertebrates. We are, however, only on the brink of a series of exciting discoveries concerning these substances, their nature and use in migration. We need to know: What precisely are the substances that are needed for the correct return? What is their biochemistry; in what concentrations are they needed; how specific are they? Is there a specific imprinting process involved as the salmon smolts start their remarkable journey, or is the correct return based upon congenital reaction patterns laid down at birth? If there is an imprinting phenomenon, is that taking place at the periphery or in the central nervous system of the fish?

The observations of the salmon in relation to the hydrographic stratification have taught us important aspects of its movements. We have learned how fishes act in a three dimensional world without reference points, and we imagine that they make use of suptile differences in gradients for navigation. But still we lack fundamental information on the sensory basis for fish orientation to understand the mysterious voyages that the salmon make, far out in the ocean and back to its home river.

Acknowledgements

The studies referred to above which were made by persons connected with the University of Oslo could not have been performed without the generous support from the Norwegian Research Council for Science and Humanities and the Norwegian Fisheries Research Council.

References

1. G. P. Arnold: 'Rheotropism in fishes'. *Biol. Rev.* **49**, 515—576 (1974).
2. G. Bertmar: 'Home range, migrations and orientation mechanisms of the River Indalsälven trout, *Salmo trutta* L'. *Inst. Freshwater Res.* (*Drottningholm*) **58**, 5—26 (1979).
3. J. R. Brett and C. Groot: 'Some aspects of olfactory and visual responses in Pacific salmon'. *J. Fish. Res. Bd. Can.* **20**, 287—303 (1963).
4. E. H. Craigie: 'A preliminary experiment upon the relation of the olfactory sense to the migration of the Sockeye salmon (*Oncorhynchus nerka* Walbaum)'. *Roy. Soc. Can. Trans.* **5**, 215—224 (1926).
5. A. C. De Lacy, L. R. Donaldson, and E. L. Brannon: 'Homing behavior of chinook salmon'. *Res. Fish Invest. Univ. Wash.* **300**, 59—60 (1968).
6. W. van Drongelen, A. Holley, and K. B. Døving: 'Convergence in the olfactory system. Quantitative aspects of odour sensitivity'. *J. Theor. Biol.* **71**, 39—48 (1978).
7. K. B. Døving: 'Studies of the relation between the frog's electro-olfactogram (EOG) and single unit activity in the olfactory bulb'. *Acta Physiol. Scand.* **60**, 150—163 (1964).
8. K. B. Døving, P. S. Enger, and H. Nordeng: 'Electrophysiological studies on the olfactory sense in char (*Salmo alpinus* L.)'. *Comp. Biochem. Physiol.* **45A**, 21—24 (1973).
9. K. B. Døving and G. Gemne: 'Electrophysiological and histological properties of the olfactory tract of the burbot (*Iota Iota* L.)'. *J. Neurophysiol.* **28**, 139—153 (1965).

10. K. B. Døving, H. Nordeng, and B. Oakley: 'Single unit discrimination of fish odours released by char (*Salmo alpinus* L.)'. *Comp. Biochem. Physiol.* **47A**, 1051—1063 (1974).
11. K. B. Døving, R. Selset, and G. Thommesen: 'Olfactory sensitivity to bile acids in salmonid fishes'. *Acta Physiol. Scand.* **108**, 123—131 (1980).
12. K. B. Døving, H. Westerberg, and P. B. Johnsen: 'Role of olfaction in the behavioral and neural responses of Atlantic salmon, *Salmo salar*, to hydrographic stratification'. *Can. J. Fish. Aquat Sci.* **42**, 1658—1667 (1985).
13. K. B. Døving, M. Dubois-Dauphin, A. Holley, and F. Jourdan: 'Functional anatomy of the olfactory organ of fish and the ciliary mechanism of water transport'. *Acta Zool. (Stockholm)* **58**, 245—255 (1977).
14. K. von Frisch: 'Über einen Schreckstoff der Fischhaut und seine biologische Bedeutung'. *Z. Vergl. Physiol.* **29**, 46—145 (1941).
15. A. B. Groves, G. B. Collins, and P. S. Trefethen: 'Roles of olfaction and vision in choice of spawning site by homing adult chinook salmon (*Oncorhynchus tshawytscha*)'. *J. Fish. Res. Bd. Can.* **25**, 867—876 (1968).
16. H. Göz: 'Über den Art- und Individualgeruch bei Fischen'. *Z. Vergl. Physiol.* **29**, 1—45 (1941).
17. T. J. Hara, K. Ueda, and A. Gorbman: 'Electroencephalographic studies on homing salmon'. *Science* **149**, 884—885 (1965).
18. F. R. Harden-Jones: *Fish Migration.* Arnold, London, p. 325 (1968).
19. A. D. Hasler: *Underwater Guideposts.* Univ. Wisconsin Press, Madison (1966).
20. A. D. Hasler and A. T. Scholtz: *Olfactory Imprinting and Homing in Salmon.* Springer Verlag, Berlin, 134 pp. (1983).
21. G. A. D. Haslewood: 'The biological importance of bile salts'. In: A. Neuberger and E. L. Tatum (eds.), *North-Holland Research Monographs, Frontiers of Biology.* North-Holland, Amsterdam, **47**, 1—206 (1978).
22. Y. Hiyama, T. Taniuchi, K. Suyama, K. Ishioka, R. Sato, T. Kajihara, and T. Maiwa: 'A preliminary experiment on the return of tagged chum salmon to the Otsuchi River, Japan'. *Bull. Jap. Soc. Sci. Fish.* **33**, 18—19 (1967).
23. N. Holmgren: 'Zur Anatomie und Histologie des Vorder- und Zwischenhirns der Knochen-fische'. *Acta Zool. (Stockholm.)* **1**, 137—315 (1920).
24. M. Ichikawa: 'Fine structure of the olfactory bulb in the goldfish, *Carassius auratus*'. *Brain Res.* **115**, 53—56 (1976).
25. M. Ichikawa and K. Ueda: 'Electron microscopic study of the termination of the centrifugal fibers in the goldfish olfactory bulb'. *Cell Tissue Res.* **197**, 256—262 (1979).
26. L. A. Jahn: 'Movements and homing of cutthroat trout (*Salmo clarki*) from open-water areas of Yellowstone Lake'. *J. Fish. Res. Bd. Can.* **26**, 1243—1261 (1969).
27. J. Jansen: 'The brain of *Myxine glutinosa*'. *J. Comp. Neurol.* **49**, 359—507 (1930).
28. P. B. Johnsen: 'A behavioral control model for homestream selection in salmonids'. In: E. L. Brannon and E. D. Salo (eds.), *Proceedings of the Salmon and Trout Migratory Behaviour Symposium.* School of Fisheries, Univ. Wash., Seattle, pp. 266—273 (1982).
29. P. B. Johnsen and A. D. Hasler: 'The use of chemical cues in the upstream migration of coho salmon, *Oncorhynchus kisutch* Walbaum'. *J. Fish Biol.* **17**, 67—73 (1980).
30. T. Kosaka and K. Hama: 'Ruffed cell: a new type of neuron with a distinctive initial unmyelinated portion of the axon in the olfactory bulb of the goldfish (*Carassius auratus*)'. *J. Comp. Neurol.* **186**, 301—320 (1979).
31. T. Kosaka and K. Hama: 'Ruffed cell: a new type of neuron with a distinctive initial unmyelinated portion of the axon in the olfactory bulb of the goldfish (*Carassius auratus*). III. Three-dimensional structure of the ruffed-cell dendrite'. *J. Comp. Neurol.* **201**, 571—587 (1981).
32. T. Kosaka and K. Hama: 'Synaptic organization in the teleost olfactory bulb'. *J. Physiol. (Paris)* **78**, 707—719 (1983).
33. G. W. Lebar: 'Movement and homing of cutthroat trout (*Salmo clarki*) in Clear and Bridge Creeks, Yellowstone National Park'. *Trans. Amer. Fish. Soc.* **100**, 41—49 (1971).
34. H. Mustaparta: 'Spatial distribution of receptor responses to stimulation with different odours'. *Acta Physiol. Scand.* **82**, 154—166 (1971).
35. H. Nordeng: 'Is the local orientation of anadromous fishes determined by pheromones?' *Nature* **233**, 411—413 (1971).

36. H. Nordeng: 'A pheromone hypothesis for homeward migration in anadromous salmonids'. *Oikos* **28**, 155—159 (1977).
37. K. Oshima, A. Gorbman, and H. Shimada: 'Memory-blocking agents: effects on olfactory discrimination in homing salmon'. *Science* **165**, 86—88 (1969).
38. K. Oshima, W. E. Hahn, and A. Gorbman: 'Olfactory discrimination of natural waters by salmon'. *J. Fish. Res. Bd. Can.* **26**, 2111—2121 (1969).
39. K. Oshima, W. E. Hahn, and A. Gorbman: 'Electroencephalographic olfactory responses in adult salmon to waters traversed in the homing migration'. *J. Fish. Res. Bd. Can.* **26**, 2123—2133 (1969).
40. M. Peters: 'Sensory mechanisms of homing in salmonids: a comment'. *Behaviour* **39**, 18—19 (1971).
41. W. Pfeiffer: 'The fright reaction of fish'. *Biol. Rev.* **37**, 495—511 (1962).
42. A. J. Pinching and T. P. S. Powell: 'The neuropil of the glomeruli of the olfactory bulb'. *J. Cell. Sci* **9**, 347—377 (1971).
43. D. A. Ramsey: 'Olfactory cues in migrating salmon'. *Science* **133**, 56—57 (1961).
44. R. Selset: 'Chemical methods for fractionation of odorants produced by char smolts and tentative suggestions for pheromone origins'. *Acta Physiol. Scand.* **108**, 97—103 (1980).
45. R. Selset and K. B. Døving: 'Behaviour of mature anadromous char (*Salmo alpinus* L.) towards odorants produced by smolts of their own population'. *Acta Physiol. Scand.* **108**, 113—122 (1980).
46. R. E. Sheldon: 'The olfactory tracts and centers in teleosts'. *J. Comp. Neurol.* **22**, 177—339 (1912).
47. P. W. Soerensen, N. E. Stacey, and J. G. Dulka: 'Effects of a potent steroidal hormone, 17-alpha, 20-beta-dihydroxy-4-pregnen-3-one, on the behaviour and physiology of goldfish'. In: Proceedings from *The Behaviour of Fishes*, Bangor, Wales, U.K. (1986).
48. O. B. Stabell: 'Homing and olfaction in salmonids: a critical review with special reference to the Atlantic salmon'. *Biol. Rev.* **59**, 333—388 (1984).
49. T. A. Stuart: 'Spawning, migration, reproduction and young stages of loch trout (*Salmo trutta* L.)'. *Freshwater and Salmon Fish. Res. Sci. Invest. Freshw. Fish. Scotland* **5**, 1—39 (1953).
50. G. Thommesen: 'The spatial distribution of odour induced potentials in the olfactory bulb of char and trout (Salmonidae)'. *Acta Physiol. Scand.* **102**, 205—217 (1978).
51. G. Thommesen: 'Specificity and distribution of receptor cells in the olfactory mucosa of char (*Salmo alpinus* L.)'. *Acta Physiol. Scand.* **115**, 47—56 (1982).
52. G. Thommesen: 'Morphology, distribution and specificity of olfactory receptor cells in salmonid fishes'. *Acta Physiol. Scand.* **117**, 241—249 (1983).
53. G. Thommesen and K. B. Døving: 'Spatial distribution of the EOG in the rat; a variation with odour quality'. *Acta Physiol. Scand.* **99**, 270—280 (1977).
54. J. H. Todd, J. Atema, and J. E. Bardach: 'Chemical communication in social behavior of a fish, the yellow bullhead (*Ictalurus natalis*)'. *Science* **158**, 672—673 (1967).
55. R. Toft: *Lukt- och synsinnets roll för lekvandrings- beteendet hos östersjölax.* Swedish Salmon Research Institute, Report no. 10 (1975).
56. P. Tytler and D. Machin: 'A comparison of the patterns of movement between indigenous and displaced brown trout (*Salmo trutta* L.) in a small shallow loch'. *Proceedings of the Royal Society of Edinburgh* **76B**, 245—268 (1978).
57. K. Ueda, T. J. Hara, and A. Gorbman: 'Electroencephalographic studies on olfactory discrimination in adult spawning salmon'. *Comp. Biochem. Physiol.* **21**, 133—144 (1967).
58. H. Westerberg: 'Ultrasonic tracking of Atlantic salmon (*Salmo salar* L.) 1/M. I. Movements in coastal regions'. *Rep. Inst. Freshwater Res. (Drottningholm)* **60**, 81—101 (1982).
59. H. Westerberg: 'Ultrasonic tracking of Atlantic salmon (*Salmo salar* L.) 1/M. II. Swimming depth and temperature stratification'. *Rep. Inst. Freshwater Res. (Drottningholm)* **60**, 102—120 (1982).
60. H. Westerberg: 'The orientation of fish and the vertical stratification at fine- and microstructure scales'. In: J. D. McCleave, W. H. Neill, J. J. Dodson, and G. P. Arnold (eds.), *Mechanisms of Migration in Fishes.* Plenum, New York, pp. 179—204 (1984).
61. W. J. Wisby and A. D. Hasler: 'Effect of olfactory occlusion on migrating silver salmon (*O. kisutch*)'. *J. Fish. Res. Bd. Can.* **11**, 472—478 (1954).
62. W. L. Wrede: 'Versuche über den Artduft der Ellritze'. *Z. Vergl. Physiol.* **17**, 510—519 (1932).

Molecules in Mammalian Communication

DIETLAND MULLER-SCHWARZE
*College of Environmental Science and Forestry, State University of New York, Syracuse,
New York, 13210 USA.*

1. Introduction

The extremely keen sense of smell of the dog and many game animals, and the
strong odors emitted by some mammals, such as musk deer, skunk, or civet, are
familiar. But we are learning only now which chemical compounds constitute the
signals that these animals produce and receive, and what functions their body
odors fulfil in their social and sexual lives. Earlier chemical studies of odoriferous
animal secretions were motivated by the perfumers' interests (e.g. [1]). For this
interdisciplinary effort I will, as a behavioral ecologist, review the main concepts of
chemical communication in mammals, and current research problems.

For other sensory modalities we employ anthropomorphic scales. 'Infrasound'
and 'ultrasound' arbitrarily denote those frequencies of sound waves beyond the
range that humans are able to hear. The first is audible to pigeons [2], the latter to
mice, dogs, bats and moths. Likewise, the narrow 'visible' (meaning to *Homo
sapiens*) section of the electromagnetic spectrum is bounded by 'infrared' and
'ultraviolet', that for instance snakes and honey bees, respectively, are able to
perceive. Odor perception of non-human animals is so alien to microsmatic Man
that no such nomenclature has ever been attempted, and for good reason. What
would 'infraodors' be? We cannot even guess how many compounds there are that
provide a clear signal to animals at concentrations that are 'odorless' to humans.
Much has been speculated about the olfactory world of the dog. Its sensitivity,
once claimed to extend all the way to concentrations of molecules in air that are
100 million times more dilute than those that are at the detection threshold for
humans [3, 4, 5], is still a matter of dispute. Animals may be so sensitive that,
for instance, the bombykol receptor cell on the antenna of the male silkworm
moth (*Bombyx mori*) responds to a single molecule of that species' female sex
attractant, making this cell a remarkable molecule counting device [6]. Tantalized
by these glimpses, we remain deeply fascinated by the olfactory worlds of other
organisms, unsmelled by humans, as the recent success of *Perfume*, Patrick
Suesskind's novel about a macrosmatic person, attests [7].

Odors serving in animal communication are termed 'semiochemicals'. These are
further subdivided into intra- and interspecific chemical signals. Semiochemicals
for intraspecific communication have commonly been labeled 'pheromones', which
are "substances secreted to the outside by an individual and received by a second

Jean Maruani (ed.), Molecules in Physics, Chemistry, and Biology, Vol. IV, 331–344.
© 1989 *by Kluwer Academic Publishers.*

individual of the same species in which they release a specific reaction, for instance a definite behavior or developmental process" [8]. Even this early definition did not assume strict species specificity. As accelerating research yields more and more information, the old terminology becomes increasingly strained. Mammals do not meet some of the criteria of the old definition. For this reason, researchers have favored different approaches that range from using the term 'pheromone' in a wide sense [9, 10] to abandoning the word 'pheromone' altogether [11]. For this review we use the term 'pheromone' as proposed by Meredith [12]: "chemicals emitted by one member of a species which, when detected by another member, result in behavioral or physiological changes that are likely to benefit both individuals". In addition to the intraspecific 'pheromones', we have to consider semiochemicals that provide signals *between* species, termed *allelochemics* [13]. If such an allelochemic is primarily of benefit to the emitter, as in the defense secretion of skunks, or in venomous animals, it is designated an *allomone*, while *kairomones* benefit the receiver, as does the odor of a prey species that a predator uses as cue [14].

This short review addresses six questions.

1. What molecules are used by mammals for communication with members of the same species?
2. What is the physiological source of these molecules?
3. What are the biological functions of odors, taking into account the ecological conditions under which various mammals live?
4. Can the choice of a molecule be predicted from its function?
5. How are the odors received by sense organs and evaluated by the central nervous system?
6. What forms and mechanisms of pheromone communication are characteristic for mammals, considering their very differentiated behavior and particularly their advanced learning capabilities?

2. Molecules Used by Mammals for Communication

In a search for semiochemicals in mammals, an evolutionary biologist would assume that already existing volatile components of excretions, secretions, and physiological byproducts would be first exploited as social or sexual cues and that the use of specific compounds manufactured for the exclusive purpose of communication represents more advanced evolutionary stages.

Unlike insects, mammals are not known for communicating via single, highly complex and species-specific compounds that by themselves or in combination with only a few other compounds have unique behavioral effects. Instead of using complex molecule structures, mammals achieve complexity by blending many compounds into one functional pheromone. Here we concentrate on those compounds that have been clearly documented to have a distinct biological effect, or at least contribute to one. Many compounds in body secretions and excretions have been isolated and identified for a great variety of other mammals, without corresponding studies of biological activity. These purely chemical studies are not considered here. They are comprehensively and lucidly covered by Albone [15].

It is usually recognized that *volatility*, hence molecular weight up to around 300—400, is a chief requirement for most airborne mammalian pheromones, even though the required degree of volatility will depend on function and therefore vary considerably. For instance, the pig pheromone steroids are of low volatility. Male guinea pigs are attracted to female urine by compounds of low volatility [16]. It should be emphasized that mammals also use as pheromones some higher-molecular compounds that are detected at short range by actual contact or licking, employing the sense of taste and/or vomeronasal organ. Two examples are the mounting pheromone in the vaginal secretion of the golden hamster, *Mesocricetus auratus*, and the puberty-accelerating pheromone in male mouse urine. The hamster mounting pheromone appears to be a protein with a molecular weight between 15 000 and 60 000 Daltons [17], and the mouse pheromone that accelerates puberty in females a peptide of about 860 Daltons [18, 19]. Non-volatile compounds can be received and evaluated by the vomeronasal organ, which they reach via a continuous stream of mucus [12], aided by the sucking action of the 'vomeronasal pump' [20] from the nasal or mouth cavities after the molecules' direct contact with these areas.

Chemically, a wide range of classes of molecules is used. Some of the identified molecules are ubiquitous and possibly of microbic origin, such as short-chain fatty acids, while others are related to steroid hormones.

To meet the second important criterion, *specificity*, a molecule must be of limited distribution, or ubiquitous compounds can be blended to become very specific chemical signals in the form of odor mixtures. Thus, in mammals we deal with single- and multicomponent pheromones of sometimes great complexity in terms of number of compounds.

Thirdly, mammalian pheromones are often found in a *matrix* of fixatives that releases them slowly and thus counteracts their volatility. Sebum on the skin or in a scent mark serves that purpose. Squalene is found in human sebum [21], in circumgenital-suprapubic secretions of tamarins (*Saguinus fuscicollis*) [22], and in the anal secretion of the sugar glider (*Petaurus breviceps*), a marsupial [23]. The effects of the substrate are extremely important for any consideration of actual concentrations of molecules available to a recipient animal.

A fourth feature is the storage of *larger, inactive molecules* that can be hydrolysed into smaller, volatile molecules that constitute all or part of the signal. Such precursor function of larger molecules has been suggested for four esters of 2-methylbutyric acid and isovaleric acid with 13-methyl-1-tetradecanol and 12-methyl-1-tetradecanol in pronghorn, *Antilocapra americana*[24]. This may also be the case for the group of 15 butyrate esters in the secretion of the circumgenital-suprapubic glands of the tamarin, *Saguinus fuscicollis* [25].

Whether pheromone molecules are *soluble* in water or in lipids is a critical characteristic for all phases of communication from production by the sender to reception by conspecifics. Water soluble compounds are found in apocrine gland secretion and will be captured by mucus in the nasal cavity, while the lipid-soluble compounds mixed into sebum provide scent marks that have to survive unpredictable environmental conditions notably precipitation. For instance, the water-soluble secretion of the (largely apocrine) metatarsal glands of the black-tailed deer, *Odocoileus hemionus columbianus*, has an airborne odor that serves as an alerting stimulus to group members [26]. The matatarsal gland is more developed in arid climates and may be even absent under humid conditions. This suggests that the function of its secretion would be de-emphasized in a humid environment [27].

Finally, whether or not mammals discriminate between *geometric isomers* and between *enantiomers* of pheromone components, has been explored very little. In black-tailed deer, *O. h. columbianus*, (Z)-4-hydroxy-6-dodecenoic acid lactone from urine is a constituent of the tarsal scent, carried on the hocks by both sexes, and elicits approach, sniffing, licking, and following by other deer. Black-tailed deer responded to the *Z* isomer, but not the *E* isomer of this compound [28], while the enantiomers of the lactone were not clearly discriminated [29].

3. Sources of Mammalian Chemosignals

Volatile compounds that serve as signals may originate in specialized glands, may be produced by microorganisms, or occur in urine, feces, saliva, breath, or birth fluids, or genital secretions. In mammals, concentrations of microscopic sebaceous and modified sweat glands (apocrine glands) in the skin constitute 'skin glands'. These glands may be elaborate organs that also contain hair that is coated with secretion or used for application of secretion to the substrate, and muscles to discharge or retain the secretion. The hair may be specialized, i.e. morphologically different from all other body hair, for the purpose of scent storage and application to the substrate. Such specialized scent hairs have been termed 'osmetrichia' [30].

Microorganisms may be producing odoriferous compounds that serve as chemosignals. The best investigated example are the odors produced in the anal glands of the red fox, *Vulpes vulpes* ([31, 32] review in [15]). In this organ, anaerobic bacteria, primarily of the genus *Clostridium*, produce volatile fatty acids whose function in communication has not yet been clearly demonstrated. Likewise, in rabbits (*Oryctolagus cuniculus*) fatty acids, especially acetic and isovaleric acids, occur in the secretion of the inguinal glands and are most likely produced by microorganisms [33]. Other rabbits respond to the odor of these acids with increased heart rates.

Urine contains chemical signals that stimulate or inhibit aggression [34] and factors that not only signal sex and breeding status to conspecifics [35] but also accelerate and delay sexual maturation in both sexes, and influence estrus cycles and male mating behavior. This is more fully described below under Priming Pheromones.

Feces appear to carry chemical signals, some due to diet-dependent metabolites, as in the maternal pheromone of the rat, *Rattus norvegicus* [36, 37], while others are added to the feces by the anal glands, as in the 'marking pellets' that rabbits, *Oryctolagus cuniculus*, deposit on 'dunghills' that serve to enhance familiarity in colony members and deter intruders (Mykytowycz, [38]).

An example of a pheromone in saliva is the sex odor in male swine, *Sus scrofa*. Two steroids, 5α-Androst-16-en-3α-ol and 5α-Androst-16-en-3-one, in the secretion of the boar's submaxillary salivary gland, are emitted during courtship and cause the sow to assume the mating stance [39, 40]. Birth fluids seem to be important in the initial attachment of a mother to her newborn young while she sniffs and licks them. Animal breeders the world over have used this information since time immemorial: an orphaned newborn lamb can be made acceptable to a strange mother if rubbed with, or wrapped in, the hide of that female's own young. The compounds that serve as signals may be diet-dependent, as in the gerbil [41] and the guinea pig [42]. Finally, mammals may communicate with odors that are derived from the environment: rats can communicate to colony members the nature of distant food sources with traces of odors that cling to their body, notably the head region [43].

4. Functions of Mammalian Pheromones

The *functions* of mammalian pheromones may be derived from two original uses of odors: an individual's body odors and/or excretions may be important for familiarization with a new area and subsequently finding familiar sites in its territory or home range. In a social context, other individuals, notably sex partners, may have used the same body odors and excretions to locate and assess other individuals in an opportunistic way, and without the presence of specific sex attractants.

We distinguish two main functional categories of pheromones: *Signaling* and *primer* pheromones. Signaling pheromones are stimuli that enable an individual to discriminate between different classes of individuals, such as familiar vs. strange (mother, or group members), dominant vs. subordinate, male vs. female, estrous vs. non-estrous, immature vs. adult, own vs. other clan, or own vs. different species. In addition to this identification of the class of the odor-emitting individual, signal pheromones also set in motion the appropriate behavior, such as sexual attraction, mounting, aggressive behavior, inhibition of aggression, scent marking, fleeing, avoiding a territory, being alarmed or alert, or caretaking by permitting a young to suckle. The numerous studies cannot be individually reviewed here. All the listed responses occur within seconds or minutes of odor perception. Therefore, signaling pheromones have their effects 'immediately'.

By contrast, primer pheromones set in motion endocrine responses that show their effects much later, i.e. after hours, days or weeks of single or repeated stimulation. The best known priming pheromones are those that affect sexual maturation and the estrous cycle, but growth may also modulated. The functions of primer pheromones include, but are not limited to acceleration and delay of puberty, blocking pregnancy, and accelerating or synchronizing estrus cycles. These pheromones are found in urine of both sexes of house mice, *Mus musculus*. Odor of male urine accelerates sexual maturity in females [44], but inhibits the development of testicles and accessory glands in young males [45]. Female urine odor, on the other hand, has different effects on young mice, depending on the physiological condition of the donor female: the urine of crowded or starving females delays puberty in other females, while urine from single, estrous females accelerates maturation [46, 47]. Likewise, urine from pregnant or lactating females accelerates puberty in other females [48]. Urine from adult females also accelerates sexual maturity in males [49]. These responses appear to be ecological adaptations to environmental conditions: if conditions are unfavorable, such as overpopulation or food shortage, the appropriate urine signals from adults relay this information to young animals whose reproduction is then delayed. This is a mechanism to bring the rate of reproduction in line with available resources [50]. If, on the other hand, favorable conditions permit reproduction to flourish, the breeding females signal this to young individuals and accelerate their puberty, thus allowing population growth at full capacity [50]. Bronson [51] suggested that this regulatory mechanism is especially important for dispersing young mice who may find themselves in varying ecological conditions.

In addition to effects on juveniles, the male urinary factors may also speed up the estrus cycle of adult females [52] and synchronize their cycles [53, 54], and the urine odor of crowded females may delay the cycle or cause pseudopregnancies [55]. Finally, a pregnant mouse may abort her fetus if exposed to the urine odor of an adult male that is genetically different from the original stud male. This is known as 'pregnancy block' [56]. As with the effects on puberty, mature mice are able to adjust their reproduction by using urine scent marks as a feedback mechanism: Female urine marks stimulate testosterone levels in males, who in turn produce more testosteron-dependent pheromone in their urine which in turn accelerates the estrus cycle in females [35].

The chemoreceptors important for many of these priming pheromone effects are located in the vomeronasal organ (reviewed in [12]). Neural projections lead from the vomeronasal organ to regions in the olfactory brain (rhinencephalon). An increase in luteinizing hormone-releasing hormone (LHRH) can be recorded in these regions of a female after exposure to only one drop of male urine [57]. Next, luteinizing hormone (LH) levels increase in the blood of males when exposed to female urine [58] and that of females exposed to male urine [59]. In prepubertal females this LH surge occurs within 30 min of exposure to male urine [60]. LH in turn stimulates subsequent release of estradiol or testosterone in females or males, respectively, with the usual consequences for reproductive processes. The sensitivity of female mice is astonishing: a dose of only 0.0001 ml male urine accelerates puberty [46, 47].

5. Can the Choice of a Molecule be Predicted from the Function of a Pheromone?

Even with our limited store of well analyzed cases we may attempt some generalizations. A few examples will follow.

a. Alarm odors that are supposed to be received by group members as rapidly as possible, and over distances from centimeters to tens, possibly hundreds of meters, should be volatile, produced in bursts and fade or decompose rapidly. *b.* Scent marks, by contrast, are designed to last for days, even weeks or months. Slowly evaporating compounds should be favored. Their release from the mark can be further slowed down by 'keeper substances', such as squalene which is present in sebum in large amounts. An animal can actually release odor molecules 'upon demand' by exhaling moist air at close range when sniffing the scent mark. The water vapor will compete with surface sites and thus liberate odor molecules. Regnier and Goodwin [61] have shown in model experiments that the scent marking compound of the Mongolian gerbil (*Meriones unguiculatus*), phenyl acetic acid, will evaporate faster both under humid conditions and when separated from its natural matrix, sebum. *c.* Discrimination of individuals by odor requires chemical 'fingerprints,' 'signatures,' or 'profiles,' and mixtures of compounds with differing ratios best serve that purpose. Such blends have been computer-analyzed and shown to be predictive for subspecies, sex, and individual differences in tamarin, *Saguinus fuscicollis* [62].

6. Chemoreception and Stimulus-Response Relationships

The sensory systems that process chemical information have been reviewed in detail periodically (e.g. [63, 12, 64]). Here we give a brief sketch to appreciate the complexity encountered in mammalian chemocommunication. In mammals, chemical stimuli can be received by five different 'olfactory organs', in addition to the sense of taste. These are the olfactory epithelium (the main sense of smell), the vomeronasal organ (VNO), located at the base of the septum between nasal and oral cavities, and three less understood structures: the trigeminal nerve, the terminal nerve, and the organ of Masera. For pheromone communication the olfactory epithelium and vomeronasal organ are most important. A certain degree of division of labor exists, with the olfactory epithelium receiving more volatile compounds, and the VNO larger molecules. It has been suggested that the VNO receives unconditional (unlearned) stimuli that are important in the regulation of reproduction, while the olfactory epithelium deals with conditioned chemical stimuli [12]. Such experiential association between inputs from the two sense organs may account for observed redundancy between these two senses, as when

mating behavior could be maintained with either input alone (see [12] for discussion of these experiments). Non-chemical stimuli, such as sounds, tactile, or
visual stimuli, may be part of a necessary stimulus configuration, or may provide
redundant signals, both simultaneously or in sequence.

Our understanding of reception and processing of naturally occurring, biologically significant, and mostly complex chemical signals is rudimentary at best.
Specific receptor sites for 5α-androst-16-en-3-one, a male sex pheromone in the
pig, have been described [65]. We are just starting to understand how the receptors
and the central nervous system processes complex odors to bring about a specific,
coordinated response. The constituents of complex odors may act on the organism
in additive, inhibitive, synergistic, or redundant fashion. There are examples of
each, and a case can be made that each different biological function calls for a
different one of these modes of action. Mammals with their extremely elaborate
decoding capabilities of the central nervous system are expected to rely very much
on such post-receptor processing of information.

Examples (or possible examples) for an *additive* relationship of blend components are all those compounds that have been identified as components of
pheromone mixtures and have been shown to account for part of the biological
activity. A case in point is *castoreum*, the scent marking secretion from the castor
sacs of the North American beaver, *Castor canadensis*. Beaver build small mud
mounds in their territories and top them with castoreum. If the odor of a strange
beaver is discovered on such a scent mound, the resident beaver respond by
sniffing, destroying, and scent marking the mound. Our bioassay exploits this
behavior: we place a scented mound in the territory of a beaver family [66, 67].
Two experiments show that a number of compounds contribute to the total effect.
First, beaver castoreum was fractionated and the fractions applied to experimental
scent mounds in territories of free-living beaver [68]. With progressive fractionation, i.e. fewer and fewer compounds in each fraction, the response diminished and
eventually disappeared. In the second experiment, single castoreum constituents
released partial responses, while a mixture of 6 such compounds was as much as
80% active (100%: entire castoreum; Muller-Schwarze *et al.*, unpublished)

Synergism: two compounds comprise the 'aggression pheromone' in the urine of
male house mice, *Mus musculus*: dehydrobrevicomin and 2-*sec* butyl-4,5-
dihydrothiazole [69]. These compounds stimulate aggression in male mice, but
only if both compounds are present together. In addition to this male-male effect,
the 'aggression pheromone' also attracts females and stimulates their estrus. It may
indeed signal the essence of 'maleness'. Therefore it should be termed 'male mouse
pheromone', analogous to the 'queen substance' of the honey bee that also has
several functions. Illuminating mammalian pheromone mechanisms is also the fact
that the two compounds are active only if dissolved in urine and not in water [70,
71].

The first example of a *redundant* mammalian pheromone was the sex pheromone of the male pig: when courting a sow, the boar produces a saliva that

contains the two steroids, 5α-androst-16-en3α-ol and 5α-androst-16-en-3-one. These compounds, 'androstenol' and 'androstenone', cause the sow to assume the mating stance. Each of these compounds separately, or both together, have all the same effect [39]. Recently, Novotny *et al.* [72] succeeded in identifying the active compounds in a mouse primer pheromone. The effects of three of these compounds provide another example of chemical redundancy. 'Crowded' female mice, i.e. living in groups of ten, produce a urine whose odor delays puberty in female mice from 27—28 to 30—31 days. The urine of adrenalectomized, but still 'crowded', females loses this puberty-inhibiting effect. Six compounds in the urine also responded to adrenalectomy by decreasing in concentration. Of these six, two acetate esters (*n*-pentyl acetate, and *cis*-2-penten-1-yl acetate), and 2,5-dimethyl-pyrazine delayed puberty. The full effect was observed whether total urine, all three compounds together, the first two, or only the pyrazine were used. Such redundancy may represent a backup mechanism that preserves the signal if the urine composition is likely to change.

7. The Mammalian Type of Pheromone Action

Given the fact that mammals use molecules for chemical communication in a much more 'open' fashion than, say insects, stimulus-response relationships cannot be assumed to be rigid in the sense that one particular molecule always releases or modulates the same response. The response to a certain pheromone depends on context, i.e. the physiological state of the receiving individual. In addition to this variability of response, it is even more important to remember that we also find variability of active stimuli: mammals may *learn* to associate certain molecules with certain contexts and link their behavioral responses to these learned odors. Mammals are not neurophysiologically 'hard-wired'. Unlike insects with their narrowly tuned pheromone receptor cells, mammals can receive many different compounds with a particular receptor cell, and a particular molecule may bind to many different receptor cells. No clear structure-activity relationships have been found. There may still exist specific receptor sites, but many different types may be located on one and the same receptor cell [73]. Such a 'multiple-site, multiple binding' system would enable a mammal to evaluate a vast number of often complex odors and assign biological significance, depending on context and experience. Such an odor recognition system does not require rigidly specific receptor characteristics. Instead, odors are possibly identified by the ratios in which different compounds bind to several receptor sites, and odor recognition would depend on comparing a received stimulus to a previously encountered one [74].

Mammals may even communicate with compounds acquired from the environment. A few experiments illustrate this flexibility: in spiny mice, *Acomys cahirinus*, a rodent from the eastern Mediterranean, young are attracted to their mothers (and *vice versa*) by a diet-dependent 'maternal pheromone'. Pups of this species even prefer a lactating female of a different species, such as the house mouse, *M. musculus*, over a conspecific female if the former was fed a 'familiar' diet, i.e. that of the biological mother, while the latter received an unfamiliar diet [75]. Porter concludes "maternal diet is more salient than are genetically determined species-specific characteristics." Nest or maternal odors that determine later sexual preferences may also be learned: male house mice prefer females with Chanel #5 odor over those with species-specific odor if they were raised in a nest artificially scented with that perfume [76]. Likewise, if male laboratory rats were suckled by a mother with citral-treated nipples, they are later attracted to females with that odor [77]. Such learning is not confined to immature animals: female house mice may learn as adults to prefer the odor of soiled bedding from males they have been exposed to during adulthood [78]. Learning may extend to other species: it was recently observed that bulls of the Asian elephant that had been housed near bulls of the other species, the African elephant, responded to secretions and their constituents of this latter species (Rasmussen, personal communication). Rats may use environmental odors to communicate with colony members. They carry odors of distant food sources on their head area which impart information to group members, leading to consumption of the indicated food [43]. Even irrelevant odors can assume signal function: male hamsters will be attracted to vanillin-treated females, but not as strongly as by the odor of genuine vaginal secretion [79]. This illuminates the important biological point that species-specific and genetic components operate in mammalian pheromone systems despite the impressive learning abilities. The evolutionary biologist accepts the challenge to analyze which environmental circumstances have favored the various possible compromises between rigid and open communication systems. Odors that serve especially in recognition of species, individuals, or breeding state are complex mixtures with varying proportions of the various components. These ratios not only change from individual to individual, but may also change over time, due to diet or hormonal changes. Such changes have to be tracked by conspecifics, and here continuous learning is important, a task that the short-lived insects, for instance, do not face. Instead of such tracking, there may be constant components in the mixture, or, alternatively, the recognition odor may be needed for only a short time [80].

On the other hand, there is evidence that male house mice are olfactorily attracted to mating partners who share all genes except for differences at the Major Histocompatibility Complex (HMC, also called H2) on chromosome 17 [81]. It is probably no coincidence that MHC is involved in both immune responses and chemical recognition by the olfactory system, and it is important to note that even here training was necessary for odor discrimination to occur [82].

If most mammals present a complex web of genetic dispositions and learned responses in their behavior, it is not surprising that despite intense efforts no simple pheromone systems have been found in humans. Suffice it to note that kin can be recognized by 'odor signatures' [83], and that axillary odors have been implicated in affecting the timing of menstrual cycles in group members [84].

In summary, molecules in the context of mammalian pheromone systems are messengers that constitute an integral part of a signal system, that has evolved for the benefit of both sender and receiver, who both belong to the same species. Selection pressures have resulted in rigid and flexible, uni- and multicomponent pheromones and the use of more or less volatile compounds, of ubiquitous or specific compounds which may be lipophilic or water soluble, depending on the function. Thus the challenge to the researcher is threefold: first, to isolate and identify in laborious bioassays the biologically active compounds; second, to determine the enormously intricate and variable stimulus-response relationships that render mammalian behavior adaptive in a variety of ecological contexts, and finally, to trace the precise evolutionary processes and pathways that resulted in the chemocommunication systems before us today.

Acknowledgements

I thank Dr. Robert M. Silverstein for critically reading this manuscript. His advice in chemical matters and uncounted inspiring suggestions over the past 20 years, part and parcel of several joint papers quoted here, are greatly appreciated. Much of the research summarized here was supported by the National Science Foundation of the U.S., the U.S. National Park Service, the New York State Department of Environmental Conservation, and the New York State Legislature.

References

1. E. Lederer: *Fortschr. Chemie organ. Naturstoffe* **6**, 87—153 (1950).
2. M. L. Kreithen and D. B. Quine: *J. comp. Physiol.* **129**, 1—4 (1979).
3. W. Neuhaus: *Z. vergl. Physiol.* **35**, 527—552 (1953).
4. W. Neuhaus: *Z. vergl. Physiol.* **38**, 238—258 (1956).
5. W. Neuhaus: *Z. vergl. Physiol.* **39**, 25—43 (1956).
6. K. E. Kaissling and E. Priesner: *Naturwissenschaften* **57**, 23 (1970).
7. P. Suesskind: *Perfume.* Alfred Knopf, New York (1986).
8. P. Karlson and M. Luescher: *Nature* **183**, 155—156 (1959).
9. D. Muller-Schwarze: 'Complex Mammalian Behavior and Pheromone Bioassay in the Field' (*Chemical Signals in Vertebrates*, v. 1, Eds. D. Muller-Schwarze and M. M. Mozell), pp. 413—433. Plenum, New York (1977).
10. D. D. Thiessen: 'Methodology and Strategies in the Laboratory' (*Chemical Signals in Vertebrates*, v. 1, Eds. D. Muller-Schwarze and M. M. Mozell), pp. 391—412. Plenum, New York (1977).
11. G. K. Beauchamp, R. L. Doty, D. G. Moulton, and R. A. Mugford: 'The Pheromone Concept in Mammalian Chemical Communication: a Critique' (*Mammalian Olfaction, Reproductive Processes, and Behavior*, Ed. R. L. Doty), pp. 143—160. Academic Press, New York (1976).
12. M. Meredith: 'Sensory Physiology of Pheromone Communication' (*Pheromones and Reproduction in Mammals*, Ed. J. G. Vandenbergh), pp. 199—252. Academic Press, New York (1983).
13. R. H. Whittacker and P. P. Feeny: *Science* **171**, 757—770 (1971).
14. W. L. Brown, T. Eisner, and R. H. Whittaker: *BioScience* **20**, 21—22 (1970).
15. E. S. Albone: *Mammalian Semiochemistry.* Wiley, Chichester (1984).
16. J. Beruter, G. K. Beauchamp, and E. L. Muetterties: *Biochem. Biophys. Res. Comm.* **53**, 264—271 (1973).

17. A. G. Singer, F. Macrides, and W. C. Agosta: 'Chemical Studies of Hamster Reproductive Pheromones' (*Chemical Signals in Vertebrates*, v. 2, Eds. D. Muller-Schwarze and R. M. Silverstein). Plenum, New York (1980).
18. J. G. Vandenbergh, J. M. Whitsett, and J. R. Lombardi: *J. Reprod. Fertil.* **43**, 515—523 (1975).
19. J. G. Vandenbergh, J. S. Finlayson, W. J. Dobrogosz, S. S. Dills, and T. A. Kost: *Biol. Reprod.* **15**, 260—265 (1976).
20. M. Meredith and R. J. O'Connell: *J. Physiol.* **286**, 301—316 (1979).
21. J. S. Strauss and P. E. Pochi: *Rec. Progr. Horm. Res.* **19**, 385—444 (1963).
22. R. G. Yarger, A. B. Smith III, G. Preti, and G. Epple: *J. Chem. Ecol.* **3**, 145 (1977).
23. H. Autrum, K. Fillies, and H. Wagner: *Naturwissenschaften* **55**, 449 (1968).
24. D. Muller-Schwarze, D. C. Muller-Schwarze, A. G. Singer, and R. M. Silverstein: *Science* **183**, 860—862 (1974).
25. G. Epple, A. M. Belcher, and A. B. Smith III: 'Chemical Signals in Callitrichid Monkeys: a Comparative Review' (*Chemical Signals in Vertebrates*, v. 4, Eds. D. Duvall, D. Muller-Schwarze, and R. M. Silverstein), Plenum, New York, pp. 653—672 (1986).
26. D. Muller-Schwarze, A. Altieri, and N. Porter: *J. Chem. Ecol.* **10**, 1707 (1984).
27. D. Muller-Schwarze: 'Evolution of Chemical Communication in Cervids' (*Symposium on Behavior and Management of Cervids*, Ed. C. Wemmer), pp. 223—234 (1987).
28. D. Muller-Schwarze, R. M. Silverstein, C. Muller-Schwarze, A. G. Singer, and N. J. Volkman: *J. Chem. Ecol.* **2**, 389—398 (1976).
29. D. Muller-Schwarze, U. Ravid, A. Claesson, A. G. Singer, R. M. Silverstein, C. Muller-Schwarze, N. J. Volkman, K. F. Zemanek, and R. G. Butler: *J. Chem. Ecol.* **4**, 247—256 (1978).
30. D. Muller-Schwarze, N. J. Volkman, and K. F. Zemanek: *J. Ultrastructure Res.* **59**, 223—230 (1977).
31. E. S. Albone and G. C. Perry: *J. Chem. Ecol.* **2**, 101—111 (1976).
32. E. S. Albone, P. E. Gosden, and G. C. Ware: 'Bacteria as a Source of Chemical Signals in Mammals' (*Chemical Signals in Vertebrates*, v. 1, Eds. D. Muller-Schwarze and M. M. Mozell), pp. 35—43. Plenum, New York (1977).
33. G. C. Merritt, B. S. Goodrich, E. R. Hesterman, and R. Mykytowycz: *J. Chem. Ecol.* **8**, 1217—1225 (1982).
34. B. Jemiolo, F. Androlini, and M. Novotny: 'Chemical and Biological Investigations of Female Mouse Pheromones' (*Chemical Signals in Vertebrates*, v. 4, Eds. D. Duvall, D. Muller-Schwarze, and R. M. Silverstein), pp. 79—85. Plenum, New York (1986).
35. F. H. Bronson and A. Coquelin: 'The Modulation of Reproduction by Priming Pheromones in House Mice: Speculations on Adaptive Function' (*Chemical Signals in Vertebrates*, v. 2, Eds. D. Muller-Schwarze and R. M. Silverstein), pp. 243—265. Plenum, New York (1980).
36. M. Leon: *Physiol Behav.* **13**, 441—453 (1974).
37. M. Leon: *Physiol. Behav.* **14**, 311—319 (1975).
38. R. Mykytowycz: *CSIRO Wildlife Res.* **11**, 11—29 (1966).
39. D. R. Melrose, H. C. B. Reed, and R. L. S. Patterson: *Brit. Vet. J.* **137**, 497—502 (1971).
40. J. P. Signoret and F. du Mesnil du Busson: *Congr. Int. Reprod. Anim. Insem. Artif.* **2**, 171—175 (1961).
41. J. Skeen and D. D. Thiessen: *Physiol. Behav.* **19**, 11—14 (1976).
42. G. K. Beauchamp: *Nature* **263**, 587—588 (1976).
43. B. G. Galef: 'Olfactory Communication Among Rats: Information Concerning Distant Diets' (*Chemical Signals in Vertebrates*, v. 4, Eds. D. Duvall, D. Muller-Schwarze, and R. M. Silverstein), pp. 487—505. Plenum, New York (1986).
44. J. G. Vandenbergh: *Endocrinology* **84**, 658—660 (1969).
45. J. G. Vandenbergh: *J. Reprod. Fertil.* **24**, 383—390 (1971).
46. L. C. Drickamer: *Dev. Psychobiol.* **15**, 433 (1982).
47. L. C. Drickamer: *Physiol. Behav.* **33**, 907—911 (1984).
48. L. C. Drickamer and J. E. Hoover: *Dev. Psychobiol.* **12**, 545 (1979).
49. K. A. Fox: *J. Reprod. Fertil.* **17**, 75—85 (1968).

50. L. C. Drickamer: 'Puberty-Influencing Chemosignals in House Mice: Ecological and Evolutionary Considerations' (*Chemical Signals in Vertebrates*, v. 4, Eds. D. Duvall, D. Muller-Schwarze, and R. M. Silverstein), pp. 441—455. Plenum, New York (1986).
51. Bronson, F. H.: *Quart. Rev. Biol.* **54**, 265 (1979).
52. W. K. Whitten: *J. Endocrinol.* **17**, 307—313 (1958).
53. W. K. Whitten: *J. Endocrinol.* **13**, 399—404 (1956).
54. H. M. Marsden and F. A. Bronson: *Science* **144**, 1469 (1964).
55. S. van der Lee and L. M. Boot: *Acta Physiol. Pharmacol. Neerl.* **4**, 442—444 (1955).
56. H. M. Bruce: *Nature* **184**, 105 (1959).
57. D. E. Dluzen, V. D. Ramirez, C. S. Carter, and L. L. Getz: *Science* **212**, 573—575 (1981).
58. F. Macrides, A. Bartke, and S. Dalterio: *Science* **189**, 1104—1106 (1975).
59. F. H. Bronson and B. Macmillan: 'Hormonal Responses to Primer Pheromones' (*Pheromones and Reproduction in Mammals*, Ed. J. G. Vandenbergh), pp. 175—197. Academic Press, New York (1983).
60. F. H. Bronson and J. A. Maruniak: *Endocrinology* **98**, 1101—1108 (1976).
61. F. E. Regnier and M. Goodwin: 'On the Chemical and Environmental Modulation of Pheromone Release from Vertebrate Scent Marks' (*Chemical Signals in Vertebrates*, v. 1, Eds. D. Muller-Schwarze and M. M. Mozell), pp. 115—133. Plenum, New York (1977).
62. A. B. Smith, A. M. Belcher, G. Epple, P. C. Jurs, and B. Lavine: *Science* **228**, 175—177 (1985).
63. P. P. C. Graziadei: 'Functional Anatomy of the Mammalian Chemoreceptor System' (*Chemical Signals in Vertebrates*, v. 1, Eds. D. Muller-Schwarze and M. M. Mozell), pp. 435—454. Plenum, New York (1977).
64. W. B. Quay: 'Olfaction in Central Neural and Neuroendocrine Systems: Integrative Review of Olfactory Representations and Interrelations' (*Chemical Signals in Vertebrates*, v. 3, Eds. D. Muller-Schwarze and R. M. Silverstein), pp. 105—118. Plenum, New York (1983).
65. J. N. Gennings, D. B. Gower, and L. H. Bannister: *Biochim. Biophys. Acta* **496**, 547—556 (1977).
66. D. Muller-Schwarze and S. Heckman: *J. Chem. Ecol.* **6**, 81 (1980).
67. D. Muller-Schwarze, S. Heckman, and B. Stagge: *Acta Zool. Fennica* **174**, 111 (1983).
68. D. Muller-Schwarze, L. Morehouse, R. Corradi, C. Zhao, and R. M. Silverstein: 'Odor Images: Responses of Beaver to Castoreum Fractions' (*Chemical Signals in Vertebrates*, v. 4, Eds. D. Duvall, D. Muller-Schwarze, and R. M. Silverstein), pp. 561—570. Plenum, New York (1986).
69. D. P. Wiesler, F. J. Schwende, M. Carmack, and M. Novotny: *J. Org. Chem.* **49**, 882—884 (1984).
70. M. Novotny, S. Harvey, B. Jemiolo, and A. Alberts: *Proc. Natl. Acad. Aci. USA* **82**, 2059—2061 (1985).
71. B. Jemiolo, J. Alberts, S. Sochinski-Wiggins, S. Harvey, and M. Novotny: *Anim. Behav.* **33**, 1114—1118 (1985).
72. M. Novotny, B. Jemiolo, S. Harvey, D. Wiesler, and A. Marchlewska-Koj: *Science* **231**, 722—725 (1986).
73. M. M. Mozell: 'Processing of Olfactory Stimuli at Peripheral Levels' (*Chemical Signals in Vertebrates*, v. 1, Eds. D. Muller-Schwarze and M. M. Mozell), pp. 465—484. Plenum, New York (1977).
74. S. G. Shirley: 'Mammalian Chemoreception' (*Mammalian Semiochemistry*, Ed. E. S. Albone), pp. 243—277. Wiley, Chichester (1984).
75. R. H. Porter: 'Chemical Signals and Kin Recognition in Spiny Mice (*Acomys cahirinus*)' (*Chemical Signals in Vertebrates*, v. 4, Eds. D. Duvall, D. Muller-Schwarze, and R. M. Silverstein), pp. 397—411. Plenum, New York (1986).
76. D. Mainardi, M. Marsan, and A. Pasquali: *Atti Soc. Scienza Naturali Museo Covico Storia Naturale Milano* **104**, 325—338 (1965).
77. T. J. Fillion and E. M. Blass: *Science* **231**, 729 (1986).
78. M. E. Albonetti and B. D. d'Udine: *Anim. Behav.* **34**, 1844—1847 (1986).
79. F. Macrides, A. N. Clancy, A. G. Singer, and W. C. Agosta: *Physiol. Behav.* **33**, 627—632 (1984).

80. Z. T. Halpin: *Adv. Study Behav.* **16**, 39—70 (1986).
81. K. Yamazaki, M. Yamaguchi, L. Baranoski, J. Bard, E. A. Boyse, and L. Thomas: *J. Exp. Med.* **150**, 755 (1979).
82. K. Yamazaki, M. Yamaguchi, E. A. Boyse, and L. Thomas: 'The Major Histocompatibility Complex as a Source of Odors Imparting Individuality among Mice' (*Chemical Signals in Vertebrates*, v. 2, Eds. D. Muller-Schwarze and R. M. Silverstein), pp. 267—273. Plenum, New York (1980).
83. R. H. Porter: *Chemical Senses* **11**, 389 (1986).
84. G. Preti, W. B. Cutler, C. M. Christensen, H. Lawley, G. R. Huggins, and C. R. Garcia: *J. Chem. Ecol.* **13**, 717—731 (1987).

Index